LF

Series Editor: Frank B. Golley
University of Georgia

PUBLISHED VOLUMES

CYCLES OF ESSENTIAL ELEMENTS / *Lawrence R. Pomeroy*
BEHAVIOR AS AN ECOLOGICAL FACTOR / *David E. Davis*
NICHE: THEORY AND APPLICATION / *Robert H. Whittaker and Simon A. Levin*
ECOLOGICAL ENERGETICS / *Richard G. Wiegert*
ECOLOGICAL SUCCESSION / *Frank B. Golley*

RELATED TITLES IN OTHER BENCHMARK SERIES
VERTEBRATE SOCIAL ORGANIZATION / *Edwin M. Banks*
EXTERNAL CONSTRUCTION BY ANIMALS / *Nicholas E. Collias and Elsie C. Collias*
SOCIAL HIERARCHY AND DOMINANCE / *Martin W. Schein*
TERRITORY / *Allen W. Stokes*
SOUND RECEPTION IN FISHES / *William N. Tavolga*
SOUND PRODUCTION IN FISHES / *William N. Tavolga*

ECOLOGICAL SUCCESSION

Edited by
FRANK B. GOLLEY
University of Georgia

Benchmark Papers in Ecology, Volume 5
Library of Congress Catalog Card Number: 76-52930
ISBN: 0-87933-256-5

79 78 77 1 2 3 4 5
Manufactured in the United States of America.

LIBRARY OF CONGRESS CATALOGING IN PUBLICATION DATA
Main entry under title:
Ecological succession.
(Benchmark papers in ecology; 5)
Includes index.
1. Ecological succession—Addresses, essays, lectures.
2. Plant succession—Addresses, essays, lectures.
I. Golley, Frank B.
QH541.E317 574.5 76-52930
ISBN 0-87933-256-5

Exclusive Distributor: **Halsted Press**
A Division of John Wiley & Sons, Inc.
ISBN: 0-470-99083-X

PREFACE

The objective of the Benchmark Series in Ecology is to trace the development of key ideas in the science of ecology and to reveal the patterns of scientific creativity. There are contrasting views on the nature of scientific creative thought. One of these stresses the spontaneous work of an individual whose insight and intuition makes possible a sudden, sometimes revolutionary, restructuring of scientific understanding. Often this individual is outside the specific field and considerable time elapses before the idea is expressed within the field, and then often by several persons independently. Another view focuses on the systematic development of ideas through detailed studies and step-by-step revision of previous interpretations of many individuals, stressing environmental factors and social conditions which set the stage for a given development. In both of these cases the idea is discussed in passing by many, while only a few systematically analyze and classify the concepts or data relating to the idea. These latter persons have the greatest effect on further trend of thought and cause the idea to be assimilated into the body of knowledge of the science. Thus, the development of a scientific concept or theory often develops as a single, isolated, crude idea, sometimes from outside the discipline, then appears in several individuals within the field, and is further developed through analytical, classificatory, and synthetic phases.

In this particular Benchmark book on ecological succession, I have attempted to reproduce this process. My major difficulty was in the interpretive or synthetic phase, since succession is still a topic that is misunderstood, or alternatively, is immature. In part, the lack of a strong unifying synthesis regarding succession may result from general characteristics of this, and other, ecological phenomenon—i.e., their complex, diverse, loosely organized character. Ecology is, however, still a young field and presumably new ideas concerning ecological succession will continue to develop, at least for a few more years.

I am grateful for the comments of Robert Whittaker, E. P. Odum, and Vincent Nabholz on the selections and discussions, and to Margaret Shedd for her editorial assistance.

FRANK B. GOLLEY

CONTENTS

CONTENTS BY AUTHOR

ECOLOGICAL SUCCESSION

INTRODUCTION

Succession is a common English word, derived from the Latin, meaning the act of coming after another in order or sequence. In ecology the term is applied to the vegetation, the animal community, or the whole ecological system and refers to an orderly process of change. According to Cowles (1911), a French biologist, Dureau de la Malle (1825) was the first to use the term succession in its ecological sense.

Ecological succession has been a major theme in ecology for the past hundred years. While the scientific study of succession is said to have begun with a description of the development of Irish bog vegetation by King (1685) in the Philosophical Transactions of the Royal Society, London, actually King's paper only reflects common knowledge. I suspect that practically everyone recognizes that the vegetation on a site may change in an orderly fashion. Indeed, from the earliest periods of human history farmers have utilized this knowledge to maintain crop production. The practice, called shifting agriculture, involves cutting down a forest plot of several hectares size and burning the dry vegetation. Next, crops are planted in the ash among the charred stumps and logs. Production continues for a few years until declining fertility, pests and weeds cause crop yields to fall, and then the farmer shifts to a new patch of forest. The plot that has been abandoned now goes through a succession, in which a tangle of vines, herbs, grass, and small trees eventually is replaced by the former forest type. After an appropriate length of time which varies from habitat to habitat, but which may be as short as eight years, the farmer returns to his original plot, recuts the second growth and begins the cycle again. Through the succession the fertility has recovered and the pests have been eliminated.

Of course, change in nature can also be disorderly or random, yet I suppose that most of us would agree with the countryman that orderly processes are dominant. Indeed, in the commonsensical world of the primitive farmer or hunter, recognition of order in the natural world is the key to developing successful strategies of life. Recognition of order is commonly expressed in art and ritual, as well as codified in taboos and folk sayings. I recall the cave drawings at Lascaux where the paleolithic hunters illustrated a diversity of animals and men in orderly sequences of gender, social groupings, and interactions. These drawings represent an order within populations meaningful to the hunter. In contrast, a ninth century Irish monk living alone in a hermitage above the sea describes the order in seasonal change in verse:

> My tidings for you, the stag bells,
> Winter snows, summer is gone.
>
> Wind high and cold, low the sun,
> Short his course, sea running high.
>
> Deep red the bracken, its shape all gone,
> The wild goose has raised his wanted cry.
>
> Cold has caught the wings of birds,
> Season of ice—these are my tidings.
>
> [De Paor and De Paor, 1958, p. 67]

Determination of the relative significance of order and disorder in ecological succession is a difficult task. It involves very careful definition of the terms used in the explanation, careful definition of the community in space and time, as well as an agreement on what features of communities are essential to the question. As we shall see, this problem has resulted in an incredibly complicated terminology, and at times, even acrimonious debate. The task is made even more difficult by the fact that the process of succession often requires a hundred or more years to run to completion and, thus, is not amenable to experimental manipulation, although several rather unsuccessful experiments have been attempted (for example Golley and Gentry, 1966).

The reader may also object to my setting order and disorder as a duality, arguing that order and disorder may occur together so that a process exhibits both tendencies. However, while I concede that this argument may be true and partly explains the diverse interpretations of succession that we will examine in the following chapters, I have used a series of dualisms in this Benchmark book as a pedagogic device. These dualistic simplifications, order-disorder, change-constancy, randomness-predictability, appear to be inherent in the way ecologists have thought and interpreted succession. They are also useful as a key to a complicated literature.

The pattern of this Benchmark emerges, therefore, as a series of questions set in a historical perspective. Part one will concern the precursors to research on succession and the question of whether natural change occurs through catastrophe or by slow, uniform processes. Part two will present a series of examples of succession in different environments. Part three will discuss the mechanisms causing these successions. Parts four and five will deal with the reasons for orderly change in ecological succession. The discussion in these chapters will involve the analogy between development of a community and an individual organism, thermodynamic principles, and systems concepts of community stability.

REFERENCES

Cowles, H. C. 1911. The Causes of Vegetative Cycles. *Bot. Gaz.* **51:**161–183.

De Paor, M., and L. De Paor. 1958. *Early Christian Ireland.* New York: Praeger.

Dureau de la Malle, A. J. C. A. 1825. Memoire sur l'alternance ou sur ce problème: la succession alternative dans la reproduction des espèces végétales vivant en société, est-elle une loi générale de la nature? *Ann. Sci. Nat. (Paris)* I **5:**353–381.

Golley, F. B., and J. B. Gentry. 1966. A Comparison of Variety and Standing Crop of Vegetation on a One-year and a Twelve-year Abandoned Field. *Oikos* **15:**185–189.

King, William. 1685. Of the Bogs and Loughs of Ireland. *R. Soc.* (London) *Philos. Trans.* **15:**948–960.

Part I

THE PRECURSORS

Editor's Comments on Paper 1

1 DAVIS

Excerpt from *The Geographical Cycle*

The observations and interpretations of science both influence and are influenced by the social-cultural environment. Thus, a scientific concept such as succession must be analysed within the context of the intellectual life of the period and people. In the modern western world of Europe and North America, this intellectual life has been strongly shaped by science. The dominant epistemology of the eighteenth and nineteenth centuries, in which the precursors of successional theory lived, has been described by two scientific metaphors (Opper, 1973): the eighteenth century by the mechanical clock and the nineteenth by the growing and developing organism. Of course we can identify other contrasting and opposing points of view in these two periods, yet each age can be defined by the general interpretation of natural events in first mechanical, and later, in biological patterns.

The theory of nature of the eighteenth century stressed the mechanistic and abstract-logical features of the observable world. Isaac Newton was one of the dominant figures of this century and a fundamental premise of Newton's cosmology was that nature is basically simple, rational, and reducible to logically quantitative abstractions. During the Enlightenment it was assumed that with the proper use of reason and enough time the laws governing all phases of natural endeavors could be discovered, bringing society to a state of perfection. This mechanistic world view had its inception in the speculation of Copernicus, with conformation by Galileo, Kepler, Descarte, and others. Acute mechanistic determinism is reached in such works as d'Holbach's (1770) *Systéme de la nature:*

> Nature is a realm of complete determinism; it knows no "order" and "disorder." We call "disorder" what disturbs or afflicts us, but all is in truth "order" in the sense that all occurs by fixed causation; order is just simply what happens. [Opper, 1973, p. 53]

In biology the Newtonian concept was expressed in the work of such scientists as Linnaeus, who arranged plants and animals into a rationally ordered system.

Gradually the fixed view of the world derived from physics and mechanics was eroded by observations of dynamic living systems. Comte de Buffon, a contemporary of Linnaeus, expresses an alternate to the static mechanical concepts of the eighteenth century in his statement, "Nature is in a state of continual flux." This new dynamic concept of the world, together with an emphasis on factual observation in nature, provided the basis for the study of ecological succession, even though the process itself was recognized and described much earlier.

Within science, there were two fundamental contributions to the development of successional studies. First, geologists such as Charles Lyell, reasoning from the actual observation of the geological record rather than abstract principles, concluded that earth processes were dynamic rather than catastrophic and mechanical. Lyell [1797–1875], a barrister, fellow of the Royal Society and a Professor of Geology at Kings College, London, showed that changes in the landscape progressed in a slow orderly fashion. He argued that the order of nature in the past was uniform with that in the present and therefore, geological phenomena could be explained by analogy with present conditions. Lyell (1850) also emphasized the ubiquity of change:

> If we turn to the present state of the animate creation and inquire whether it has now become fixed and stationary, we discover that, on the contrary, it is in a state of continual flux—that there are many causes in action which tend to the extinction of species, and which are conclusive against the doctrine of their unlimited durability. [p. 175]

The significance of Lyell's interpretation to ecological succession was that it focused on the inherent processes on the earth rather than on chance and the unusual event. In other words, Lyell provided sound scientific evidence for orderly change as opposed to mechanical catastrophe.

The next step pertinent to ecological succession was made by William Morris Davis who developed a theory of the orderly development of landscape forms. Davis [1850–1934] was a geologist, meteorologist, and physical geographer and served as Professor of Physical Geography and a Professor of Geology at Harvard University. Davis termed his theory "the Geographical Cycle," quoting Webster in defense of his use of the word cycle. His definition follows:

> An interval of time in which a certain succession of events returns again and again, uniformly and continuously in the same order, a periodical space of time marked by the recurrence of something peculiar. [Davis, 1899, Eighth International Geographical Congress]

While some geographers objected to the term "cycle," since the first and last member of a series may be unlike, Davis responded:

> As far as plains and plateaus are concerned, this objection does not hold, for they begin and end their ideal cycle of changes as low, featureless expanses. [Davis, 1899, Eighth International Geographical Congress]

Davis's Geographical Cycle is described in the selection that follows. This excerpt from his paper describes the cycle in terms of an idealized landscape starting with an uplift of mountains and ending with a rolling lowland. Davis's analogy between the orderly development in the ideal cycle and the development in the life history of an organism illustrates the biocentric attitude of the period.

The second underlying set of premises of the study of succession came from the biologists. Widespread exploration and travel had provided field naturalists with experience in a variety of natural communities and with many new organisms. Linnaeus had organized these taxa into a fixed scheme, but even Linnaeus began to encounter difficulties in making precise distinctions between species because of overlaps in their characteristics. The transition from a static to a dynamic concept of organic nature was characterized by evolutionary speculation in cosmology, geology, biology and embryology, and by ideas of progress in philosophy, political science and history. Charles Darwin is the central figure in this development. Darwin's theory was based on careful field observation and stressed that random variations in a given species were responsible for those individuals adapted to the environment, as expressed in this excerpt from *The Origin of the Species* (Darwin, 1909):

> . . . A great amount of variability, under which term individual differences are always included, will evidently be favourable. A large number of individuals, by giving a better chance within any given period for the appearance of profitable variations, will compensate for a lesser amount of variability in each individual, and is, I believe, a highly important element of success. Though Nature grants long periods of time for the work of natural selection, she does not grant an indefinite period; for as all organic beings are striving to seize on each place in the economy of nature, if any one species does not become modified and improved in a corresponding degree with its

> competitors, it will be exterminated. Unless favourable variations be inherited by some, at least, of the offspring, nothing can be affected by natural selection. The tendency to reversion may often check or prevent the work; but as this tendency has not prevented man from forming, by selection, numerous domestic races, why should it prevail against natural selection? [p. 114]

It is this metaphor of a growing, developing and evolving biological organism that dominated the age.

Finally, theories of succession rest on yet another set of biological observations: the studies of the ecological community. Everyone recognizes the differences between forests and fields, and those who depend upon these natural systems, such as hunter-gatherers, foresters or range managers, recognize a variety of types of each. These types are termed communities and they are characterized as the complex of plants, animals, and microorganisms living together at a defined time and place.

Two fundamental problems of community ecology relate to studies of ecological succession. First, ecologists have asked if communities are distinct, real entities in nature or are they mental constructs? This problem is exceedingly important since if communities are neither distinct nor definable then their succession is of little interest. Successional studies have contributed to our understanding of the nature of the community and this problem will be discussed later when we consider the interpretation of successional data.

The second problem grows out of the first. If communities are real, how are they related to other natural systems? It was recognized early in the nineteenth century that the individual organism was composed of subsystems, such as organs, tissues and cells. Further, the cell theory, developed from the observations of Schleiden with respect to plants and Schwann with respect to animals, has been recognized as an extension of the atomic theory of matter in general. Alfred North Whitehead (1925) writes:

> The living cell is to biology what the electron and the proton are to physics. Apart from cells and from aggregates of cells there are no biological phenomena. The cell theory was introduced into biology contemporaneously with, and independently of, Dalton's atomic theory. The two theories are independent exemplifications of the same idea of "atomism." [p. 94]

E. P. Odum (1971) popularized for ecologists the expansion of the "aggregates of cells" into a biological hierarchy of organization in which cells are aggregated into tissues, organs, individuals, populations, and communities. Thus, in Odum's hierarchy, communities are related to other biological systems and share characteristics with them. If this hierarchical set of systems is accepted, then the student of succession may search for explanation of the orderly sequences he/she observes in the dynamics of individuals, populations and/or cells.

It has been suggested that a dominant metaphor of the nineteenth century was the dynamic evolving organism derived from the observations and theory of biologists, supported by geological observations. All phases of life were interpreted through this metaphor. One more example will suffice to make the point. Frank Lloyd Wright [1869–1959] developing his concept of organic architecture stated, "an organic form grows its structure out of conditions as a plant grows out of soil . . . both unfold similarly from within," and further, "An inner-life principle is a gift to every seed. An inner-life principle is also necessary for every idea of a good building." It is not surprising that some students of communities interpreted the community as an organism and interpreted succession of communities as analogous to evolution of species and development of individuals. The ecologist may well ask if the study of succession is the study of a metaphor or if it is the study of a real phenomenon.

We have already mentioned that the first published account of succession is that of King (1685), who reported on the origin of bog vegetation derived from floating mats in Irish ponds. Stephen Spurr (1952) traces the history of successional studies in forests through Hundeshagen (1830), Gand (1840) and Cochon (1846) in Europe and cities Dawson's (1847) study of forests in the Maritime provinces as the first detailed North American account. Spurr also points out that Henry David Thoreau (1860) studied succession in central New England.

In the next part we will examine some of these sequences.

REFERENCES

Cochon, R. 1846. Alternance des essences dans les forêts. *Ann. For.* **5:**1–13.

Darwin, C. 1909. *The Origin of Species.* New York: Collier & Son.

Davis, W. M. 1899. Complications of the Geographical Cycle. *Eighth International Geographic Congress,* pp. 150–163.

Dawson, J. W. 1847. On the Destruction and Partial Reproduction of Forests in British North America. *Edinburg New Philos. J.* **42:**259–271.

Gand, G. 1940. *Memoire sur l'Alternance des essences forestières.* Paris: Académie des sciences.

d'Holbach. 1770. *The Eighteenth Century Background.* Paraphrased by B. Wiley, 1940. New York: Colombia University Press.

Hundeshagen, J. C. 1830. Ueber die natürliche Unwandlung der Wälder, order die sogenannte Wanderung der Pflanzen. *Forstl. Ber. Misc.* **1:**36–51.

King, W. 1685. Of the Bogs and Loughs of Ireland. *R. Soc. (London) Philos. Trans.* **15:**948–960.

Lyell, C. 1850. *Principles of Geology*. 8th ed. London: John Murray.
Odum, E. P. 1971. *Fundamentals of Ecology*. 3rd ed. Philadelphia: W. B. Saunders.
Opper, J. 1973. *Science and the Arts, a Study in Relationships from 1600 to 1900*. Rutherford, N. J.: Fairleigh Dickson University Press.
Spurr, S. H. 1952. Origin of the Concept of Forest Succession. *Ecology* **33:**426–427.
Thoreau, H. D. 1960. *Succession of Forest Trees.* Mass. Board Agric. Rep. VIII.
Whitehead, A. N. 1925. *Science and the Modern World*. New York: Macmillan.

1

Reprinted from pp. 254–256 of *Geographical Essays,* D. W. Johnson, ed., Ginn & Co., 1909, 777 pp.

THE GEOGRAPHICAL CYCLE

W. M. Davis

[*Editor's Note:* In the original, material precedes this excerpt.]

The Ideal Geographical Cycle. The sequence in the developmental changes of land forms is, in its own way, as systematic as the sequence of changes found in the more evident development of organic forms. Indeed, it is chiefly for this reason that the study of the origin of land forms — or geomorphogeny, as some call it — becomes a practical aid, helpful to the geographer at every turn. This will be made clearer by the specific consideration of an ideal case, and here a graphic form of expression will be found of assistance.

In Fig. 4 the base line $\alpha\omega$ represents the passage of time, while verticals above the base line measure altitude above sea-level. At the epoch 1 let a region of whatever structure and form be uplifted, *B* representing the average altitude of its higher parts and *A* that of its lower parts, *AB* thus measuring its average initial relief. The surface rocks are attacked by the weather. Rain falls on the weathered surface and washes some of the loosened waste down the initial slopes to the trough lines, where two converging slopes meet; there the streams are formed, flowing in directions consequent upon the descent of the trough lines. The machinery of the destructive processes is thus put in motion, and the destructive development of the region is begun. The larger rivers, whose channels initially had an altitude

A, quickly deepen their valleys, and at the epoch 2 have reduced their main channels to a moderate altitude represented by *C*. The higher parts of the interstream uplands, acted on only by the weather without the concentration of water in streams, waste away much more slowly, and at epoch 2 are reduced in height only to *D*. The relief of the surface has thus been increased from *AB* to *CD*. The main rivers then deepen their channels very slowly for the rest of their lives, as shown by the curve *CEGJ*, and the wasting of the uplands, much dissected by branch streams, comes to be more rapid than the deepening of the main valleys, as shown by comparing the curves *DFHK* and *CEGJ*. The period 3–4 is the time of the most rapid consumption of the uplands, and thus stands in strong contrast with the period 1–2, when there was the most rapid deepening of the main valleys. In the earlier period the relief was rapidly increasing in value,

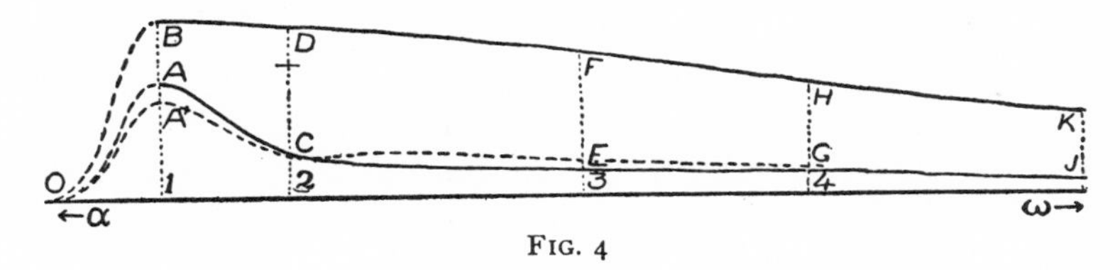

FIG. 4

as steep-sided valleys were cut beneath the initial troughs. Through the period 2–3 the maximum value of relief is reached, and the variety of form is greatly increased by the headward growth of side valleys. During the period 3–4 relief is decreasing faster than at any other time, and the slope of the valley sides is becoming much gentler than before; but these changes advance much more slowly than those of the first period. From epoch 4 onward the remaining relief is gradually reduced to smaller and smaller measures, and the slopes become fainter and fainter, so that some time after the latest stage of the diagram the region is only a rolling lowland, whatever may have been its original height. So slowly do the later changes advance that the reduction of the reduced *JK* to half of its value might well require as much time as all that which has already elapsed; and from the gentle slopes that would then remain, the further removal of waste must indeed be exceedingly slow.

The frequency of torrential floods and of landslides in young and in mature mountains, in contrast to the quiescence of the sluggish streams and the slow movement of the soil on lowlands of denudation, suffices to show that rate of denudation is a matter of strictly geographical as well as of geological interest.

It follows from this brief analysis that a geographical cycle may be subdivided into parts of unequal duration, each one of which will be characterized by the degree and variety of the relief, and by the rate of change, as well as by the amount of change that has been accomplished since the initiation of the cycle. There will be a brief youth of rapidly increasing relief, a maturity of strongest relief and greatest variety of form, a transition period of most rapidly yet slowly decreasing relief, and an indefinitely long old age of faint relief, in which further changes are exceedingly slow. There are, of course, no breaks between these subdivisions or stages; each one merges into its successor, yet each one is in the main distinctly characterized by features found at no other time.

[*Editor's Note:* Material has been omitted at this point.]

Part II

THE PATTERN

Editor's Comments on Papers 2 Through 6

2 **COWLES**
Excerpt from *The Physiographic Ecology of Chicago and Vicinity; A Study of the Origin, Development, and Classification of Plant Societies*

3 **COOPER**
The Recent Ecological History of Glacier Bay, Alaska: II. The Present Vegetation Cycle

4 **CRAFTON and WELLS**
The Old Field Prisere: An Ecological Study

5 **SHELFORD**
Ecological Succession. II. Pond Fishes

6 **ODUM**
Organic Production and Turnover in Old Field Succession

In part one we introduced the observation that a sequence or succession of communities appears after a disturbance. We suggested that these sequences seemed to be repeatable over space and time. Here we will examine several examples of successions.

The development of stages of vegetation are especially obvious in certain habitats. These habitats often provide a bare surface, created by physical forces, on which plants invade, grow and eventually form characteristic communities. Sand dunes, glacial moraines, lake margins, river sandbars, peat bogs and abandoned agricultural fields all have been the focus of successional studies. Among the literally thousands of papers and books describing succession, I have chosen a series which illustrates the patterns on a variety of habitats and which also provides examples of major advances in successional studies. These patterns have been divided into those occurring on new substrates that have never supported life (termed primary succession) and those cases in which the community is disturbed or destroyed, but at least some soil formed by it remains as a substrate for succession (secondary succession).

The classification of succession into primary and secondary types is widely used and derives from Clements (1916):

> The whole course of succession rests upon the nature of the bare area which initiates it. We have already seen that the essential nature of a bare area is expressed in the amount and kind of water. Hence, in attempting to group naturally all the foregoing areas, i.e., from the standpoint of succession, it is necessary to recognize that water areas and rock areas constitute the two primary groups. While these are opposed in water-content and density, they agree in presenting extreme conditions in which development is necessarily slow and of long duration. The denudation of either area in the course of succession results in the sudden reappearance of earlier conditions, which cause the repetition of certain stages. If denudation consists of the destruction of the vegetation alone, the soil factors are changed relatively little. The sere thus initiated is relatively short, consisting of fewer stages and reaching the climax in a short time. If the soil is much disturbed, however, the conditions produced approach much nearer the original extreme, and the resulting sere is correspondingly longer and more complex. The degree of disturbance may be so great as to bring back the original extreme conditions, in which case the normal course of development is repeated. This amounts to the production of a new area, both with respect to the extreme condition and the lack of germules. Hence, all bare areas fall into a second basic grouping into primary and secondary areas. Primary bare areas present extreme conditions as to water-content, possess no viable germules of other than pioneer species, require long-continued reaction before they are ready for climax states, and hence give rise to long and complex seres. Secondary bare areas present less extreme conditons, normally possess viable germules of more than one stage, often in large number, retain more or less of the preceding reactions, and consequently give rise to relatively short and simple seres. From the standpoint of succession, secondary areas are related to primary ones. In consequence, the most natural classification of all bare areas seems to be into primary and secondary with a subdivision into water, rock, and soil. [p. 60]

Since succession concerns a complex natural phenomenon there are numerous ways to view the events that occur, and depending upon one's background and scientific interests different emphasis in a classification can be placed on climate, soil, plants, animals, and so on.

Cowles (1911) represents a viewpoint quite distinct from Clements. Indeed, he felt that Clements's

> . . . classification seems not to be of fundamental value, since it separates such closely related phenomena as those of erosion and deposition, and places together such unlike things as human agencies and the subsidence of land. [p. 167]

Cowles's alternative is to distinguish between physiographic and biotic agencies causing succession. He further divides physiographic influences into regional and topographic factors. Regional successions are those responding to long-term and very slow changes in regional climates such as the post-glacial invasion of southern forms of life into boreal regions following retreat of the glacial ice. Topographical successions are associated with running water, wind, ice, gravity and volcanism, which usually involve erosion and deposition. Finally, biotic successions are those under the influence of plants and animals. Most classifications include portions of one or the other of these two schemes. The issue is never solved, of course. Pierre Dansereau (1974) recently discussed types of succession in the *Handbook of Vegetation Science.*

Henry Chandler Cowles [1869–1939] published a lengthy paper in 1899 on the sand dune vegetation in northwestern Indiana (Cowles, 1899). He was Professor of Botany at the University of Chicago and ecologists familiar with the Chicago School of Ecology know that the dunes on the southeastern shore of Lake Michigan provide an unusually dynamic environment for the study of ecology. Cowles gave us the first clear description of the sequence of vegetative successions on the dune habitat—a topic investigated by European biologists as far back as Kerner—which provided a sequence widely used as an example in teaching but also used as the basis for many other studies, up to and including the important reanalysis of dune ecology by Olson (1958). The following excerpt (Paper 2) describing dune succession was taken from Cowles's 1901 paper on the physiographic ecology of Chicago. It provides a useful summary of the original 114-page report. Note that in this very dynamic topographic environment where the slope and shape of the dunes changes relatively rapidly, there may be a variety of vegetational units growing on the dune habitat. Eventually when the moving dunes are far enough from the lake so that wind speed is lowered, vegetation may cover the sand and a forest becomes established which stabilizes the substrate. Note also that Cowles predicts that eventually this habitat might be covered by a forest similar to that growing in the region of Chicago—a mesophytic white oak, red oak, hickory type, followed by a beech-maple forest.

Sand dunes are a harsh environment for plant and animal life. A similar habitat also fruitful for the study of successional sequences is the glacial moraine. William S. Cooper, former Professor of Botany at the University of Minnesota, made a series of detailed studies of succession at Glacier Bay, Alaska, and his papers provided another major influence in successional ecology. Glacier Bay is a very interesting fiord, with a relatively well known geological history. The last glacial advance culminated a century and a half to two centuries ago, when the glacier overwhelmed the mature forest communities dominated by Sitka Spruce (*Picea sitchensis*) and hemlock (*Tsuga heterophylla* or *T. mertensiana*).

Cooper's successional studies are based on permanent quadrats and many separate observations on sites free of ice for varying periods of time (Cooper, 1923 and Paper 3). The report selected to be reprinted in this volume focuses on vegetation on glacial debris and rock surfaces. The paper is interesting for several reasons. First, Cooper divides the vegetation into three communities on the basis of physiography and species composition, yet recognizes that individuals typical of any community may appear in any other. Egler (1954) later emphasizes this point very strongly—seldom do we observe complete replacement of floras and faunas, although the physiography changes from meadow to shrub to forest. Second, Cooper shows the very strong influence of substrate and local climate on development of the vegetation. Finally, the role of plant reproduction is emphasized. For example, pioneer species usually have very mobile disseminules—spores in the case of the moss *Rhacomitrium*. In contrast, spruce reproduces vigorously by layering—a process described as involving daughter branches which leave the main trunk at ground level and then become erect and form tree-like shoots.

These two examples of primary succession can be contrasted to an example of secondary succession. One of the dominant forms of secondary succession is that occurring on abandoned farm land. In the southeastern United States, the combined impact of crop pests, urbanization and economic conditions has caused a drastic reduction in farm acreage. For example, in the state of Georgia, the percentage of the land area in harvested crop land declined from 29 percent in 1920 to 9.7 percent in 1969, according to the U.S. census. In many instances small farms were abandoned during this time and the fields reverted to natural vegetation. Crafton and Wells's study of old-field succession on the Piedmont of North Carolina (Paper 4) is one of the earliest examples of this sere. I have chosen this paper not only for its historical interest, but, more importantly, because it sets the stage for a long sequence of investigations at Duke University. Henry Oosting's 126-page report on successional communities on the North Carolina Piedmont (Oosting, 1942) was a high point in this sequence and was followed by a series of studies of the mechanisms of succession by Oosting and his students. Ecologists at the University of Georgia also carried out a series of investigations on old-field succession, but in this case focused on production, species diversity, and similar functional-structural attributes of the community (see Paper 6).

Crafton and Wells describe the normal pattern of vegetation developing upon abandoned land as a pioneer stage, a tall weed stage, followed by broomsedge, then pine, and finally, hardwood forest. They show that the initial vegetation may vary depending upon a variety of environmental and, as we will see in part three, biological factors.

As is typical of papers from this period, there are a number of terms that require definition for the modern reader. For example, *ecesis* means germination, growth and development of a plant in a new location. *Ecads* are organisms showing somatic adaptation to a specific environment. *Associes* are temporary developmental communities, while a *consocies* is a morphological part of an associes, characterized by the presence of a single dominant. These terms originate with Clements; definitions are given in Hanson's *Dictionary of Ecology* (Hanson, 1962).

Thus far, the examples have largely been concerned with terrestrial habitats. Aquatic habitats, especially bogs, ponds and small lakes, have also been studied from the successional point of view. Victór E. Shelford's 1911 paper on pond fishes (Paper 5) was chosen as an early example of this type of investigation. Several points in the paper need emphasis. Shelford [1877–1968] was a zoologist and served as Professor of Zoology at the University of Illinois. He collaborated with Henry Cowles and was strongly influenced by Cowles's ideas on succession. But Shelford went beyond Cowles in emphasizing the physiological phenomena underlying the observed successional sequence. Indeed, Shelford points out that a given species may even change physiologically or behaviorally in a series of communities of different ages.

However, the key idea that should be noted in Shelford's paper is that the organisms influence the environment in such a way as to change it and make it undesirable for the species which are present. In a sense, one set of species is thought to prepare the habitat for a new set, which replaces them. These two ideas—the impact of the biota on the environment creating a changed environment and the replacement of one set of species by another—are encountered again and again in ecological literature.

In large aquatic systems, the impact of the biota is less obvious. Nevertheless, in lakes it has been widely held that a succession may occur, with the eventual conversion of the lake into a terrestrial community. Lakes may be classified into two broad types: eutrophic and oligotrophic. These words refer to the chemical nutritive conditions of the lake waters; eutrophic lakes are rich in nutrients, while oligotrophic lakes are relatively poor. These conditions are a function of nutrient input and retention, as influenced by watershed characteristics and drainage, basin shape, depth, outflow and other factors. Eutrophic lakes are relatively shallow, with well-developed submerged beaches and rich littoral vegetation. The water contains relatively high quantities of nitrogen and phosphorus. The plankton is richly developed and the bottom sediments are anaerobic during major portions of the year. In contrast, oligotrophic lakes typically have a deep basin, with steep slopes, little beach development or littoral vegetation. The plankton are of a different type and the sediments typically are aerobic.

It has been proposed that succession in lakes proceeds from the oligotrophic form to the eutrophic. For example, Ström (1928) states:

> The natural process of the maturing of a lake is that of eutrophication. The original state of all lakes must be assumed to be oligotroph, but during the course of time there will always be a surplus of organic sediments accruing from the life processes of a lake, and the originally *oligotroph* lake is changed to a *eutroph*. The quantities of plankton, oxygen curves, and average depths are the first features to be changed; the bottom fauna the last. . . . When a lake has reached the real eutroph stage, the changes toward a greater degree of eutrophy are very speedily effected, indeed; the character of a lake can be materially changed within a generation. . . . Finally, the *eutroph* lake disappears as a lake when the littoral vegetation has gained foothold throughout its bottom, and in the very last stage it is transformed into low moor. [p. 109]

Ström points out that agricultural fertilization and sewage influences this process and that the rate of change varies greatly depending upon the sedimentation rates and erosion of the watershed of the lake. Indeed, certain lakes were originally of low depth and through geological processes have become progressively deeper and apparently have changed in the reverse direction from eutrophic to oligotrophic conditions.

This classification of lakes is useful to the modern day limnologist, but the idea of lake types being related in a succession is of less value—in modern limnology textbooks *succession* seldom appears even in the index. Rather, limnologists are concerned today with the interaction of lakes and their watersheds and the impacts of man on lake conditions.

In the Benchmark papers thus far, the method of analysis has involved comparison of a sequence of habitats that represent different ages of materials—sand dunes, glacial surfaces, and small ponds. Another method of studying succession is to follow a set of habitats over time observing the sequence of changes on one given surface. This method is more certain since no assumptions about age of habitat need to be made, but it also requires a long period of time and an area that can be protected. Eugene P. Odum, Fuller E. Callaway Professor of Ecology at the University of Georgia, utilized this latter approach in studies of succession on abandoned agricultural fields (locally called old-fields) on the United States Atomic Energy Commissions' (now Energy Research and Development Administration) Savannah River Plant. Odum began observations in 1952, and they have been continued in one form or another ever since. His paper on old-field succession (Paper 6) is noteworthy for several reasons besides the approach or methodology.

First, Odum began studies of both the structure and function of the vegetation. Structure is defined as the static features of the vegetation—species composition, biomass, and height, while function refers to kinetic features such as production, decomposition and energy flow. Odum emphasized productivity in his studies and thus developed a very different description of succession than his predecessors.

Second, Odum demonstrated that during the first six years (the so-called weed stage), structural and functional parameters showed several trends. For example, production began at a high rate, then declined and stabilized. In contrast, species numbers were low at the onset, increased and became more constant. These trends were also repeated in the next stage of succession on the same fields (the grass stage), as shown in one of the editor's papers (Golley, 1965). This was one of the first demonstrations of a possible relationship between species numbers and production of the vegetation—a topic that we will examine in a later section.

REFERENCES

Clements, F. E. 1916. *Plant Succession: An Analysis of the Development of Vegetation*. Carnegie Institution of Washington Publication No. 242.

Cooper, W. S. 1923. The Recent Ecological History of Glacier Bay, Alaska. III. Permanent Quadrats at Glacier Bay. An Initial Report on a Long-Period Study. *Ecology* **4:**355–365.

Cowles, H. C. 1899. The Ecological Relations of the Vegetation on the Sand Dunes of Lake Michigan. *Bot. Gaz.* **27:**95–117, 167–202, 281–308, 361–391.

———. 1911. The Causes of Vegetative Cycles. *Bot. Gaz.* **51:**161–183.

Dansereau, P. 1974. Types of Succession. In *Vegetation Dynamics,* ed. R. Knapp, pp. 123–135. The Hague, The Netherlands: W. Junk Publ.

Egler, F. E. 1954. Vegetation Science Concepts. I. Initial Floristic Composition; a Factor in Old-Field Vegetation Development. *Vegetatio* **4:**412–417.

Golley, F. B. 1965. Structure and Function of an Old-Friend Broomsedge Community. *Ecol. Monogr.* **35:**113–131.

Hanson, H. C. 1962. *Dictionary of Ecology*. New York: Philosophical Library.

Olson, J. S. 1958. Rates of Succession and Soil Changes on Southern Lake Michigan Dunes. *Bot. Gaz.* **119:**125–170.

Oosting, H. J. 1942. An Ecological Analysis of the Plant Communities of Piedmont, North Carolina. *Am. Midl. Nat.* **28**(1):1–126.

Ström, K. M. 1928. Recent advances in Limnology. *Linn. Soc. London Proc.* **140:**96–110.

2

Reprinted from pp. 170–177 of *Bot. Gaz.* **31**(3):145–182 (1901)

THE PHYSIOGRAPHIC ECOLOGY OF CHICAGO AND VICINITY

H. C. Cowles

[*Editor's Note:* In the original, material precedes this excerpt.]

2. THE BEACH-DUNE-SANDHILL SERIES.

A. *The beach.*—The author has previously discussed in considerable detail the dynamics of the dune societies,[25] and it will not be necessary to do more here than to summarize the chief conclusions, and add a few new data. Before long it is expected that a paper will appear giving the changes that have taken place since the first observations were made in 1896.

The beach in the Chicago area is xerophytic throughout. There is nothing analogous to the salt marshes of the Atlantic coast, nor to the hydrophytic shores farther north along Lake Michigan. The lower portion of the beach is exposed to alternate washing by the waves and desiccation in the sun, and is devoid of life. The middle beach, which is washed by winter waves, though not by those of summer, has in consequence a vegetation of xerophytic annuals, the most prominent of which is *Cakile Americana*. The upper beach is beyond present wave action, and is tenanted by biennials and perennials in addition to the annuals. *Fig. 33* shows a beach of this type, the lower beach being smooth and even, the middle beach covered with débris, while the upper beach has a scattered perennial vegetation.

The beach at the base of cliffs shows similar subdivisions, though the zones are much narrower as a rule. The vegetation, too, is much the same, though some forms, as Strophostyles, have not been seen as yet on the beaches of the dune district. At the foot of cliffs there often occur alluvial fans of sand, which have been deposited by the torrents during and following rain storms. These fans have a comparatively rich vegetation and species sometimes occur here that are not found elsewhere on the beach.

[25] COWLES, H. C.: The ecological relations of the vegetation on the sand dunes of Lake Michigan. BOT. GAZ. **27**: 95–117, 167–202, 281–308, 361–391. 1899.

B. *The embryonic or stationary beach dunes.*—Wherever plants occur on a beach that is swept by sand-laden winds, deposition of sand must take place, since the plants offer obstacles to the progress of the wind. If these plants are extreme xerophytes and are able to endure covering or uncovering without injury, they may cause the formation of beach dunes. Among the

FIG. 33.—Beach at Dune park, showing the smooth and naked lower beach, the middle beach with its line ot débris, the upper beach with scattered shrubs, and the dunes.

dune-forming plants of this type are *Ammophila arundinacea*, *Salix glaucophylla* and *S. adenophylla*, *Prunus pumila*, and *Populus monilifera*. The shapes of these beach dunes vary with the characteristics of these dune-forming plants. Ammophila dunes are extensive but low, because of strong horizontal rhizome propagation. Prunus and Populus dunes are smaller but higher, because of the relative lack of horizontal propagation and the presence of great vertical growth capacity. Dunes are formed more slowly in protected places, and here the dune-forming species may be plants that are ill adapted to the severest beach conditions, such as the creeping juniper.

C. *The active or wandering dunes. The dune complex.*—The stationary embryonic dunes on the beach begin to wander as soon as the conditions become too severe for the dune-forming plants. The first result of this change is seen in the reshaping of the dune to correspond with the contour of a purely wind-made form. The rapidity of this process is largely determined by the success or failure of the dune-formers as dune-holders. The best dune-holders are Calamagrostis, Ammophila, and Prunus.

There are all gradations between a simple moving dune and a moving landscape; the latter may be called a dune-complex. The complex is a restless maze, advancing as a whole in one direction, but with individual portions advancing in all directions. It shows all stages of dune development and is forever changing. The windward slopes are gentle and are furrowed by the wind, as it sweeps along; the lee slopes are much steeper. The only plant that flourishes everywhere on the complex is the succulent annual, *Corispermum hyssopifolium*, although *Populus monilifera* is frequent. The scanty flora is not due to the lack of water in the soil, but to the instability of the soil and to the xerophytic air.

The influence of an encroaching dune upon a preexisting flora varies with the rate of advance, the height of the dune above the country on which it encroaches, and the nature of the vegetation. The burial of forests is a common phenomenon. The dominant forest trees in the path of advancing dunes are *Pinus Banksiana* and *Quercus coccinea tinctoria*. These trees are destroyed long before they are completely buried. The dead trees may be uncovered later, as the dune passes on beyond.

In the Dune park region there are a number of swamps upon which dunes are advancing. While most of the vegetation is destroyed at once, *Salix glaucophylla*, *S. adenophylla*, and *Cornus stolonifera* are able to adapt themselves to the new conditions, by elongating their stems and sending out roots from the buried portions. Thus hydrophytic shrubs are better able to meet the dune's advance successfully than any other plants. The water relations of these plants, however, are not rapidly altered in the new conditions. It may be, too, that these shrubs have adapted themselves to an essentially xerophytic life through living in undrained swamps. Again it may be true that inhabitants of undrained swamps are better able to withstand a partial burial than are other plants.

Vegetation appears to be unable to capture a rapidly moving dune. While many plants can grow even on rapidly advancing slopes, they do not succeed in stopping the dune. The movement of a dune is checked chiefly by a decrease in the available wind energy, due to increasing distance from the lake or to barriers. A slowly advancing slope is soon captured by plants, because they have a power of vertical growth greater than the vertical component of advance. Vegetation commonly gets its first foothold at the base of lee slopes about the outer margin of the complex, because of soil moisture and protection from the wind. The plants tend to creep up the slopes by vegetative propagation. Antecedent and subsequent vegetation work together toward the common end. Where there is no antecedent vegetation, Ammophila and other herbs first appear, and then a dense shrub growth of Cornus, Salix, *Vitis cordifolia*, and *Prunus Virginiana*. Capture may also begin within the complex, especially in protected depressions, where *Salix longifolia* is often abundant.

D. *The established dunes.*— No order of succession in this entire region is so hard to decipher as is that of the established dunes. There are at least three types of these dunes so far as the vegetation is concerned, and it is not yet possible to figure out their relationships. The continuation of the conditions as outlined in the preceding paragraph results in a forest society on the lee slope, in which is found the basswood, together with a most remarkable collection of mesophytic trees, shrubs, and climbers, which have developed xerophytic structures. These dunes are evidently but recently established, as is shown by the absence of a vegetation carpet; furthermore the slopes are almost always steep.

Again, there are forest societies in which the pines dominate, either *Pinus Banksiana* or *P. Strobus*. These arise from a heath, composed in the main of Arctostaphylos and Juniperus. The heath appears to originate on fossil beaches or on secondary embryonic dunes or other places where the danger of burial is not great. It will be noted that both the heath and the pine forest are dominated by evergreens. These societies commonly occur near the lake or on lakeward slopes, which are northern slopes as well. On these coniferous dune slopes there is to be found another notable collection of northern plants, resembling ecologically the peat bog plants already mentioned. Heaths and coniferous forests also occur on sterile barrens and in depressions where the conditions are unfavorable for deciduous forests. A slight change in the physical conditions may bring about the rejuvenation of the coniferous dunes, because of their exposed situation. This rejuvenation commonly begins by the formation of a wind sweep, and the vegetation on either hand is forced to succumb to sand-blast action and gravity.

A third type of established dune is that in which the oaks predominate, and especially *Quercus coccinea tinctoria*. The oak dunes are more common inland and on southern slopes. Probably the oaks follow the pines, but the evidence on which this is based is not voluminous. The pines certainly have a wider range of habitat than the oaks, occurring in wetter and in drier soil and also in more exposed situations. The mutual relations of the pines and oaks are certainly interesting and deserve some very careful study. Pine forests prevail on the north or lakeward slopes and oak forests on the south or inland slopes. With the pines are other northern evergreen forms, such as Arctostaphylos, while with the oaks are Opuntia, Euphorbia, and other more southern types. The density of the vegetation on the north side is also in contrast with the sparser and more open vegetation of the south side. The cause for this radical difference on the two slopes is doubtless complex, but it is obvious that the north slope has greater moisture, shade, and cold, and probably more wind. Which of these is the more important is not certain, but the presence of the northern species seems in favor of cold or wind as the chief factor.

There are a number of interesting sand hills and ridges at

some distance from the lake. Some of these are fifteen miles from the present lake shore, while others are found at various intervals nearer and nearer the lake. It has been found that these can be grouped for the most part into three series, representing three beach lines of Lake Chicago, as the glacial extension of Lake Michigan has been called. The upper and oldest

FIG. 34.— Portion of an ancient beach line (Calumet beach) at Summit, showing the characteristic oak vegetation, in this case chiefly bur oaks (*Quercus macrocarpa*).

of these ridges has been termed the Glenwood beach, the intermediate ridge the Calumet beach, and the lower and younger ridge the Tolleston beach. The geographic relations of these beaches is well discussed by Leverett[26] and also by Salisbury and Alden,[27] and nothing need be said here except as to the vegetation. In general these ridges and hills have a xerophytic forest flora, dominated by the bur, black, and white oaks (*Quercus macrocarpa*, *Q. coccinea tinctoria*, *Q. alba*). The proportions between these trees varies strikingly, though the bur or

[26] *Op. cit.* 55–85. [27] *Op. cit.* 31–51.

black oak is usually the chief character tree. No satisfactory reason can yet be given for these variations, though the bur oak appears to be more abundant on the lower and less drained ridges, while the black oak is more abundant on the higher ridges. The shrub undergrowth is commonly sparse, and the most frequent members of this stratum are the hazel (*Corylus*

FIG. 35.—Portion of an ancient beach (Glenwood beach) near Thornton. The trees here are chiefly black oaks (*Quercus coccinea tinctoria*); the beach is higher, and the trees more luxuriant than usual.

Americana), Rosa, the New Jersey tea (*Ceanothus Americanus*); *Salix humilis*, the low blueberry (*Vaccinium Pennsylvanicum*), and the huckleberry (*Gaylussacia resinosa*). Among the commoner herbs are *Silene stellata*, *Antennaria plantaginifolia*, *Heuchera hispida*, *Rumex Acetosella*, *Carex Pennsylvanica*, *Potentilla argentea*, *Poa compressa*, *Pteris aquilina*, *Ceratodon purpureus*. In open places there are often almost pure growths of Poa or Potentilla. *Figs. 34* and *35* show portions of these ancient beaches in which the oaks dominate; *fig. 34* shows, perhaps, the more common condition, *i. e.*, a rather low beach with a sparse tree growth.

The future of the vegetation on the established dunes and beaches is somewhat problematical. From analogy with other plant societies in this region, and from established dunes in Michigan, we should expect a mesophytic forest, probably of the white oak-red oak-hickory type at first and then followed by a beech-maple forest. There are evidences that some such changes are now taking place. On many of the oak dunes, especially where protected from exposure, there is already a considerable accumulation of humus. Herbaceous ravine mesophytes like Hepatica, Arisaema, and Trillium are already present, and with them mesophytic shrubs and trees, including the sugar maple itself, though the beech has not been found on the dunes of our area, as it has in Michigan. One might expect that the flora of the older Glenwood beach would have advanced more toward the mesophytic stage than has the flora of the younger Tolleston beach. Such, indeed, seems to be the case, especially at Glenwood, where the white oaks are more numerous, and the black oaks much larger and more luxuriant. The humus is richer and most things look as if the age of this beach were notably greater than that of the Calumet or Tolleston beaches. This subject, however, needs much further investigation. In any event, one character of the sand hill stands out in bold relief, viz., its great resistance to physiographic change. Not only is its erosion slower than that of the clay hill, but the advance of its vegetation is vastly slower at all points along the line. The slowness of humus accumulation accounts for this, perhaps, more than all else.

III. Summary and conclusion.

In the present paper the author has endeavored to show the need for a classification of plant societies which shall form a logical and connected whole. Warming's classification, based on the water content of the soil, is doubtless the best possible classification, if but one factor is used. Graebner's classification, based on soil characteristics, includes the advantages of Warming's scheme, and adds desirable new features.

[*Editor's Note:* Material has been omitted at this point.]

3

Reprinted from *Ecology* **4**(3):223–246 (1923)

THE RECENT ECOLOGICAL HISTORY OF GLACIER BAY, ALASKA: II. THE PRESENT VEGETATION CYCLE

William Skinner Cooper

University of Minnesota

Contents

I. Introduction

In the first paper of this series, after presenting an introductory section dealing with the climate, topography, geology, and especially the glacial history of the Glacier Bay region, I described the remnants of an older vegetation cycle associated with a period of shrunken ice fields preceding the last glacial advance. The shores of the bay at that time, and the surrounding mountains, were clothed with a forest which had attained almost, and in some places, quite to the climax state, and which was identical in every way with the forest of similar habitat and stage of development in southeastern Alaska today. The peaceful order of its life was suddenly disturbed by the invasion of ice tongues which descended from the higher mountains, coalesced into broad piedmont expanses, and finally united to form one huge glacier which moved down the main bay almost to its mouth. The forest upon the upper mountain sides was swept clean away, and that upon the lower slopes and the lowlands was buried beneath hundreds of feet of sediments deposited by glacial streams during the advance. Over all poured the great ice flood, thousands of feet deep. Along the mountainous shores of the lower bay, where the ice did not rise so high, an interrupted strip of forest between the glacier surface and timberline, steadily widening southward, escaped destruction, and remains to this day a sample of the growth which once clothed all the slopes.

The ice margin began to recede, and, as the masses shrank, an ever-widening expanse of bare ground was laid open to renewed invasion by plants. The new vegetation cycle, thus initiated in the region formerly covered by the ice, and now in active process of development, forms the subject-matter of this paper.

The present cycle is closely linked with the old. The pioneer and shrub vegetation, though it must have been largely destroyed by the advance of the ice into the dense forest, nevertheless persisted in places, as is suggested by the series of incipient successions associated with the accumulation of the gravels (Cooper, '23, p. 122). When the retreat began, disseminules from the survivors existing along the edge of the ice began to repopulate the fresh surfaces with pioneer species and shrubs. An additional source of the former class, more important later in the retreat, was the region above timberline, from which seeds, fruits, and whole plants would be easily carried down by water and avalanching snow. The mature forest above and without the ice-covered area furnished invaders for the reestablishment of the climax. As an example of this, we have already noted (*loc. cit.*, pp. 123–125) the creeping down of the young forest from the sharply cut lower limit of the old. A final link of a different nature is of no slight importance—the organic matter in the gravels, derived from the plants of the ancient forest and now available for utilization by those of the newer cycle.

For the physical background of the present study the reader is referred to the first paper of the series, where the topography, geology, and glacial history are outlined. The plan of the present paper is to describe first the vegetation of the present day, or, borrowing a word from the geologists, the "areal" ecology of the region. The main portion deals with the process of revegetation of the denuded area, and this is prefaced by an analysis of the environment in which the pioneers find themselves.

II. **Areal Ecology**

Such a locality as Glacier Bay, with its constant exposure of fresh surfaces for plant invasion, emphasizes the importance of development as the fundamental fact in vegetation study, and the indefiniteness of vegetation units where change is at all rapid. It is useful, however, and even essential, to make a preliminary survey of the plant population as it exists today before passing to a dynamic study. Certain few communities do stand out, but their distinctness is more apparent than real, being due to increasing dominance of a new growth-form, as when the low-growing pioneers give place to tall shrubs, and these in turn to trees. We may thus, for the sake of a vivid presentation, divide our vegetation into three communities, or in dynamic terms, stages: the pioneer community, the willow-alder thicket, and the conifer forest; with the understanding that these are mere cross-sections of a continuous stream of development, made with reference to certain points in this stream where change is most evident.

The map in the first paper (*loc. cit.*, fig. 1) gives the distribution of the communities as they existed along the coasts of Glacier Bay in the year 1921. In the light of the rapid rate of vegetational change that has prevailed for the last century, we can not expect this map to remain true for many years, but it will serve as a basis for comparison with the results of future observations. Coast chart 8306 of the U. S. Coast and Geodetic Survey served as a base.

FIG. 1. Station 19, with Muir ice cliff a mile and a half distant. Shelter containing hydro-thermograph, and atmometer to the right of it.

The 1,000-foot contours put in by the makers of the map are approximate only, but do give a generally true picture of the topography. The mapping of the vegetation was done from the water and from points of vantage on the shores, and the results must be accepted with the reservations inseparable from such a method. Their reasonable accuracy, however, is assured by the conveniently open and abrupt character of most of the bordering slopes. Certain low areas, especially west of the entrance, were invisible from all stations occupied, and therefore many boundaries in this region have been necessarily left indefinite.

The Pioneer Community

This type (figs. 1 and 3) is altogether too scattered and indefinite to map. It occurs on all slopes of the bay, not occupied by other communities, where present conditions of the habitat permit its existence. It varies in density from a state of widely scattered individuals to a fairly close cover. The majority of the plants are of low stature by habit; the trees and high shrubs, erect and often tall in more favorable habitats, are here more or less depressed, or often as thoroughly prostrate as the more characteristic forms. A classified list of the principal species is given below with the percentage of the number of localities studied (twenty-two) in which they were found. Forty-four additional species were noted in one or two localities.

	Percent of Localities where Found
Lichen	
Stereocaulon alpinum Th. Fr.	18.2
Mosses	
Rhacomitrium canescens Brid.	13.6
Rhacomitrium lanuginosum (Hedw.) Brid.	9.1
Perennial Herbs	
Epilobium latifolium L. Broad-leaved willow-herb	91.0
Dryas drummondii Rich. Dryas	72.8
Equisetum variegatum Schleich. Variegated horsetail	63.7
Equisetum arvense L. Common horsetail	27.3
Poa alpina L.	27.3
Limnorchis sp. Rein-orchis	27.3
Saxifraga oppositifolia L. Purple saxifrage	22.7
Euphrasia mollis (Ledeb.) Wettst. Eyebright	18.2
Trisetum spicatum (L.) Richter	18.2
Sagina saginoides (L.) Britton. Pearlwort	18.2
Pyrola secunda L. Shinleaf	18.2
Silene acaulis L. Alpine pink	13.6
Prostrate Shrubs	
Salix arctica Pall. Arctic willow	86.4
Salix stolonifera Coville	45.4
Salix reticulata L.	31.9
Arctous alpina (L.) Niedenzu. Alpine bear-berry	13.6
Erect Shrubs	
Salix sitchensis Sans. Sitka willow	86.4
Salix alaxensis (And.) Coville. Alaska willow	72.8
Salix barclayi And.	54.6
Alnus tenuifolia Nutt. Alder	27.6
Salix commutata Bebb	22.7
Salix glauca L.	22.7
Salix scouleriana Barrett	13.6
Lepargyraea canadensis (L.) Greene. Buffalo-berry	13.6

TREES

Populus trichocarpa T. and G. Cottonwood.................... 59.2
Picea sitchensis Carr. Sitka spruce.......................... 22.7

THE WILLOW-ALDER COMMUNITY

The distribution of this type is indicated on the map (*loc. cit.*, fig. 1), and the general appearance illustrated in figures 6 and 7. In general, it occupies the slopes around the middle portion of the bay, the lower part being inhabited by the climax forest and the upper portions in an incomplete manner by the pioneers. Traveling southward in the main bay, or outward in its branches, we first encounter the willow-alder thickets as isolated patches, mainly upon the terrace gravels and ground moraine, but also upon ledges of the steep mountain slopes. The edges are frequently sharply defined, because the favorable spots are definitely bounded, and sometimes because of cutting away of the gravels by erosion. As one goes farther down the bay, the patches become larger and finally coalesce into a dense cover that extends from the water's edge to the normal timberline at an altitude of two thousand feet and more.

The alder, *Alnus tenuifolia,* is nearly everywhere dominant, and is easily recognizable from its bright olive green tone. In the mature thicket its height averages twenty feet. Two willows are always present in some abundance, *Salix alaxensis,* the more important, and *S. sitchensis.* In stature they equal the alder, and appear from a distance as gray-green patches in the olive. The cottonwood, *Populus trichocarpa,* is commonly present, not numerous in individuals, but conspicuous because of its greater height.

The undergrowth of the alder thicket includes such mesophytes as the ferns, *Aspidium spinulosum* (O. F. Müller) Sw., *Polystichum braunii* (Spenner) Fée, and *Asplenium filix-foemina* (L.) Bernh.; ground pine, *Lycopodium selago* L., coral-root, *Corallorhiza mertensiana* Bong., and colt's foot, *Petasites frigida* (L.) Fries; and the mosses, *Rhytidiadelphus triquetrus* (L.) Warnst. and *R. squarrosus* (L.) Warnst. Along the shores of the main bay the alder thickets are overtopped by an occasional young spruce (fig. 7). Southward the spruces become more and more abundant and of continually greater stature, until the alder thicket merges insensibly into the third community, the conifer forest.

THE CONIFER FOREST

The young conifer forest which occupies the older portions of the ice-invaded region is a nearly pure stand of Sitka spruce, *Picea sitchensis.* It grades into the alder thicket of the central region and extends southward along both shores of the bay to the entrance. It covers the Beardslee Islands and the broad foreland east of the bay in solid formation. The trees average fifty to sixty feet in height and stand densely together. Their symmetrical form, the presence of leafy branches almost to the ground, and the absence of standing dead trunks testify to the youth of the forest. Ring counts confirm it, the oldest tree in a number examined at Bear Track Cove (*loc. cit.*, fig. 15) in 1916 being but seventy-one years of age. Occasional slender, narrow-topped cottonwoods accompany the spruces, and infrequent individuals of coast hemlock, *Tsuga heterophylla* Sarg., and mountain hemlock, *Tsuga mertensiana* Carr., also occur, usually in the understory. Ancient alders and willows, *Salix sitchensis* and *S. alaxensis*, obviously having a hard time of it, sprawl beneath the trees in the more open places. Another group of shrubs is more at home: salmon-berry, *Rubus spectabilis* Pursh; red-berried elder, *Sambucus racemosa* L.; devil's club, *Fatsia horrida* (Sm.) B. and H.; and blue currant, *Ribes bracteosum* Dougl. The herbs are those of the great Alaska forest: *Aspidium spinulosum, Polystichum braunii, Asplenium filix-foemina;* twayblade, *Ophrys cordata* L.; baneberry, *Actaea arguta* Nutt.; shinleaf, *Pyrola secunda;* and one-flowered shinleaf, *Moneses uniflora* (L.) Gray. The most important element in the lower vegetation is the moss carpet, which covers the gravel and boulders many inches deep. The bulk of it is made up of four species: *Hylocomium proliferum* (L.) Lindb., *Rhytidiadelphus triquetrus* (L.) Warnst., *R. loreus* (L.) Warnst., and *R. squarrosus* (L.) Warnst. *Mnium insigne* Mitt., *Dicranum rugosum* (Hoffm.) Brid., *D. scoparium* (L.) Hedw., *Drepanocladus aduncus* (L.) Warnst., and *Brachythecium albicans occidentale* R. & C. are common, and *Ulota crispa* Brid., *Mnium glabrescens* Kindb., *Antitrichia curtipendula gigantea* Sull. & Lesq., *Pseudoleskea stenophylla* Ren. & Card., and *Plagiothecium piliferum* (Sw.) B. S. & G. are rather partial to trunks and branches, especially of defunct and dying alders and willows. This list makes no pretensions to completeness, the bryophyte flora being exceedingly rich. In brief, this forest differs from the average coastal climax of southeastern Alaska only in the youthfulness of the spruces and the small proportion of hemlock.

III. **The Present Vegetation Cycle**

The Primitive Habitats

The ice, in its retreat, is exposing three types of land surface: bare rock, moraine, and glacio-fluvial deposits. The rock surfaces react in varying degree to the attacks of the elements. The argillites and slates furnish with little delay a favorable substratum for the pioneer plants, but continued weathering aided by gravity destroys many of the early arrivals, and on very steep slopes notably retards the progress of succession by constantly producing fresh surfaces. Where the slope is reasonably gentle, however, this type of bedrock affords the best physical conditions of any for the rapid establishment of vegetation. The diorites, quickly breaking into large blocks and slabs, are decidedly less favorable. The limestone and marble surfaces are worst of all, retaining their smoothly glaciated contours indefinitely, weathering by solution rather than disintegration. The amount of vegetation upon similar slopes of these three types, of equal subaerial age, is plainly proportional to their comparative rate of weathering.

True glacial till, unmodified by water, is scarce except near the mouth of the bay; elsewhere it either was sparingly deposited during the rapid retreat of the ice or has been removed by postglacial stream activity. A thin layer of ground moraine is seen to cover the interglacial gravels when they emerge from beneath the ice (*loc. cit.,* fig. 4); a sparse scattering of the same material partially conceals many rock surfaces (fig. 3).

The gravels which are so important around Muir Inlet have been described in the first paper. They vary greatly in character, but as a class present no constant efficient differences from the till, so far as relation to vegetation is concerned. It is therefore possible to classify the primitive bare areas into two principal types: rock surfaces and depositional accumulations. Another of opposite nature must be added: the pools which occur here and there, especially where the ground moraine is relatively important.

Moisture in the large can not be considered a critical factor in southeastern Alaska, and yet there are places where it is manifestly a limiting influence. The primitive soils of Glacier Bay differ greatly in dominant size of particles; their water-retaining capacities differ accordingly; and there are periods during the growing season when the coarser superficial layers of soil become relatively dry. For well-established plants this is of little importance, since the deeper layers must always retain an adequate supply, but for recently germinated seedlings the water deficit must frequently prove fatal. The effect of such differences is evident in the denser covering of pioneer vegetation upon the finer textured soils, the subaerial age being the same. Soil temperature must be exceedingly low in freshly exposed areas, and the progressive course of amelioration as the glacier recedes should parallel that of the atmospheric temperature about to be presented. The general lack of organic matter is an extremely important characteristic of the freshly exposed soils.

As to the atmospheric factors of the primitive habitat, I have instrumental data to present, scanty but significant. During my visit of 1921 a Friez hygro-thermograph and a Livingston white spherical porous-cup atmometer were operated for a period of about ten days at station 19 (fig. 1), which is a low knob of rock littered with ground moraine, one and one-half miles south of the Muir Glacier front and half that distance from the Plateau Glacier cliff to the west. It was uncovered about 1907, and the melting edge of the Cushing Plateau is still but a quarter of a mile distant. It is, in fact, nearly surrounded by ice, for the masses of the old Muir field extend down both sides of the Inlet for several miles. In every way it is a typical sample of ground very recently laid open to invasion by plants. The present vegetation consists of a few scattered mosses, perennial herbs, and creeping willows.

TABLE 1. *Atmospheric conditions near the Muir ice front (station 19)*

	Clear				Partly Cloudy			Rainy		
August..........	21	22	23	24	25	26	27	28	29	30
Temperature, Fahr.										
Mean daily.....		44.4°	42.4°	39.4°	38.3°	39.0°	39.6°	36.7°	37.7°	
Maximum daily.	53	56	52	48	43	46	47	41	42	39
Minimum daily.		35	36	34	35	35	35	35	35	33
Relative humidity										
Mean daily.....		79.2%	81.9%	88.6%	94.3%	88.6%	92.8%	100%	100%	
Maximum daily.		93	93	99	98	100	94	100	100	100
Minimum daily.	77	65	71	78	85	76	82	100	100	

	Mean Temperature	Mean Relative Humidity		Average Daily Evaporation
8 days...............	39.7°	90.7%	Aug. 21–30............	11.93 c.c.
3 clear days......	42.1°	83.2	Aug. 21–27............ (4 clear, 2 cloudy)	16.53
3 cloudy days.....	38.9°	91.8		
2 rainy days......	37.2°	100.0	Aug. 27–30............ (1 cloudy, 2 rainy)	2.73

The instrumental data obtained at station 19 are brought together in Table 1. The period August 22–29, 1921, provided a very fair sample of coastal Alaska weather. There were three clear, three partly cloudy, and two rainy days. The mean temperature for the eight complete days was 39.7° F. For three clear days it was 42.1°, and for two rainy days 37.2°—a difference of slight importance. The absolute maximum was 56° and the absolute minimum 33°. The extreme lowness of the temperature for a period of at least average sunniness is noteworthy. I have attempted to translate these meager data into the terms of Livingston's physiological temperature indices (Livingston, '16). Assuming that the eight-day period studied represents an average for the whole frostless season, and that the frostless season coincides in length with that of Sitka (about 160 days), we obtain a physiological summation index of 145. Other Pacific coast stations, according to Livingston, have the following indices:

Seattle	3692	San Francisco...............	4122
Portland....................	4780	Los Angeles.................	8451

With so minute an efficiency it is remarkable that any growth is possible.

Atmospheric moisture conditions are, on the other hand, extremely favorable. The mean relative humidity for eight days was 90.7 per cent, for three clear days 83.2 per cent, and during the rainy period continuously at the saturation point. The average daily evaporation for ten days was 11.93 c.c., and the difference in rate during the dry and wet weather was striking—16.53 c.c. and 2.73 c.c.

Considering the two factors in combination, it is clear that the close proximity of a large body of ice produces temperature conditions so extremely unfavorable that the best moisture conditions can hardly compensate. Nevertheless, in various places in Alaska there are plain indications that low temperature is by no means absolutely prohibitive. Vegetation, even climax forest, is frequently found extending to the very edge of the ice, and even beyond it where it is stagnant and thickly mantled with moraine.

It was my intention to operate two sets of instruments for the comparison of a locality close to the ice with one more remote. Because of the non-arrival of one of the hygro-thermographs it was necessary to substitute a pair of maximum-minimum thermometers at the second locality. These, with an atmometer, were in operation at station 4, on Strawberry Island (fig. 2), from

Fig. 2. Station 4, on Strawberry Island. Maximum-minimum thermometers and atmometer in position.

August 20 to August 31. Strawberry Island, an outlier of the Beardslee group near the mouth of the bay, is distant thirty-one miles from the tidal front of the Muir Glacier. It is composed of terrace gravels, and the vegetation is made up mainly of low willows, with infrequent alders and invading young spruces. The instrument station was near the northern edge of the island, about one hundred feet above the water. A comparison of atmospheric conditions at stations 19 and 4 is presented in Table 2.

TABLE 2. *Comparison of conditions near the ice front with those in the lower bay*

	Station 19 (Muir Glacier) Aug. 21-30	Station 4 (Strawberry-Island) Aug. 20-31
Maximum temperature	56°	67°
Minimum temperature	33°	58°
Average daily evaporation	11.93 c.c.	6.45 c.c.

Differences in both temperature and evaporation are striking. The minimum temperature on Strawberry Island is higher than the maximum at the Muir Glacier. The small range at the former station is noteworthy. The evaporation rate at the Muir is nearly double that upon Strawberry Island. At first thought this seems surprising, but a plausible explanation suggests itself: that the constant air movement down the glacier and Muir Inlet carries away the moisture from around the porous cup as fast as it enters the atmosphere, while the stillness so characteristic of the lower bay permits it to accumulate. It is quite evident that conditions both as to temperature and moisture are notably more favorable for vegetation upon Strawberry Island than close to the Muir Glacier. The fact may be stated in another way, more suitable to a developmental study: the environmental conditions for plants, unfavorable and almost prohibitive while the ice sheet is close by, become notably ameliorated as the ice edge retreats from the vicinity, both as to atmospheric (and therefore soil) temperature and evaporation.

The Process of Development

The rate and manner of invasion are determined by three sets of factors: the number and mobility of available disseminules, the rate of exposure of new territory by retreat of the ice, and the character of the habitat thus provided.

The most important pioneers have very mobile disseminules: spores in the case of *Rhacomitrium* and *Equisetum,* plumed seeds or fruits in *Epilobium, Dryas,* and the willows. They all have the power to travel far, but it will still be true that the greatest numbers of seeds and spores will fall in the vicinity of the parent plants.

The prospective habitat for the early invaders has just been characterized: temperature, both of soil and atmosphere, exceedingly unfavorable; soil moisture subject to depletion in the surface layers for short periods, and evaporation low.

The manner of invasion is by isolated outposts—here and there a lone plant fighting for its life. Those that survive become centers of colonization, new invaders continue to arrive, and so in time a solid cover of vegetation is established. If the rate of glacial retreat be slow, the invaders will from the first be more closely placed, being nearer to the parent plants; consolidation will proceed with relative rapidity, and the advance will partake more or less of the nature of mass extension, its edge closely following the receding ice front. If the glacier retreat rapidly, disseminules in smaller numbers will be distributed over a vaster area, so that the general effect for miles will be of utter barrenness, and this appearance will persist for a long period, since consolidation must go on with corresponding slowness. In any case the rate of vegetational development in a given spot will gradually accelerate, because of the amelioration of temperature conditions as the ice recedes, and the continual accumulation of organic matter due to the activities of the plants themselves.

As to the species which are active in invasion, almost any plant of the region may be found among the vanguard, even occasional individuals of the climax dominants. This naturally results in pronounced telescoping of stages. It is, however, true that the species differ in ability to withstand severe conditions, and thus several groups may be distinguished whose successive arrivals point off weakly marked stages in the process.

At this point it is necessary to distinguish three lines of development, correlated with three varieties of primitive habitat: the rock surfaces, the moraines and terrace gravels, and the ponds.

Development on Rock Surfaces

Vegetational development upon rock surfaces is similar throughout northern regions. Its rapidity depends upon the slope of the surface, its initial character, and its response to the continued assaults of the agencies of weathering, including the plants themselves. It has already been stated that upon

FIG. 3. Glaciated rock surfaces and thin ground moraine with pioneer vegetation. Station 26.

the slates and argillites development may be rapid. The rock breaks down so quickly that where the degree of slope allows the products of disintegration to accumulate, conditions soon approximate those of the moraines and gravels. Upon steep slopes the removal of the loose materials by gravity results in repeated reinvasion by pioneers and their quick destruction, and this state continues until physiographic stability is attained, after which the normal progress of succession is possible. Upon the limestone development is slowest, since the smooth, rounded surfaces resulting from glaciation are preserved indefinitely.

If the slope permit, mosses start wherever a foothold is possible—behind small débris, in shallow cavities, and similar situations. The crevices, however, are the all-important starting points for the pioneers. The most abundant crevice herbs in the localities studied were the grass, *Poa hispidula* Vasey; alum-root, *Heuchera glabra* Willd.; purple saxifrage, *Saxifraga oppositifolia;* dryas, *Dryas drummondii;* broad-leaved willow-herb, *Epilobium latifolium;* eyebright, *Euphrasia mollis;* yarrow, *Achillea borealis* Bong.; and pearly everlasting, *Anaphalis margaritacea* Benth. A complete list would probably include a large proportion of the plants of the region, for the crevice, though limited in extent, is a relatively favorable habitat. Growing with the herbs, and vastly more important than they, are found creeping shrubs: the willows, *Salix arctica* and *S. reticulata,* and the bear-berry, *Arctostaphylos uva-ursi* Spreng. These, anchored in the crevices, form dense mats which spread over the adjoining rock surfaces, merging with the moss patches and forming the foundation of the thick layer of organic soil that will eventually rest upon the still uneroded rock surface. Another class of pioneers comprises the erect shrubs, which are at first confined to the crevices, but which soon establish themselves upon the vegetation mat itself. These are *Salix barclayi, S. alaxensis* and the other willows, and alder. The trees are not far behind, for cottonwood and Sitka spruce are found in the crevices with the pioneers.

The more favorable spots, such as level or depressed areas, or surfaces with many crevices, soon become covered with a luxuriant turf-like growth, in which all the species listed above are represented, and many others as well: the grasses, *Trisetum spicatum, Poa alpina,* and *P. laxa* Haenke; stonecrop, *Sedum roseum* L., *Sibbaldia procumbens* L., *Hedysarum americanum* (Michx.) Britton, *Astragalus alpinus* L.; shinleaf, *Pyrola secunda, P. asarifolia* Michx.; and alpine bear-berry, *Arctous alpina.* By increase of the shrubby species such areas are rapidly converted into thickets in which alder and willows are dominant, while the adjacent steeper and smoother surfaces are still bare of plants. Such is the condition today upon the limestone islands of the lower bay, Drake and Willoughby. The spruces, thickly scattered upon the meadow and thicket areas, indicate the future course of development.

Enormous stretches of mountain slope throughout southeastern Alaska have attained to the climax through such a process of development as I have described. The mass of humus which supports the present forest vegetation rests upon a glaciated rock surface comparatively unmodified, and only here and there are cliff faces still visible. The steepness of certain mountain slopes which bear full developed climax forest is one of the remarkable features of the region.

This and the following lines of succession do not differ materially in the species involved nor in the general sequence of stages. The distinction lies in the necessity, in the case of the rock surfaces, of building up an entire soil which is wholly organic, contrasted with the immediate availability of a soil (inorganic) in the case of the moraines and gravels. There is further contrast in the relative uniformity in rate of development upon the gravels as against the pronounced unevenness upon the rock surfaces, due to varying degrees of steepness and to alternation of creviced and uncreviced areas.

Development upon Moraine and Gravels

This series is of particular interest at Glacier Bay because of the unusual extent of the gravel deposits, which, moreover, provide optimum conditions for invasion, so far as physical soil character is concerned. The rate of development is limited only by the rate of glacial retreat, with its effect upon temperature, and on the ability of the plants to accumulate humus. Telescoping is here especially pronounced, since the initial soil is relatively so favorable.

For the honor of being the hardiest pioneer of all there are three candidates: the closely related mosses, *Rhacomitrium canescens* and *R. lanuginosum,* and the perennial herb, *Epilobium latifolium.* These are present wherever there are any plants at all, and in the rawest and coldest localities they are the only vegetation. Station 29, when studied in 1916, was as uncongenial a situation as can be imagined: uncovered since 1907 and three miles from the present terminus of the Grand Pacific Glacier; at the bottom of a deep trough surrounded by ice-laden peaks. The soil consisted of a thin layer of detritus resting upon a polished rock surface. A careful search revealed some scattered protonema of a moss, probably *Rhacomitrium,* with a few small sporophytes, and about twenty diminutive plants of *Epilobium,* mostly starved seedlings less than an inch high. One branched specimen was four inches high, with no flowers, and there were several dead plants of similar size. There are many square miles of mountain slopes and gravel terraces surrounding the upper bay and Muir Inlet which support vegetation little better in quality and variety.

Rhacomitrium and *Epilobium* retain a position of importance as long as the vegetation remains scattered. Their luxuriance is a fair indicator of the quality of the habitat. I have a series of specimens of *Rhacomitrium canescens* ranging from the short-stemmed often silvery form that covers the stones of the primitive habitat with a thin layer to the type of more favorable situations that makes dense green tufts composed of stems as much as four inches long. Its range of habitat includes even the more open places in the climax forest. *Epilobium* is equally variable, ranging from the dwarfs of station 29 to great clumps of thrifty stems two feet in height and loaded with flowers. These two species are important in the contribution of humus which they make. The amount is not large, but is of the utmost consequence in a soil practically devoid of organic constituents.

The next two species in order of arrival are variegated horsetail, *Equisetum variegatum,* and *Dryas drummondii.* A divergence in habitat preference is here apparent. *Dryas* characterizes the coarser and therefore drier soils, and *Equisetum* frequents the accumulations of fine silt, especially in wet depressions and around the margins of ponds. The importance of *Dryas* lies in the fact that it is the first of the mat-forming plants, which are the most effective of the early arrivals in advancing the process of development. It produces a large number of short, stocky subterranean or surface branches, each of which bears a dense tuft of leaves, the whole mass thoroughly dominating the area and spreading with great rapidity.

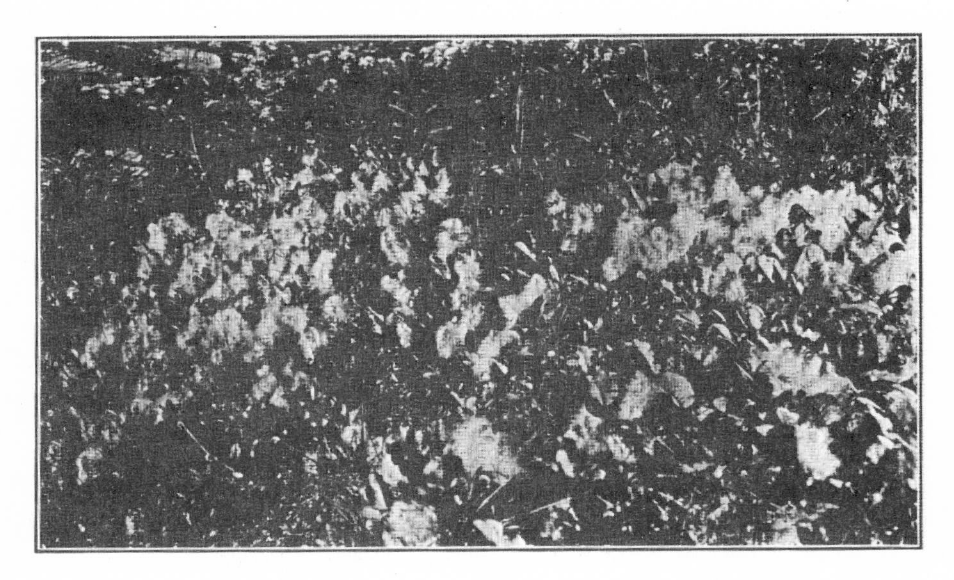

FIG. 4. Arctic willow, *Salix arctica,* one of the important mat-forming pioneers.

These five species are universally important; several others deserve a few words of special notice. Where the soil is constantly saturated by reason of seepage, a definite group of species quickly forms a thick solid turf. The common horsetail, *Equisetum arvense,* and the mosses, *Philonotis fontana* Brid. and *Drepanocladus kneiffii* (Schimp.) Warnst., make the bulk of this, and the latter become so saturated that the water may be wrung from them as from a sponge. An excellent instance was found at station 15, interesting also because of interglacial forest remnants. The stiff basal clay is overlaid by a three-inch layer of hard peat (the ancient forest floor), and this by a heavy accumulation of coarse gravel. The over-riding glacier beveled the surface, cutting obliquely across the peat layer, so that today the water draining from the superficial gravels issues from the bank at the level of the old forest floor and flows down the gentle slope below, keeping the surface layers constantly saturated. The peat makes the dividing line between two absolutely different vegetation types, the solid carpet of mosses and horsetail below and the scattered community of creeping willows above. It is possible that organic material derived from the remains of the old forest floor makes the lower area still more favorable.

With the perennials so far treated come occasional individuals of the next group, the prostrate willows. These increase in number and with *Dryas* are the most important of the early arrivals. They produce thick woody stems which lie closely appressed to the ground and are attached to it by abundant adventitious roots. Their rôle is the formation and retention of large amounts of humus; they initiate the layer of humus-filled soil which is a prime requisite for the establishment of the climax forest. Three species occur at Glacier Bay: *Salix arctica, S. reticulata,* and *S. stolonifera.* The first is of greater consequence than the other two combined, both in abundance and in bulk (fig. 4). It forms dense solid mats a yard or more across, and its numerous roots, thick stems, and large leaves and catkins furnish much organic matter to the soil. The willows grow well upon soil of any texture, but are especially conspicuous, with *Dryas,* upon the coarser deposits because there are fewer individuals of other species thereon.

Very rarely do the prostrate willows form anything approaching a solid ground cover, for the next group of plants arrive with them, increasing more slowly than they, but soon surpassing them in importance. These are the erect willows, and the following six species occur at Glacier Bay: *Salix alaxensis, S. barclayi, S. commutata, S. glauca, S. scouleriana,* and *S. sitchensis.* In coarse soils and subject to unfavorable conditions they are at first far from erect, often sprawling and even prostrate. In such condition they are similar in appearance to the truly prostrate species, but differ in that they are not so closely appressed, and therefore not so effective in retention of humus. Later, when the root system is well established in the deeper soil layers, the depressed stems send up erect branches which become the dominant part of the plants (fig. 5).

It is impossible to select any of the above list that actually precede the others. Certain rapidly attain a quasi-dominance which does not endure long, these being the species of lower stature, *Salix barclayi, S. commutata, S. glauca,* and *S. scouleriana.* The other two, *S. alaxensis* and *S. sitchensis,* in the minority and therefore less prominent while all are young, are potentially small trees, and later dominate by reason of their size, while the others, being obligate shrubs, steadily decrease in importance. *Salix alaxensis* is of especially rapid growth, and therefore in favorable situations is conspicuous from the first. At station 26 an erect stem eight feet high and two inches in diameter at the base was found to be but eight years old.

FIG. 5. Alaska willow, *Salix alaxensis,* in a pioneer situation. At first prostrate, the plant has begun to produce erect shoots.

The increase of the shrubby willows and the addition of the alder, which, like the other shrubs, first enters with the pioneers, results in the development of a thicket, first evident in the more favorable areas, gradually spreading over all. In a few places the willows alone make a recognizable community. The highest part of Strawberry Island (station 4) is covered by a low thicket growth of various species of *Salix,* the erect ones being dominant (fig. 2). *Salix arctica,* however, retains a position of importance, carpeting large areas with a luxuriant mat. The ground is thoroughly covered and humus is abundant. Alders and spruces are invading in company from more sheltered ground to the south. Station 22, though farther north, shows a later stage. The principal species in 1916 were *Salix barclayi, S. commutata, S. sitchensis,* and *S. alaxensis,* averaging six feet in height, and *S. arctica* flourishing in the more open spots. An occasional cottonwood and frequent spruces rose above the general level. Even the mountain hemlock was present in small numbers.

The herbaceous and low shrubby growth of the willow thicket is made up of relict pioneers and new arrivals. The latter group includes baked-apple berry, *Rubus chamaemorus* L.; strawberry, *Fragaria chiloensis* (L.) Duch.;

shinleaf, *Pyrola asarifolia;* bear-berry, *Arctostaphylos uva-ursi;* yarrow, *Achillea borealis;* and *Arnica chamissonis* Less.—a decidely more mesophytic assortment than we have previously met. *Fragaria* covers large areas in the more open portions and bears enormous quantities of delicious fruit. The incoming mosses are also mesophytic: *Rhytidiadelphus triquetrus* and *R. squarrosus.* The former grows thickly upon mats of *Salix arctica* and perhaps is instrumental in bringing about its elimination.

FIG. 6. Alder thicket upon a gravel terrace. West side of Muir Inlet near Station 21.

Thickets of pure willow are rare. It is far commoner for the alder to develop so rapidly that by the time the thicket stage is reached it is more abundant than the willows (fig. 6). In such cases the alders average about sixty per cent of the stand, and the two tall-growing willows, *Salix sitchensis* and *S. alaxensis,* divide the remaining forty, the latter retaining its share of importance longest. Cottonwoods and spruces are always present, rising above the general level, the first as relicts, the second mainly as invaders. A mature thicket of this kind is difficult to penetrate because of the interlacing assurgent stems of the closely placed shrubs, which may be several inches in diameter.

The humus, mixed at first in large quantity with the mineral soil, now, in addition, forms a thick carpet over the gravels and cobbles of the old terrace surface. For the first time, too, in the progress of the succession, decrease of light intensity becomes an important factor; this and the increase of humus, with accompanying stabilization of the water content at a high point, are reflected in the character of the lower vegetation. The pioneers are practically gone, and the shrubby willows, too, except those which make a part of the dominant stratum. The new arrivals are mesophytic species that will maintain themselves into the climax forest and become permanent members of it. A brief list has been given in a previous section.

The cottonwood (*Populus trichocarpa*) deserves a few words of special comment. It is one of the very earliest colonizers, being found in thirteen of the twenty-two pioneer stations. In five of these cottonwoods were the oldest of the woody plants whose ages were determined. In the barest habitats it goes through a hard struggle, the effects of which are seen in its low contorted form and early yellowing leaves. At station 14 (ice limit of about 1903) a semi-prostrate individual was found in 1921, seventeen years old. Nevertheless, many individuals manage to survive through the pioneer stages and to surpass the willows and alders in rate of growth, so that isolated projecting narrow-topped cottonwoods are characteristic of the thickets, and, in ever-decreasing numbers, of the young conifer forest as well. The species being intolerant of shade, new arrivals are excluded after the thicket cover becomes complete.

Finally, even the climax trees make their first appearance with the pioneers. Of the twenty-two stations, the spruce was found in five, and the two hemlocks in one. Like the cottonwood, they at first undergo adverse conditions, which are often reflected in their form and condition. At station 12, which, though it has had a subaerial history of more than forty years, is still largely in pioneer condition because of exposure to glacier winds, a matted spruce was found, two feet in height, with a trunk three inches thick, thirty-nine years old. Several ring counts indicate that growth, which is very slow for a few years, ordinarily augments in rate as the root system becomes well established and humus increases in amount. While the willows and alders are progressing toward dominance, the conifers are also increasing in number and size, so that by the time the willow-alder thicket is mature it is commonly overtopped here and there by slender spruces. New seedlings start beneath the alders, probably derived in large part from seeds borne by trees already established in the vicinity. They are nowhere abundant, but their multiplication is sufficient to bring about a gradual, steady increase of the species, uniform in rate over wide areas.

FIG. 7. Willow thicket at left, alder thicket at right; Sitka spruce invading both. Station 22.

The establishment of the climax does not depend upon previous dominance of alder, for in the areas of pure willow thicket the spruces were found to be invading with equal vigor (fig. 7). In such areas this tree reproduces vigorously by layering (Cooper, '11). On Strawberry Island (station 4) apparent groups of young spruces were in reality made up of a single parent tree surrounded by layered shoots of almost equal height (fig. 8). The branches which produce the daughter trees leave the main trunk at the surface of the ground or a few inches above it, and are in whole or in part buried by humus accumulations. At a distance of several feet they turn erect, suddenly increase in diameter, and form tree-like shoots, which obtain most of their sustenance from large adventitious roots originating at the point of bending. Aside from the possibility of multiplication of individuals by the death of the connecting branch, the habit is of importance in that it results in rapid increase of the area controlled by the conifer element.

FIG. 8. Sitka spruce: a group composed of a parent trunk and several layered branches. Station 4, Strawberry Island.

The individual spruces that first arrive probably seldom form a part of the ultimate dominant stand, being at a decided disadvantage in competition with the later arrivals, which have higher temperatures, shelter, and an ample humus supply from the very first. The pioneers persist in a suppressed condition beneath the more fortunate later arrivals. At station 6, on Bear Track Cove (Cooper, '23, fig. 15, p. 123), an area of young climax, or more accurately, of subclimax state, age counts of trees close together gave the following interesting results:

	Diameter, Inches	Age
Dominant stand	24	64
	24	50
	20	53
	20	50
Average	22	54.25
Suppressed	7	57 (?)
	6	67
	5	64
	4	71
Average	5.5	64.75

The suppressed trees, with one fourth the diameter of the dominants, average ten years older, and the oldest tree of all is the smallest.

After the development of a solid stand of spruce, the elimination of the alders and willows is a mere matter of time. At station 5 there was a dense tangle of dead alder and willow trunks beneath the spruce, loaded with mosses, and an occasional one still living, sometimes prostrate, half buried in moss, but sending up erect leafy shoots at intervals along the trunk. Alder, Sitka willow, and Alaska willow were found in such condition, the last bearing enormous leaves, measuring eight inches in length and three in width—more than twice the normal size.

With the establishment of the spruces in solid stand comes a rapid increase of the moss contingent, producing the already described luxuriant carpet of many species. The three species of *Rhytidiadelphus* are dominant at first, and *Hylocomium proliferum* is the last to arrive. Great abundance of this species is a sure indication that successional maturity is close at hand.

We have traced the course of development to the farthest point yet reached upon the areas which were ice-covered during the last great advance, to a state of early maturity, characterized by nearly pure dominance of Sitka spruce. The final step, which will bring the condition of full maturity that exists throughout the great forests of southeastern Alaska, will consist in the increase of the two hemlocks until they are at least of equal importance with the spruce. The hemlocks, though they appear in small numbers early in the succession, play no rôle in its development up to this point. Their future increase is plainly indicated by the frequent youthful individuals in the understory of the forests near the mouth of the bay.

Development in Ponds

It remains to add a few words concerning the vegetational development in ponds upon the till and gravels. Such habitats are rather infrequent and mostly small and shallow. Of plants growing mainly or wholly submerged there are very few, green algae and *Ruppia maritima* L. being the only ones found in the pioneer regions. *Myriophyllum spicatum* L. (?) grew in a pond surrounded by willow-alder thicket. Most important are the amphibious species, including some that flourish as well on the surrounding gravels, especially the horsetails, *Equisetum variegatum* and *E. arvense*. The former in

Fig. 9. Shallow pool in a ground moraine, with a dense growth of horsetail, *Equisetum variegatum*. Station 26.

such places has a very characteristic manner of growth, forming dense masses of tall, stout, erect culms (fig. 9), while upon the gravels it occurs in scattered fashion, producing from one to several reclining or weakly ascending stems from each root. The preference of *Equisetum arvense* for very wet places has already been noted. Certain species are confined to such localities: cotton-grass, *Eriophorum scheuchzeri* Hoppe; sedge, *Carex vulgaris* Fries; and rush, *Juncus haenkei* E. Mey. The mosses *Philonotis fontana* Brid., *Drepanocladus kneiffii* (Schimp.) Warnst., and *D. revolvens* (Sw.) Warnst. are frequently abundant. These species of more or less amphibious habit combine to form a dense growth of vegetation which completely fills the shallow pools and forms a marginal zone around the deeper ones, extending out over the surrounding gravels for some distance. There is usually a ring of thrifty willows associated with this fringe of amphibious plants. The filling-in process goes on in the orthodox way, but in the larger depressions it is so slow that many ponds become surrounded by fully developed climax forest.

In figure 10 the successional rôles of eighteen important species are indicated. For each stage the percentage of the total number of stations in which the species was found is given, and thus the presentation is faulty in that it

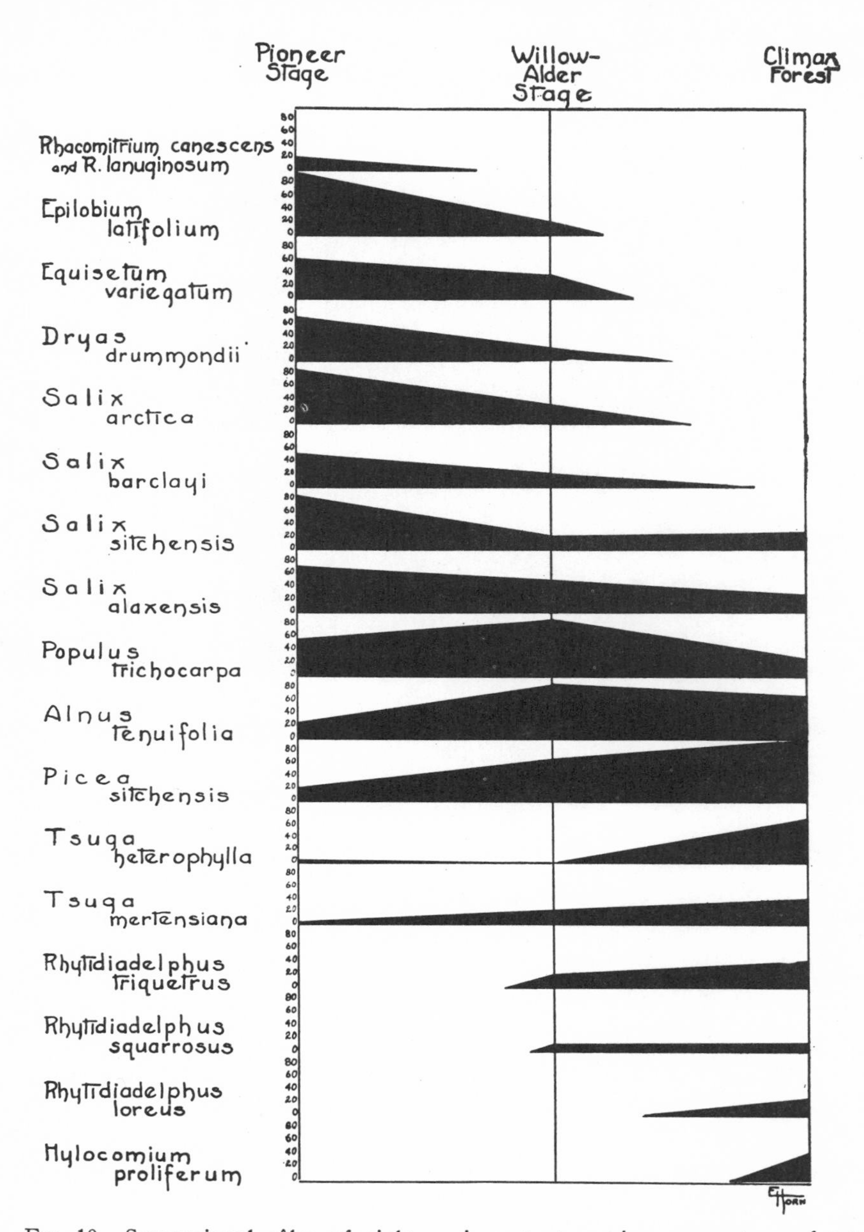

FIG. 10. Successional rôles of eighteen important species; percentage of total number of stations in which each was found is shown on the left of the diagram, and indicated by the thickness of the black areas.

provides no indication of relative abundance. Nevertheless, certain facts are clearly demonstrated: the persistence of many species through two or more stages, resulting in the blurring or extinction of community limits; and the immaturity of the climax forest, indicated by the continued presence of species of the preceding stage.

IV. Summary

1. *Areal Ecology.*—The vegetation of the area which has been recently laid bare by the final retreat of the glaciers may be roughly marked off into three communities: the *pioneer community,* upon the areas most recently vacated, characterized by one or two xerophytic mosses, certain perennial herbs, and prostrate willows; the *willow-alder thicket,* occupying the slopes around the middle portion of the bay; and the *conifer forest,* upon the shores near the entrance, characterized by a nearly pure growth of Sitka spruce. Because of the rapidity of vegetational change these communities lack sharpness of definition, and must be considered merely as cross-sections of a continuous and rapid stream of successional development.

2. *The Primitive Habitats.*—The ice, in its retreat, is exposing two general types of land surface: rock exposure and depositional accumulations, the latter including moraine and glacio-fluvial deposits. Occasional pools provide a habitat of opposite character. The primitive soils are characterized by low temperature, abundance of water in general, but occasional desiccation in the surface layers, and lack of organic matter. Atmospheric conditions are favorable in comparatively low rate of evaporation, and unfavorable in general lowness of temperature. As the ice margin recedes from a given spot, the temperature rises and the evaporation rate drops still lower.

3. *The Process of Development.*—The rate and manner of invasion are determined by the number and mobility of available disseminules, the rate of exposure of new territory by retreat of ice, and the character of the habitat thus provided. The manner of invasion is by isolated outposts, followed by gradual consolidation. If the rate of glacial retreat be slow, consolidation will proceed with relative rapidity; if rapid, the general effect of barrenness will persist for a long period, since consolidation must progress slowly under such conditions.

4. *Development on Rock Surfaces.*—The rapidity of invasion upon rock surfaces depends on the slope of the surface, its initial character and its response to weathering, and accordingly varies widely. Crevices are the most important starting points for the pioneers, which include herbs, shrubs, and even trees. Turf-like growths, anchored at the crevices, come to cover the level or depressed areas, and these rapidly develop into thickets and even into forest while the steeper and smoother surfaces are still bare. This line of succession does not greatly differ from the next in general character, but is slower, due to the necessity of building up an entire soil which is wholly

organic; and is less uniform in rate, because of the varying degree of favorableness of the primitive surfaces.

5. *Development upon Moraine and Gravels.*—The hardiest pioneers are the mosses of the genus *Rhacomitrium* and the broad-leaved willow-herb, *Epilobium latifolium.* Horsetail, *Equisetum variegatum,* and *Dryas drummondii,* a mat-builder, are next, the former preferring the wetter places, the latter the coarser soils. The prostrate willows complete the list of pioneers, and are by far the most important because of their bulk and ground-covering ability. Pronounced telescoping of the successional stages is indicated by the frequent establishment of individuals of the tall shrubs and even of the climax trees in distinctly pioneer habitats. By increase of the erect willows and the addition of the alder a thicket develops, first evident in the more favorable areas, gradually spreading over all. The Sitka spruce, already present here and there, grows in importance, finally forming a practically pure forest, in which the shrubs languish and die. Occasional hemlocks in the understory indicate the approach of the mature climax state.

6. *Development in Ponds.*—These habitats, which are rather infrequent and mostly small and shallow, come to support a growth of horsetail, rushes, cotton-grass, and aquatic mosses. The filling-in process is so slow that the larger ponds, still retaining their pioneer character, may become surrounded by fully developed climax forest.

Literature Cited

Cooper, W. S.

'11 Reproduction by Layering among Conifers. Bot. Gaz. **52**: 369–379.

'23 The Recent Ecological History of Glacier Bay, Alaska. I. The Interglacial Forests of Glacier Bay. Ecology **4**: 93–128.

Livingston, B. E.

'16 Physiological Temperature Indices for the Study of Plant Growth in Relation to Climatic Conditions. Physiol. Res. **1**: 399–420.

4

Reprinted from *Elisha Mitchell Sci. Soc. J.* **50**:225–246 (1934)

THE OLD FIELD PRISERE: AN ECOLOGICAL STUDY

By W. M. Crafton and B. W. Wells

INTRODUCTION

In the eastern mesic forest region the old field vegetation is an ecological unit strikingly distinct from that of the climatic climax vegetation. The primary stages of the developmental vegetation, the prisere, consist of an herbaceous flora entirely. This is followed by the tree flora which constitutes the dominant natural vegetation of the eastern United States.

The herbaceous flora, composed of grasses and weeds, ranges throughout old fields and waste places in a more or less regular series. In the southern upland bare areas the most frequent pioneer is the crab grass (*Syntherisma sanguinale*). It is particularly abundant in recently cultivated fields. It enters during cultivation and maintains dominance rarely after the second year. The general seeding and growth of invaders result in the exclusion of this annual. These invaders—tall weeds (*Eupatorium*, *Aster*, *Solidago*, and others)—play the dominant rôle for a time. A few years later the old field vegetation changes again and the third phase is initiated. This time broom-sedge (*Andropogon* spp.) invades—reaches its extreme expression—then passes out with the ecesis of pines (*Pinus taeda*, *P. echinata*, and *P. virginiana*).

Where the tree vegetation is open in stand in the old pine fields, broom-sedge is well represented. On the other hand, where the pines form dense consocies by mass invasion this grass soon becomes only a relict of the final prisere stage.

The rate of establishment of the elements of the prisere are profoundly influenced by the edaphic and aerial conditions. Further, the sequence, in part at least, bears a direct relation to the ecesis of the final prisere element. This study deals primarily with the habitat in relation to the distribution of the vegetation. With this approach, certain conclusions can be drawn to account for the definite stages in the prisere.

It is not intended to discuss or even mention all of the plants in the old fields. The species under observation proved of interest because of the aspects they present, importance in the habitat analyses, and, of course, their direct relation to the prisere.

METHOD

The studies of this problem were confined to the fields about Raleigh, North Carolina. For the measure of plant invasion and competition meter quadrats were marked off in several communities of different soil habitat. The vegetation in the quadrats was recorded late in 1930 and early spring, summer, and fall of 1931. Bare areas were made in the spring of 1931 by spading up meter square plots. Records of plant growth were obtained as in the other quadrat areas. In many of the quadrats, 20 cm. intra-quadrats were marked off. This permitted closer checking of the effect of competition. Further, a rather detailed study of a terraced slope was made.

The water supplying power of the soil was obtained only for comparative purposes. In certain striking situations texture analyses of the soil were made. Further, relative light values were obtained. All of these data were gathered from the various representative communities.

Information in regard to seedling growth and response to drought proved significant in the interpretation of the stages of the prisere development. This information was obtained from studies in both the green house and the field.

Close observation of the old field vegetation coupled with the data gathered from the above sources comprised the mode of attack on this problem.

THE VEGETATION

Like all primary vegetation, that of the old fields is somewhat xeric in nature. The upland areas tend to display a great many species adapted to the xeric habitats; the lowlands contain those species adapted to the more mesic conditions.

Usually the vegetation in either situation is so variegated that the casual observer sees nothing but a heterogenous mass of grasses and weeds. However, close and frequent observations will disclose a rather graded series of species in the revegetation of old fields. The appearance of this series is related to the nature of the soil and the species involved.

Successional stages of the old field associes

The pioneer consocies. It is a common observation that cultivated fields become covered with crab grass. Bermuda grass (*Cynodon*

Dactylon) is less common as a dominant. Here and there occasionally appears *Paspalum* sp., but its occurrence is rare. In local moist areas cockle bur (*Xanthium commune*) and smart weed (*Polygonum* spp.) often become subdominants or even dominants. However, in upland fields they are always minor elements. In eroded fields where the subsoil is exposed, the pioneer stage is represented by ragweed (*Ambrosia artemisiifolia*), button weed (*Diodia teres*), and poverty grass (*Aristida* sp.). These exposed clay areas and the *Cynodon* consocies have the effect of holding up the usual rapid development of the prisere.

The annual crab grass often retains its dominance for a second growing season. At this time horse weed (*Leptilon canadense*) is found in addition to the minor plants of the previous year (pl. 18, fig. 1). This is the

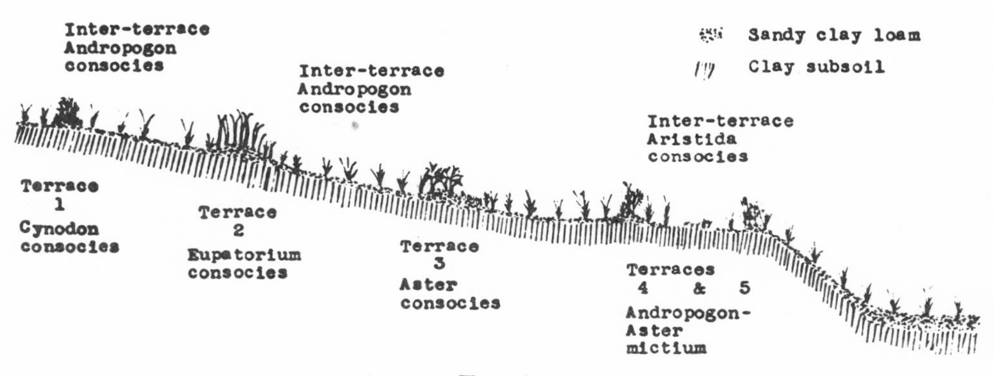

FIG. 1

most conspicuous plant of the associes because of its great height. Aster or some other tall weed may become well represented at this time, or at least the year following. Orchards, cultivated three or four times during the growing season, display crab grass and scattered aster. Foxtail (*Chaetochloa* spp.) is often present. Fields from which small grain crops have been harvested in the early summer, are given over to these grasses. Other fields which have reached the weed stage prior to planting to small grain sometimes skip the crab grass stage. Such fields are usually disked. After harvest the weeds become dominant again. Crab grass and foxtail may be found scattered throughout these areas representing remnants of the true pioneer stage. In sections of the upper piedmont it is a fairly common practice to prepare the fields in the summer for fall sowing. The plowing is usually done in July. With a few rains the fields become covered with crab grass.

Other plants in recently cultivated areas are mostly annuals. Occasionally a few perennial weeds become established, but they are never abundant. Even the regular minor elements are more or less widely scattered.

The most striking aspects in this stage are contributed by toad-flax (*Linaria canadensis*) and tall sheep sorrel (*Rumex stricta*). In the sandy regions of the state they present beautiful vernal aspects of lavender and crimson. Fields cultivated the previous year are often covered by either or both of these plants. However, in the piedmont they are less common. Ox-eye daisy (*Chrysanthemum Leucanthemum*) and ragwort (*Senecio Smallii*) appear in the pioneer stage only in Bermuda grass areas which have been established for several years (fig. 2). Further, these species are much more common in the piedmont and mountain areas.

List of the Characteristic Plants of the Pioneer Consocies

Dominants

Ambrosia artemisiifolia
Aristida sp.
Cynodon Dactylon
Paspalum sp.
Syntherisma sanguinale

Subdominants

Chaetochloa glauca
Chaetochloa viridis
Leptilon canadense

Minor plants

Allium sp.
Amaranthus retroflexus
Amaranthus spinosus
Antennaria plantaginifolia
Bidens spp.
Brassica sp.
Chenopodium album
Chenopodium Botrys
Chrysanthemum Leucanthemum
Convolvulus arvenis
Coreopsis sp.
Datura sp.
Diodia teres
Erigeron spp.
Eragrostis sp.
Hypericum gentianoides
Linaria canadensis
Oenothera laciniata
Oxalis stricta
Panicum capillare
Panicum virgatum
Rubus sp.
Scleranthus annuus
Sida spinosa
Solanum carolinense
Xanthium commune

The tall weed consocies. This is the intermediate stage, composed mainly of aster (*Aster* spp.), golden rod (*Solidago* spp.), dog fennel (*Eupatorium* spp.), sneeze weed (*Helenium tenuifolium*), and Carolina golden rod (*Euthamia caroliniana*). The last three are often dominant in neglected pastures. With grazing these weeds retain prominence for years. However, when this factor is withdrawn, broom-sedge assumes dominance. This grass, though long seeded, is apparently held in check by grazing.

Frequently one observes succession within a weed consocies. In the coarser soils dog fennel may be followed by *Aster*. However, the former may not be entirely eliminated until the final stage is initiated.

In the open weed communities the dominants of the pioneer stage strongly resist the invaders. These grasses are best represented in short weed consocies, that is, communities of Carolina golden rod and sneeze weed. On the other hand, where the plant cover is heavy there is only occasional evidence of the retiring species.

A longer time is required for the suppression of Bermuda grass than any other of the pioneers. In many cases it is represented throughout this stage as an underplant community. It is entirely eliminated only in dense stands of weeds. Otherwise, it may persist long after broom-sedge has taken possession of contiguous areas.

Of all these weeds, the dog fennels (*E. hyssopifolium* and *E. capillifolium*) are probably the most important ones which extend far into the coastal plain. *Solidago odora* appears in the lower sand regions where clay is near the surface. *Aster ericoides*, most common in the piedmont, also disappears in the sandy regions of the state.

In open weed communities, daisy fleabane (*Erigeron* spp.), rabbit foot clover (*Trifolium arvense*), and hop clover (*T. agrarium*) represent the spring aspects. However, where the weed stand is dense these plants are apt to be eliminated.

LIST OF THE CHARACTERISTIC PLANTS OF THE TALL WEED CONSOCIES

Dominants

Aster ericoides
Aster Tradescanti
Eupatorium capillifolium
Solidago odora
Solidago rugosa

Subdominants

Achillea millefolium
Eupatorium hyssopifolium
Eupatorium sp.
Euthamia caroliniana
Helenium tenuifolium

Minor plants

Antennaria plantaginifolia
Bidens spp.
Daucus carota
Desmodium acuminata
Desmodium sp.
Erigeron spp.
Lactuca canadensis
Lespedeza striata
Lespedeza procumbens
Trifolium agrarium
Trifolium arvense
Rubus spp.
Senecio Smallii
Solanum carolinense

The broom-sedge consocies. This final consocies of the prisere may be found in various stages of expression. In different fields it is often observed sharing an area with tall weeds, short weeds, pines, or almost by itself (figs. 3, 4, and 5). With tall weeds it represents the part of an invader. Where the short weeds are found a more advanced stage is represented. The extreme expression of dominance is reached when even the short weeds are excluded. This latter situation, however, is of rare occurrence.

A few differences in the distribution of the tall weeds in the piedmont and coastal plain have been pointed out. The broom-sedge stage in these two sections of the state is represented mainly by *Andropogon virginicus, A. Elliottii,* and *A. argyraeus.* However, in the coastal plain *A. glaucopsis* becomes prominent in addition to *A. virginicus.*

In connection with this stage it is well to mention the contributors to the spring aspects. Fleabane (*Erigeron* spp.) and ragwort (*Senecio Smallii*) appear locally in broom-sedge areas. The appearance of rabbit foot clover and hop clover is more or less limited.

LIST OF CHARACTERISTIC PLANTS OF THE BROOM-SEDGE CONSOCIES

Dominants

Andropogon Elliottii
Andropogon virginicus

Subdominants

Andropogon glaucopsis
Eupatorium sp.
Euthamia caroliniana
Solidago sp.

Minor plants

Andropogon argyraeus
Antennaria plantaginifolia
Aster grandiflorus
Chamaecrista nictitans
Chrysopsis graminifolia
Daucus carota
Desmodium acuminatum
Desmodium sp.
Erigeron spp.
Eragrostis sp.
Gnaphalium purpurea
Helianthus sp.
Hypericum gentianoides
Lactuca canadensis
Lespedeza augustifolia
Lespedeza procumbens
Physalis sp.
Senecio Smallii
Sieglingia seslerioides
Smilax rotundifolia
Smilax Bona-nox

Seedling growth

Seedling growth is of great importance in regard to the rôle of plants in any sere. While the juvenile growth alone is not an adequate index of ecesis, knowledge of this phase in vegetational studies is paramount.

In the early period a slight shift to the dry soil condition may easily prove fatal. With rapidly growing species, however, the importance of such a change is less significant. Likewise, plants growing in shade may be affected to a less extent at such times.

Experiments. Seedlings of aster, golden rod, broom-sedge, and crab grass were placed in competition in several pots of loam soil and subjected to artificial drought. The tall weed and broom-sedge seedlings were 1 month old. The crab grass was less than 2 weeks old. After 1 week with no water the weeds could not be revived. Only 2 of the 6 broom-sedge seedlings survived sufficiently to start growing again. One of 2 crab grass seedlings lived. At the end of the week without water, the leaves of crab grass were only rolled; those of the other plants were rolled and dried. Seedlings in other pots subjected to the same drought treatment, but removed from the direct sunlight, rolled after 2 weeks.

Two pots of 14 day old broom-sedge seedlings were placed in the shade of potted plants in the greenhouse. Two other pots were not shaded. The shaded plants lived 3 days without water. The seedlings left exposed to the direct sunlight died the first day.

Bermuda grass and broom-sedge seedlings were grown in the same pot. The plants were not watered through the day as is necessary for the perpetuation of broom-sedge in the open. The Bermuda grass showed no indication of a drought condition, but the broom-sedge died.

The apparently slow growth of the weeds and broom-sedge is of importance. After 6 weeks of growth the leaves of broom-sedge averaged 2 cm. in length. Tillering was very pronounced at this time. Weeds of the same age were twice as tall. Crab grass in the period of 6 weeks flowered at the height of 25 cm.

Field observations. Seedling crab grass and weeds are rarely found before the middle of May. Broom-sedge, on the other hand, may be found in abundance by the first of the month. The oldest seedlings at this time may be about 3 weeks of age. By the middle of May there are very few small seedlings surviving in the bare areas. The rains become more and more infrequent and the critical periods—a few days without rain—come on and only the oldest seedlings are able to endure.

Broom-sedge seedlings literally cover the ground in lowland areas. The frequency with which they occur here is much greater than in the adjoining upland regions. In the latter situation the leaves may roll after a few days with no rain. However, under the same conditions those seedlings which grow near clumps of weeds or under the leaves of broom-sedge do not appear to be affected. Search for seedlings of this plant in Bermuda grass consocies revealed not a trace of their presence.

THE HABITAT IN RELATION TO THE VEGETATION

The bare sunny areas give rise to those species which are usually photophilous. However, from the standpoint of xerism the exposed subsoil and sandy clay loam soils are occupied by very different pioneers, and changes in the vegetation progress at different rates. Where the surface soil is eroded, leaving the red subsoil exposed, changes in soil conditions must be brought about before the broom-sedge can successfully ecize; or more correctly, the habitat must be made favorable for the successful invasion of the tall weeds. Usually in these areas the nutrient-water complex operates. The only change produced in the habitat of the loam and sandy soils is that induced by the tall weeds. The soil water conditions are much more favorable at this time for seedling establishment. This can be shown to be of vital importance to broom-sedge. However, the influence of texture becomes a factor even in this regard.

Variations in soil and aerial habitats are often clearly shown on slopes. And the vegetation of the old fields shows marked response to the water, texture, and nutrient conditions often present in such areas.

The prisere stages are always altered by the depth of the surface soil. In deep sandy soils the advance is rarely as rapid as in those more shallow.

Quadrat studies

Bare area quadrats. Meter square areas were spaded up in various field communities in the early spring of 1931. The vegetation in these quadrats was recorded twice (early summer and fall) during the year.

On a sandy clay loam upland area, Bermuda grass became the dominant element, while ragweed played the rôle of the subdominant. A plot was located in an area dominated by the grass just mentioned. In June 3 ragweed seedlings were growing well in the 20 cm. intraquadrat. By fall only 1 attenuated plant remained, while 8 grass plants had appeared.

In another Bermuda grass consocies, bared plots gave rise to a luxuriant growth of crab grass. Smart weed and cockle bur were minor elements. This area was of sandy clay, but in lowground. Crab grass became the dominant element in an upland sandy loam area. Yellow foxtail played the minor part.

The bared plot in an *Andropogon* consocies gave rise to 43 ragweed and 24 Carolina golden rod plants by June. In the fall 34 ragweeds remained. These ranged from 15–60 cm. in height. At the same time the golden rod numbered 23. Some showed definite suppression. Not all of these flowered. The height of these plants ranged from 10–40 cm. Crab grass was represented by 53 hardy plants. *Lespedeza procumbens,* though reduced in numbers, was still present in the fall. In the 20 cm. intra-quadrat ragweed and *Lespedeza* paid the heaviest toll to competition. Of 6 *Lespedeza* seedlings in June, 2 survived until fall; and of 4 ragweed only 1 survived. All 6 crab grass plants flowered.

The open plot in poverty grass was in contrast to the above mentioned area. This quadrat was located higher on the same gentle west-facing slope. Other than the few inches of sandy clay, the soil is of heavy red clay. The soil habitat is in direct contrast to the lower part of the slope which is of deep sandy clay loam.

By late May, 13 plants grew in the quadrat. One week later there were 114 seedlings. Though *Lespedeza* and ragweed had increased numerically, bracted plantain (*Plantago aristida*) was dominant. In contrast to this there were 400 seedlings in the lower quadrat. By fall *Lespedeza* was reduced from 29 to 6 plants and ragweed from 19 to 10. The mortality of all seedlings was greater in the stony sandy clay soil as compared to those in the loam soil. Crab grass was represented by only a few plants.

Crab grass became the dominant element in the bared areas in *Eupatorium* and *Helenium* consocies. Bermuda grass and sneeze weed were the minor elements. The soil in these contiguous consocies was of the sandy clay loam type.

It is commonly observed that where the Cecil clay subsoil appears at the surface the vegetation is scant. Poverty grass and button weed usually dominate such areas. And Carolina golden rod is often scattered over the clay surfaces. For experimental purposes soil from a *Cynodon* consocies was removed to an "outcrop" of the red clay. Where fertilizer was applied with the transported soil, ragweed and crab grass grew. There were but 3 plants of the crab grass. These plants did not appear elsewhere on the galled area.

Aristida consocies. This particular area had been dominated by poverty grass for a number of years. The soil to the 6 inch depth is of sandy clay. Beneath this depth a heavy red clay sub-soil is found. The entire area is so stony that it was almost impossible to obtain soil point data.

The spring vegetation consisted of the usual hop clover and bracted plantain. This was followed by rabbit foot clover.

In the fall of 1930 four asters flowered. The following spring 16 rosettes appeared, but these were reduced to 8 by the middle of June. At this time the average height of these plants was 10 cm. Only 4 of the 8 plants flowered the next fall. It may be well to call attention to the fact that in deep sandy loam soils aster had reached the height of 25 cm. by the first of May. This contrast is no more striking than that of the rabbit foot clover in various localities. It should be stated that this clover grew to the average height of 20 cm. in this consocies. Where manure had been applied the plants averaged 45 cm. in height.

Cynodon consocies. This upland area of sandy clay loam had been uncultivated for 4 years. In the quadrat 3 asters flowered in the fall of 1930. Early the next spring 34 rosettes of this plant were growing. By June this number had increased to 38. However, only 7 of this number survived and flowered in the fall. The height of these plants ranged from 25–40 cm. This species of aster (*A. ericoides*) normally grows much higher.

Aster consocies. This area apparently had been in Bermuda grass a few years previously. It indicates directly what will happen in the *Cynodon* consocies in the next few years. This *Aster* consocies was on a north-facing slope of shallow sandy loam.

The vegetation in the fall of 1930 was represented predominantly by aster (*A. ericoides*) and Bermuda grass. By May of the following year rabbit foot and hop clover covered the area. A month later the vegetation was changing to aster; Canada lettuce, ragwort, and wild carrot were the minor elements. Bermuda grass was reduced from 23 to 4 plants in the course of a year. During the same period aster was eliminated to almost half of the 1930 growth. Of 57 plants in June, only 14 flowered in the fall. On the other hand, broom-sedge increased from 6 plants to 8 in the course of the year. The tussocks did not develop to great size.

Solidago consocies. The soil condition is the same as that in the *Aster* consocies. The quadrat was approximately in the same location.

By the middle of May the plants in this consocies had reached the average height of 25 cm. A month later it was definitely affecting field sorrel (*Rumex acetosella*) and daisy fleabane (*Erigeron ramosa*). Of the 69 weeds present in June, 51 flowered in the fall. Sensitive pea (*Chamaecrista nictitans*), Bermuda grass, and beggar-ticks (*Desmodium acuminata*) were much suppressed. Only the *Desmodium* flowered.

Syntherisma-Helenium associes. These plants were the main elements in a pasture. The soil is a sandy clay loam.

In spite of the evenly scattered sneeze weed, crab grass was still dominant in the fall of 1931. The minor element was represented by Bermuda grass. Both dog fennel and broom-sedge were invading this area. However, the former was most aggressive.

Eupatorium consocies. This community occupied the deep sandy loam area near the *Syntherisma-Helenium* associes.

There was no change in this area over the one year period. However, it was of interest to note the prevalence of *Lespedeza,* but the absence of crab grass. Sneeze weed was sparingly represented. Broom-sedge was much better represented in this community than in the previously mentioned associes.

Solidago-Andropogon mictium. The soil type of this area is a deep sandy loam.

A record of the 1930 growth was not obtained from this area. Of the 38 golden rod plants which appeared in June of 1931, only 21 remained until fall. The broom-sedge maintained its representation of 16 tussocks.

Euthamia-Andropogon associes. The sandy clay loam soil of this area is 60 cm. deep.

Other than the changing spring vegetation of hop clover and rabbit foot clover this area remained the same throughout the growing season. Many of the golden rod plants reached only 20 cm. in height. On the other hand, broom-sedge grew to the maximum height of 70 cm.

Andropogon consocies. This community was on an upland area of red Cecil clay.

Of the 25 tussocks in the quadrat only 10 flowered in 1930. About half of this number flowered the following year. By June of 1931 several aster plants appeared. These were entirely eliminated during the summer.

Summary of quadrat studies

The greatest number of a single species that appeared in the bared areas was crab grass. It grew abundantly on lowland and upland loam soils. On the clay soils this grass was meagerly represented. These areas of thin soils gave rise to ragweed and poverty grass.

The persistence of crab grass was observed to be most pronounced in the short weed communities. In areas dominated by tall weeds it was rarely found. Conversely, broom-sedge was best represented in the tall weed communities. The study of the *Cynodon* consocies indicates the slow ecesis of *Aster*. A further stage of this was demonstrated in the *Aster* consocies in which Bermuda grass was definitely suppressed. The increased representation of broom-sedge in the same quadrat points to the trend of the succession. The establishment of broom-sedge with resultant suppression of weeds was indicated in the *Andropogon-Solidago* quadrat. The competition of tall weeds with broom-sedge seemed to be more intense than that in weed consocies.

From the standpoint of soil conditions, it will be sufficient to point out here, that the succession rate seems to be retarded most on the exposed subsoil. The pioneers on these areas are of the xeric type. On the rather shallow or deep loam soils the usual pioneers are present.

Terraced slope transect

This particular study will bring out three major influences which retard the prisere. However, the same factors which operate to produce this effect are general and important for all the situations as they exist in the ordinary development of the prisere.

Except for a few areas of exposed subsoil, the field is sandy clay loam from 2 feet to 6 inches in depth. The greatest depth of the surface layer is on the terraces and at the foot of the slope. Below this range is the red subsoil.

The inter-terraces were rather uniformly covered by broom-sedge; the only exceptions being the galled areas on which bracted plantain and poverty grass grew (text fig. 1). Boneset (*Eupatorium* sp.), golden rod (*Solidago* sp.), and cudweed (*Antennaria plantaginifolia*), which usually accompany broom-sedge, were well represented throughout the field. Other minors were beggar-ticks, sensitive pea, tickle grass, Carolina golden rod, aster, golden rod, and dog fennel. At the base of the slope the mesic *Andropogon glaucopsis* was co-dominant with *A. virginicus*. But on the slope this more mesic species thinned out to only occasional tussocks of reduced size. Trailing bramble and other species of *Rubus* were well represented in the low ground. These ranged half-way up the slope and were most prominent on the terraces.

The series of terraces presented a peculiar distribution of old field vegetation. The first of the 5 terraces considered was occupied by Bermuda grass as dominant, with aster (*A. Tradescanti*) subdominant. The upper part of the second terrace was crowned by dog fennel (*E. capillifolium*) (fig. 6). In a small area golden rod (*Solidago odora*) was dominant. On the lower end of this same terrace, which gradually disappeared, was aster. Broom-sedge was dominant where the terrace completely faded out. On the third terrace aster played the dominant rôle in a local area, but broom-sedge was invading this consocies. Again, broom-sedge was dominant where this terrace leveled to the slope. Broom-sedge, aster, and Bermuda grass occupied the two remaining ridges with bramble, golden rod, and Carolina golden rod as minors.

Thus, where the terraces were still prominent, Bermuda grass and weed associes registered dominance. However, where the ridges faded out broom-sedge—tall weed mictia were found. The ecotone was abrupt on the upper side of each terrace; on the lower side weeds and broom-sedge mingled (fig. 6).

In the spring the terraces were densely covered with rabbit foot clover. This plant was practically confined to the ridges. On the lower side of the "aster terrace" (No. 3) vetch was the spring dominant. The ecotone between this species and broom-sedge was very definite. By fall this area was given over to Bermuda grass with aster only a minor element (fig. 7).

Several changes occurred the following year ('32). Bermuda grass no longer was dominant. It was replaced by aster. The dog fennel on the second terrace gave way to aster also (figs. 8 and 9).

Water relations

The importance of water in plant ecesis cannot be overemphasized. The expression of this factor is closely related to texture. And texture plays a very important part in nutrient relations as well.

Water supplying power of the soil. For the measurement of this factor Livingston soil points were used. The points were inserted at the 15 cm. depth and removed at the end of an hour. The following data were obtained from the several communities on the slope. The readings are given in milligrams.

	May 29 (5 days after rain)
Eupatorium consocies	600
Andropogon consocies	710
Cynodon consocies	980
Andropogon-Aster associes	930
Aristida consocies	540

Texture. The rôle this factor plays in water relations is of major importance. This is particularly true for plants in the juvenile stage. Samples for analyses were taken from the surface and the 15 cm. depth.

TABLE 1

	DOG FENNEL	ASTER	BROOM-SEDGE		ASTER	BERMUDA GRASS	ANDROPOGON-ASTER ASSOCIES	BASE OF SLOPE
Coarse	18.10	13.27*	14.15†	10.00	12.80		9.76	
Medium	14.00	5.41	2.94	5.45	4.00		7.33	
Fine	12.12	3.50	3.80	7.05	6.42		8.10	
Very fine	42.17	48.00	53.70	55.00	52.00		40.15	
Clay	13.40	29.56	24.00	22.20	24.00		34.21	
	99.79	99.74	98.59	99.70	99.22		99.55	
Coarse	18.30	13.35	12.90	7.00	7.95	9.30	6.60	7.00
Medium	0.00	4.00	3.30	0.00	3.30	3.00	2.15	6.80
Fine	3.50	7.20	4.39	6.00	4.60	4.00	9.40	12.50
Very fine	49.00	53.50	51.80	49.60	55.13	55.00	52.10	46.00
Clay	28.60	21.85	27.10	37.15	28.80	28.50	29.34	27.45
	99.40	99.90	99.49	99.75	99.78	99.80	99.59	99.75

* Near terminus of terrace.
† At terminus of same terrace.

TABLE 2

	ANDROPOGON CONSOCIES	EUPATORIUM CONSOCIES
Coarse	13.27	13.40
Medium	6.20	7.25
Fine	8.54	9.75
Very fine	42.31	42.53
Clay	29.45	26.65
	99.77	99.58
Coarse	7.85	9.25
Medium	5.82	8.73
Fine	8.43	11.70
Very fine	38.65	40.64
Clay	38.40	29.22
	99.15	99.54

The tables give in percentages the various grades found. The first set of figures is the surface soil analyses (table 1).

The texture data (table 2) further emphasize the relation of texture to the ecesis of broom-sedge. In the *Andropogon* consocies, dog fennel was still subdominant. Joining this area was the *Eupatorium* consocies with a few broom-sedge tussocks. These communities were not on the terraced slope. Soil for analyses was taken from the surface and 15 cm. depths. The first set of figures (table 1) is the analyses of the surface soil.

Light

A stop-watch photometer was used to determine this factor. The method of procedure and calculation recommended by Clements[1] was used. A series of exposures was made on a clear day at noon. This gave the standard, or the most intense light values. Exposures were then made in the various communities. All data except that from the *Vicia* socies were obtained in September at one-half the height of the vegetation. The exposure in pine was made by directing the aperture of the photometer so that the light which filtered through the branches would strike the photographic paper.

The relative average values are given below.

	per cent
Dog fennel	9.5
Aster	12.2
Bermuda grass	8.5
Vetch	1.7
Pine	2.5

Summary of habitat studies

In the coarser soils capillary activity is comparatively less than in the fine soils. Wells and Shunk[2] demonstrated this to be the case in their studies of the sand hill region of the lower coastal plain. That the water supply is tied up with texture is fairly well demonstrated by the results in the Raleigh region.

The *Eupatorium* consocies gave the lowest water supplying power on the basis of texture. The soil texture is of a coarser grade than that of the other communities. In the *Cynodon* consocies and the *Andropogon-Aster* associes the soil point readings approached equal values. The texture is approximately the same for these two communities. The

[1] Clements, Frederic E. *Plant Physiology and Ecology.* Henry Holt and Company, New York, 1907.

[2] Wells, B. W., and Shunk, I. V. The Vegetation and Habitat Factors of the Coarser Sands of the North Carolina Coastal Plain. Ecological Monographs 1: 465–570. Oct., 1931.

Aristida consocies gave the lowest water supply value. In this case the soil type was compact red clay. The water either penetrates to only a slight extent or it is held more firmly by the soil particles than in other soils. The low reading in broom-sedge may be accounted for by the fact that the soil points were on the border line between the surface sandy clay loam and the red clay layers. In addition to this, all of the communities except the *Aristida* and *Andropogon* consocies were heavily covered by spring plants. In the contiguous *Andropogon* and *Eupatorium* consocies (not on the slope) we notice only slight differences in the texture. But the finest textured soil is in the broom-sedge area.

The significant facts brought out by these data are: the relation of texture to water as indicated in the two *Eupatorium* consocies, and the low water supplying power of the raw red clay soils in the *Aristida*.

The greatest response to the light factor was found in the dense stand of *Aster* on the third terrace. Bermuda grass seemed to be unable to grow to any extent in this situation. However, it is of interest to note that the dense shade of the early vetch had no effect upon the later development of this grass. In open consocies of Bermuda grass growth begins early in the spring. On the other hand, in the vetch community by the latter part of May there were but few plants arising from the underground stems. Apparently after the death of this legume, Bermuda grass grew rapidly.

The aster, dog fennel, and Bermuda grass communities gave approximately the same light values. Nevertheless, the shade produced by the former seemed sufficient to reduce the grass considerably.

The light value under pine is relatively low as it filters through the branches. This effect is minimized by the reflected light which enters under the branches. Broom-sedge shows little response to this factor where the tree vegetation is not crowded.

The density of broom-sedge tussocks is dependent upon the water supply. Since they are a foot or more apart, light could never become a factor. Those plants which ecize in an *Andropogon* consocies require a maximum amount of light, but their water requirement is probably low. Cudweed, beggar-ticks, sensitive pea, etc., are to be found in the otherwise bare spaces between the broom-sedge tussocks. Especially to be mentioned are the pines which constitute the next stage in the sere.

Ecads

The response of plants to several known factors can often be readily determined by field observations. With the soil water and the light factors playing a decisive part in the establishment and maintenance of the principal old field elements, it seems desirable to present data concerning the certain responses to factor changes.

Broom-sedge. The distribution of the leaves on the flowering stalks were not as abundant nor as large in the *Eupatorium-Andropogon* complex as in the *Andropogon* consocies. However, the coloration was good. The flowering stalks were of the average height but very slender. Further, the number of stalks produced were comparatively few. All of these conditions indicate insufficient water and low light. Water was probably the most influential. The texture studies in this area are helpful in the interpretation of the water-texture relationship. This is especially important in relation to seedlings.

Broom-sedge under pines showed severe suppression. In such instances, it was probably due to root competition since the plants showed no evident elongation response to the lessened light.

Dog fennel. The ecads of this plant showed definite suppression on the inter-terrace where broom-sedge was dominant. Not only were the stems reduced in size, but those which grew from the rosettes were fewer in number. In many cases the number of stems was decreased and the size reduced over the number at the same rosette of 1930. Where three stalks of the 1930 growth could be counted only one grew in 1931. Suppression is most noticeable where the dog fennel grows very near the broom-sedge tussocks. In areas where broom-sedge has become well represented, a few minutes' search will disclose many such situations.

Aster. The effect of competition for water was very noticeable where this representative grew with broom-sedge. The leaves were not so numerous nor large as in the associes. Occasionally several stems grew from a single rosette. This happened only in the more open spaces between the broom-sedge tussocks. Suppression was greatest where the two species grew closely together. Golden rod made a similar response in competition with broom-sedge.

Bermuda grass. By contrasting the plants in the consocies with those in the *Aster-Cynodon* mictium the effects of competition become very noticeable. Line transects of 40 cm. lengths in the consocies gave an average count of 8 plants. In contrast to this, transects in open aster gave an average count of only 4 plants. In the dense stand of aster an occasional spear of grass could be seen. These plants had fewer leaves, and the stems were unusually long.

DISCUSSION

Inhabitants of an area become established because of adaptations to the conditions present in that particular habitat. The habitats of bare and covered areas are, of course, very different so far as the water and light conditions are concerned. Evaporation of water from the soil surface on a bare area is much greater than that where vegetation has developed. And the height and density of the vegetation become factors in the maintenance of surface soil moisture. The evaporating power of the air in such instances is decreased due to the check of wind velocity and solar radiation.

The above facts form a basis for the concepts to account for the vegetational sequence in the prisere. A crab grass community is best represented on bare areas. In competition with open communities of annuals it will maintain dominance. We notice that crab grass holds on for considerable time in the short weed community of *Helenium*. And when opened to light, the small-grain fields often become vegetated with this grass. However, in areas where perennials are dominant, though the light is favorable, this grass is not found. Low water during drought periods is probably the limiting factor, because dominant plants are always as fully represented as the habitat and the time permit.

The rapid growth of crab grass is responsible for its ready appearance in cultivated fields. The hydrophilic condition of colloidal particles in the large water-holding cells may contribute a great deal to its maintenance in open ground. The weeds grow more slowly and are more susceptible to drought in the seedling stage. Further, cultivation may hinder the growth of these plants should they appear. Just why tall weeds do not always become dominant the season following abandonment is not certainly known. The general seeding is probably such that they should become dominant at this time. Sometimes this is the case. In other situations the number of tall weeds that ecize the first season are few. This has been observed on both thin stony soil and lowland loam. A second season after abandonment a weed community is usually found. The number of the previous year is increased both by seeding and by the appearance of plants from the stolons of those already established. The resulting shade eliminates the crab grass.

Bermuda grass loses out to tall weeds, because of reduced light, also. This was clearly seen in the transect study. In the *Aster* consocies, which shaded into the *Cynodon* consocies, there was a decrease in the representation of the grass. A further stage was shown on the fourth terrace in which broom-sedge and aster mingled and Bermuda grass was a minor underplant. Invasion into Bermuda grass by weeds requires a much longer time than entrance into one of crab grass. This is related to the growth habits of the grass species. The annual crab grass has no chance with a perennial species. Bermuda grass, a perennial, competes successfully with weeds until considerable shade is produced. It is then suppressed.

The invasion of broom-sedge into old fields rests upon the establishment of more mesic conditions induced by the tall weed flora. Pot experiments have shown broom-sedge seedlings to have a slow growth rate and to be very sensitive to a slightly dry soil condition. In the field it has been observed to germinate and grow in the open, but in a few days without rain the seedlings disappeared. However, shaded seedlings withstood the adverse conditions better. The concept advanced here is that the soil in closed tall weed communities does not dry out so rapidly between the spring rains so that broom-sedge can become established. The various situations on the terraced slope intensify this concept. In the finer soils where aster had suppressed the Bermuda grass, broom-sedge was ecizing. On the coarser soils (terrace 2) the broom-sedge was scarcely represented. It can never enter the *Eupatorium* consocies until there is almost daily rainfall through the spring. The coarse, surface soils dry out much too rapidly for the slow-growing grass to become established. This characteristic is probably responsible for the minor representation of *Andropogon virginicus* in the "sand hills." Similar response to texture difference was noted in another *Eupatorium* consocies and the joining *Andropogon* consocies. The water relations as influenced by texture could have been the only factor operating against the uniform establishment of the broom-sedge.

The ecotone between broom-sedge and Bermuda grass is definite where these two grasses inhabit contiguous areas. The former is rarely found with Bermuda grass. The ecotone between a field of crab grass and one of broom-sedge is just as sharp. If the invasion of broom-sedge was due solely to its apparently slow growth in the juvenile stage, the tall weeds and this grass would appear simultaneously. This does not occur. Not until the weeds have become well established does broom-sedge invade. And we notice that it rarely enters communities of annual weeds. This can be explained by the comparative growth of the annual and perennial weeds in relation to their influence on the soil and aerial habitats during the spring season. Annuals do not grow to any extent until the middle of May. Summer annuals probably do not germinate before that time. Thus, the shade produced by these plants is not sufficient to keep the surface soil from drying out before the intermittent drought periods of early summer come on. On the other hand, the tall weeds which arise from perennial roots and reach an average height of 25 cm. by the first of May, create a different moisture condition in the surface soils. The surface soil in plant communities of this height does not dry out appreciably in the short time between rains. The rainfall from April 26 to June 15 in 1931 averaged 1 rain every 3 days. The growth of broom-sedge under the weed cover is usually possible throughout May at least. The increasing infrequency of rains after this time would tend to reduce seedling establishment. The shade of the weeds introduces another factor which may be advantageous to the broom-sedge. Transpiration from the seedling grass may be reduced considerably. The low light value in shaded communities is apparently of little consequence to the broom-sedge.

Once established under these conditions, the grass is able to compete with the weeds. The root system so thoroughly ramifies the soil that the weeds are finally eliminated. The keen competition broom-sedge offers was indicated in the reduction of the weeds in the aster quadrat. Wherever broom-sedge is found with other herbaceous vegetation the severity of competition may be seen. And it is only under the influence of the tree vegetation that broom-sedge, once established, shows marked suppression.

The exclusion of the many minor species from the tall weed communities is probably due to their intolerance to low light. Many of the same plants that are found in open fields, appear again in broom-sedge communities. Their representation is always of minor importance. Cud weed, boneset, sensitive pea, and Carolina golden rod are never found in closed communities. While not all of these appear in open fields they often are present in broom-sedge fields. That Bermuda grass is a strong competitor for the fullest expression in an area, is shown by the absence of these plants. The habit of growth is probably the limiting factor. Many of them are annuals. The early growth of Bermuda grass might be sufficient to eliminate these plants. Of the perennials, the Carolina golden rod is known to begin growth comparatively late in the spring. Its possibilities of survival are best in broom-sedge, which seems to develop less rapidly than the Bermuda grass. The growth habit and the water factor probably account for the appearance of these plants in the one open community and their absence in the other.

Another influence in the distribution of vegetation, often present in fields, is the galled areas. The nutrient-containing soil is eroded in such instances, and the soil which remains is such that very little water penetrates. Consequently a major part of the water that falls on these red clay surfaces quickly runs off. Hence, the plants growing on the galled areas must be very xeric in nature.

Where poverty grass or button weed is found in abundance it is safe to conclude that the soil habitat is of an unusually dry, sterile type. These plants play the dominant rôle year after year. They are freed almost entirely from competition with broom-sedge, and only gradually are weeds able to ecize. Even then, the number of these invaders does not increase to any extent over a period of years. When low nutrient and low water factors are involved jointly, the change in vegetation is brought about by improvement of the texture and the nutrient conditions. This, of course, requires a considerable lapse of time.

It should be mentioned that where the subsoil appears on the level, revegetation is more rapid. Nutrient accumulates more rapidly, and more favorable water conditions exist in these situations than on slopes.

SUMMARY

1. There are three definite stages in the revegetation of abandoned fields in the vicinity of Raleigh, North Carolina. While there are a few variations in the species' representation throughout the state, the general trend is the same. The pioneer stage may involve several species. On the thin soil, poverty grass, ragweed, and button weed become the dominants. The more fertile soils become vegetated commonly with crab grass. Bermuda grass is less common. The intermediate stage is always represented by the tall weeds. The several species occupy most any soil habitat. It seems that dog fennel is most particular and favors the sandy soils. Broom-sedge becomes established following the ecesis of weeds. Usually the pines follow this grass.

2. Studies of seedling growth showed very clearly why the three stages in the prisere exist.

3. Quadrats established in various plant communities gave an index to the nature of competition and the resultant successional trend in the prisere. A terraced slope transect introduced a number of differentials which were unusually helpful in carrying out the analysis of priseral development. The texture-water complex proved most important here.

4. Soil point data showed no great difference in the water supplying power of the soil in the several communities. The very coarse soil gave a low reading. The soil point data from red clay soils show clearly why the more xeric plants become the pioneers.

5. The light values were relative. There was a considerable response to reduced light in the *Aster-Cynodon* mictium. And where this factor was not involved in annual short weed communities, crab grass held on.

6. Ecads proved valuable as indicators of small variations in the habitats.

7. This study leads to several definite conclusions: The various soil habitats produce no important changes in the priseral sequence. The rapidity with which crab grass grows, and its ability to withstand longer drought periods than the other species, is responsible for its position as the pioneer. The slower-growing weeds can not become established before the year following abandonment. Slow seedling growth and susceptibility to prolonged drought are probably most influential. The modification in the surface soil moisture brought about by the tall weeds makes possible the invasion of broom-sedge. While the series may be retarded for a time, as soon as the tall weeds ecize abundantly enough to decrease the water loss from the surface soil, broom-sedge will become established. Broom-sedge does not enter bare areas, nor areas with short vegetation, because of the sensitiveness of the seedlings to even slightly dry soil conditions. Both crab grass and Bermuda grass are affected by reduced light. Hence, the establishment of weeds means the elimination of these pioneers. The suppression of the weeds is due to the competition set up by the perennial root system of the broom-sedge.

[*Editor's Note:* Plates 18, 19, and 20 are not reproduced here.]

5

Reprinted from *Biol. Bull.* **21**(3):127–151 (1911)

ECOLOGICAL SUCCESSION.

II. Pond Fishes.

VICTOR E. SHELFORD.

I. Introduction.

In the first paper of this series ("Stream Fishes and Physiographic Analysis"), we pointed out the reasons for undertaking this investigation and stated the purposes around which the work has centered. There we were concerned with the value of the principles of physiography in a method of locating the animal in the environment and of determining something of its character as a whole. Here we are to discuss the value of plant succession in a similar way and to have a better opportunity to show the validity of the principle of succession as applied to animals. Furthermore, as we pointed out in the other paper, ecological succession is to be differentiated from geological succession. Ecological succession is succession of ecological types regardless of species, while geological succession is the succession of species. The data presented here affords an excellent opportunity to bring out the differences and relations of these two types of succession.

We noted also in the first paper that the first recognition of plant and animal succession came with the development of genetic physiography. It was mainly the successsion which accompanies physiographic change. Cowles ('01) also clearly recognized succession due to the action of plants themselves. This latter idea has been elaborated by Clements ('05) and essentially demonstrated by Schantz ('06) and Dacknowski ('08). Animals must obviously play an important rôle in this type of succession, but unfortunately this has not been investigated.

The succession with which we will deal in this paper is that resulting fron the action of organisms on their own environment. For all practical purposes the area selected for this study has been in a condition of physiographic stability for a considerable period. The selection and analysis of the place of study is the most important step in the whole investigation. Indeed there are only a few suitable localities in North America.

II. Area of Study.

Owing to the fact that succession is always either dependent upon, or modified by changes in conditions, a correct interpretation of this phenomenon depends largely upon accurate knowledge of the area under consideration.

1. *Location and General Character.*—The ponds which are the subject of this study lie in the sand area at the south end of Lake Michigan, within the corporate limits of the city of Gary, Ind. They may be reached from stations known as Pine, Buffington, or Clark Junction. This locality is characterized by a large series of sand ridges, for the most part nearly parallel with the lake shore (Map I.). Their average width is about 100 feet. They are separated by ponds which are somewhat narrower (Map II., p. 131). Most of these ponds are several miles long. They vary in depth during the spring high water, from a few inches to four or five feet. They describe an arc somewhat longer than the lake shore (Map I.), and are farthest from it about midway of their lengths.

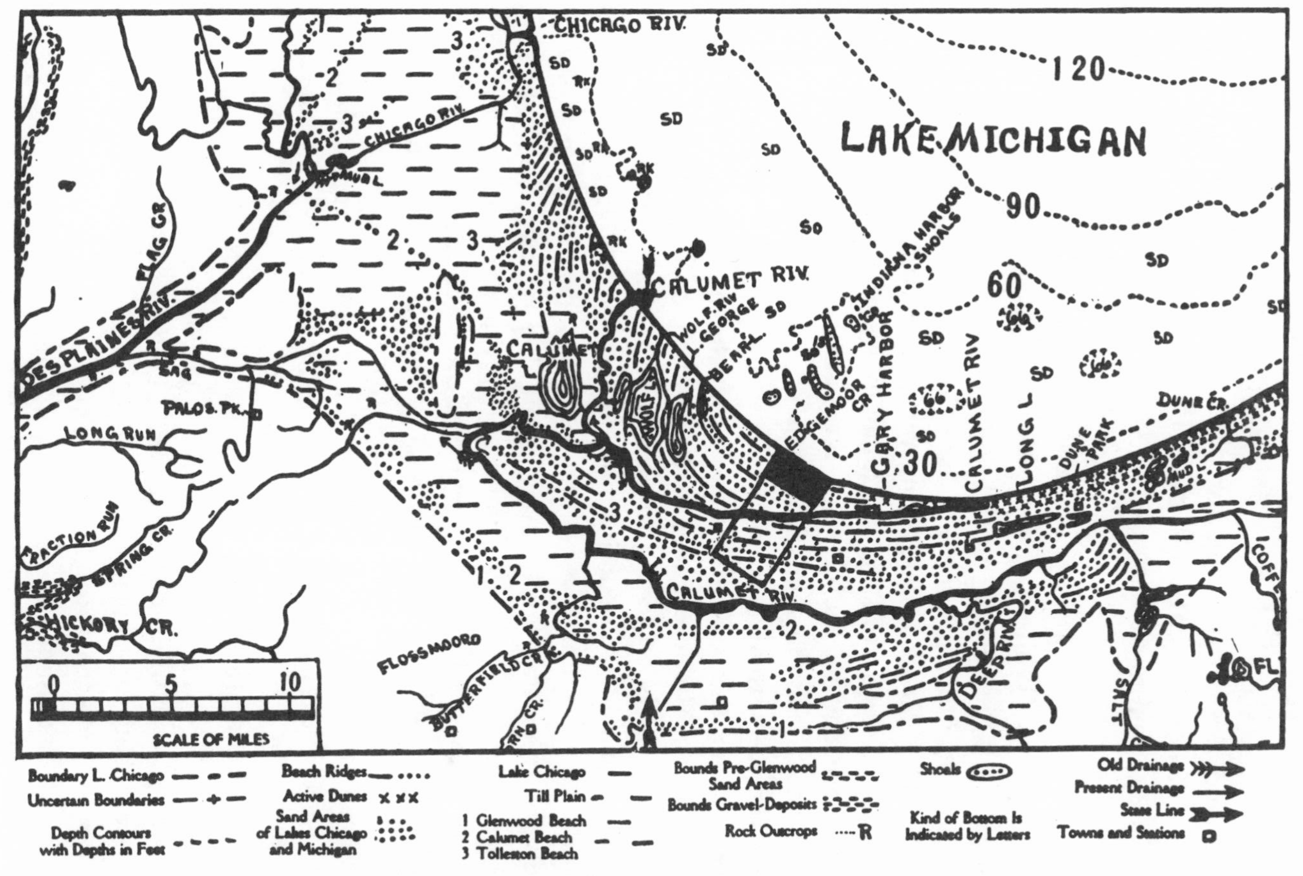

MAP. I. This shows the arrangement of deposits at the south end of Lake Michigan. The general direction of the beach ridges is indicated and accordingly the general directions of the ponds. The number of such ponds and ridges cannot be indicated on a map of this scale. The former extent of Lake Chicago and subsequent lakes is indicated. The area in which the fish have been studied is represented by the larger rectangle. The black rectangle represents the area covered by Map. II.

Originally, there were probably a number of outlets to the system of ponds which were joined to one another by these outlets and through various low places in the sand ridges. In the area of our concentrated study there is an outlet (Map II.) which has served to connect all the younger ponds in an intimate fashion.

The building of sewers associated with the growth of northern Indiana towns (Whiting, East Chicago, Hammond and Gary) has drained large portions of the ponds, while roads and railroads have isolated other portions.

2. *Origin of the Ponds.*—During the final retreat of the North American ice sheet, its Lake Michigan lobe stood for a time with its southern end at the crest of the Valparaiso moraine which lives concentrically around the southern end of Lake Michigan. When the ice retreated from this position, water occupied the space between this lobe of the sheet and the moraine. This was Lake Chicago, the forerunner of Lake Michigan. After having stood respectively 60 feet, 40 feet and 20 feet above its present level, long enough to deposit conspicuous beaches, it took up a position at a 12-foot level. The waters appear to have fallen gradually from this level. At present one or two ridges and depressions similar to those found above the water on old Lake Chicago plain, are below the surface of the water along the shore at the south end of Lake Michigan. The retreat of the water has evidently exposed such ridges as fast as they were formed. This has left a *series of parallel ridges and ponds arranged in the order of their age—the oldest farthest from the lake, the youngest nearest the lake.* [1]

These ponds are not all of the same size. The largest ones were selected for study and will be referred to in the paper by number as the entire series is counted inland from the lake shore. There are between seventy-five and ninety of these ponds or depressions between the lake shore and the 20-foot beach level. This is the maximum number. Map I. shows that as we pass in either direction from the area of study their number decreases.

[1] For a treatment of this subject, see articles cited in the bibliography but not specifically referred to here. Professor R. D. Salisbury tells me that there is no question concerning the relative age, from physiographic evidence alone.

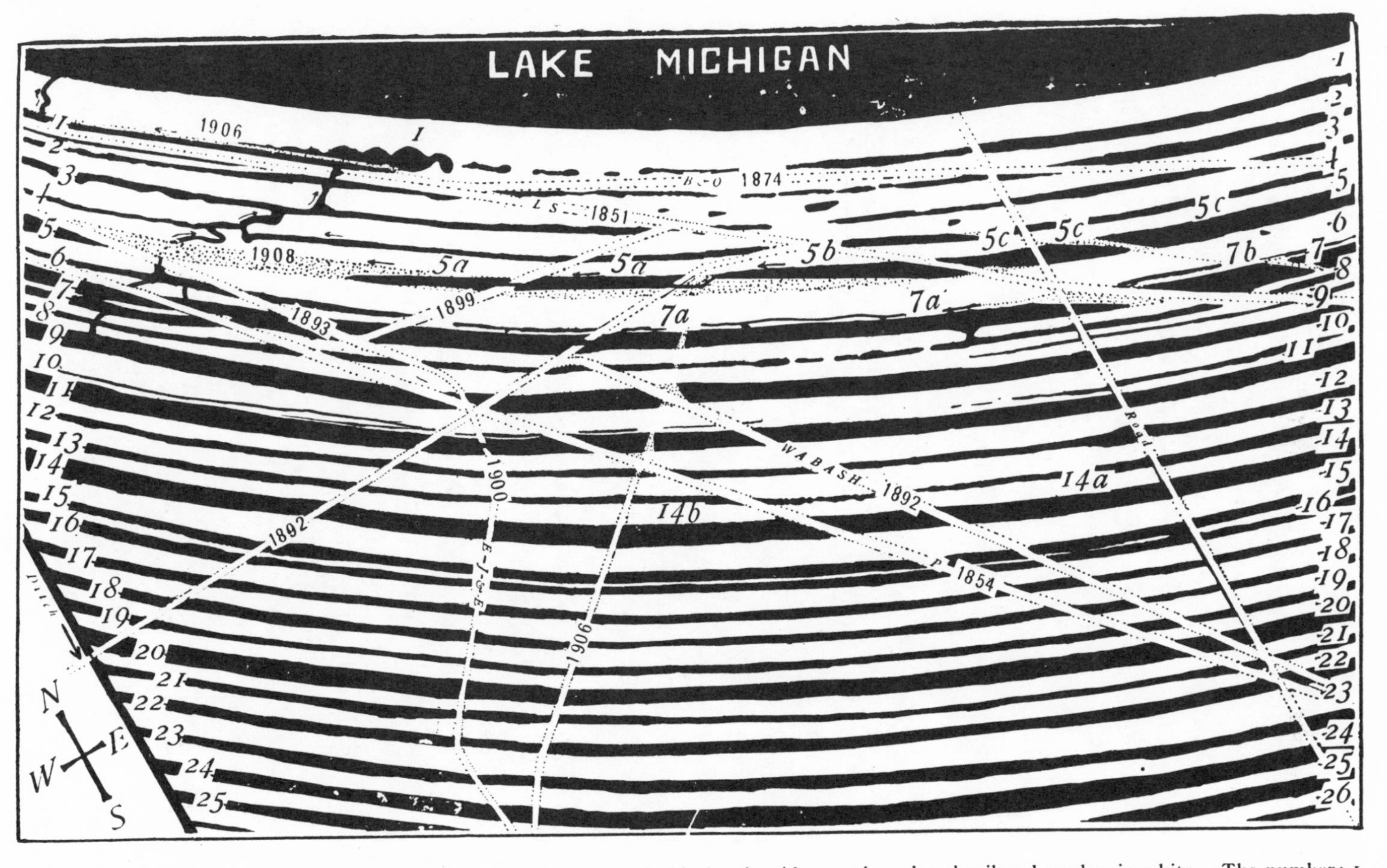

Map II. Shows the area of special study. The water is shown in black; the ridges and road and railroad grades, in white. The numbers 1, 5, 7, 14, etc., refer to the numbers of the ponds and ridges opposite which they are placed, as these ponds and ridges are counted from the lake shore. The dates associated with the indicated railroad grades refer to the approximate date of building of the grade, and probable damming of the long ponds where sluices were not maintained. The dotted lines are to bound the railroad grades; stippled areas represent recent filling.

Map II. shows the relations of the first twenty-four of these ponds. The recent changes of this smaller area are indicated on this map by the dates of the building of the various road and railroad grades.

III. The Data.

All methods of collecting have been employed. The dip net has been found most effective, but the drought of 1908 and draining by sewers have yielded cruical results in Ponds 1, the outlet, Pond 5*a*, 5*c*, 14*a* and 7*b*.

With the exception of *Amia calva*, the records are for adults, or for young where the pond is known to be *isolated*, and the presence of adults is therefore necessarily implied.

The record of *Ameiurus melas* in Pond 56 is based upon the presence of young and adults; identification of the adults depended upon identification of the young, because the adults were too badly macerated to make identification practicable.

1. *Distribution of the Species of Fish.*—The distribution of the fish in the various isolated parts of the ponds is of much importance. Turning to Map II., we note that the extension of the ponds to the east and west is not shown. Toward the east Ponds 1 to 17 become shallow near the Gary Steel Plant and are *not* connected with other bodies of water. Parts of Ponds 1 to 7 have been directly connected with the lake by the outlet (Map II.) until within the past few years. Excepting in the case of pond 7, other connections with the lake, to the northwest of the area shown on our map, have probably been closed for a much longer period. We may concern ourselves largely with the parts of the ponds to the east of the outlet, which have been isolated by railroad grades, and with their relation to the outlet.

Table I., taken with Map II., presents the exact data on the distribution of the fish in the various parts of the ponds. All fish nomenclature is after Forbes and Richardson, '08.

In Table I. we note first that the large-mouthed black bass, the sunfishes, the pumpkin seed, and the warmouth, are all confined to the outermost pond, or Pond 1. The passage from pond 1 to the outlet and to pond 5 was open until 1906. The study began in 1907. In the autumn of 1908 the drought reduced the water in the outlet to a minimum, but none of these species was found there, though the passage between the outlet and pond 1 had been open two years before. Why have the fish vacated or avoided the outlet? One can only suggest a behavior reaction as the cause.

TABLE I.

DISTRIBUTION OF THE FISH.

The letters and numbers at the heads of the columns refer to the various isolated parts of ponds and, excepting 0 which refers to the outlet, may be located on Map II. The last column indicates the occurrence of fish in the older ponds of the series, which are not found on the map; the numbers refer to the number of the pond in which the fish were found. ? indicates an incomplete identification.

Common Name.	Scientific Name.	Ponds.									
		1	0	5*a*	5*b*	5*c*	7*a*	7*b*	14*a*	14*b*	
Large-mouthed black bass	*Micropterus salmoides*	*									
Blue gill	*Lepomis pallidus*	*									
Blue-spotted sun fish	*Lepomis cyanellus*	*									56
Pumpkin seed	*Eupomotis gibbosus*	*									
Warmouth bass	*Chænobryttus gulosus*	*									
Yellow perch	*Perca flavescens*	*				*					
Chub sucker	*Erimyzon sucetta*	*	*		*	*		*			
Spotted bullhead	*Ameiurus nebulosus*	*	*		*	*	*	*	?		
Tadpole cat	*Schilbeodes gyrinus*	*	*		*	*	*	*	?		
Pickerel	*Esox vermiculatus*	*	*	*	*	*	*	*	*		58
Mud minnow	*Umbra limi*	*	*	*	*	*	*	*	*		58
Golden shiner	*Abramis crysoleucas*		*	*	*	*	*	*	?		
Yellow bullhead	*Ameiurus natalis*						*				
Black bullhead	*Ameiurus melas*						*	*	*	*	58
Dog fish	*Amia calva*										15

The perch is distributed in a still more peculiar way. It is found in Ponds 1 and 5*c*, but *not* in the outlet or in Ponds 5*a* and 5*b*. 5*a* and 5*b* were connected with the outlet until 1907 but 5*c* has probably been separated since 1851, when the L. S. & M. S. R. R. was built. I have no evidence that a sluice was ever maintained under this railroad in Pond 5. The perch is then distributed in such a way as to necessitate the conclusion, either that it was artificially introduced or that it was once in the outlet and in Pond 5*a* and 5*b*, because these make the only passage to Pond 5*c* where it now occurs.

The chub sucker is in all the ponds up to 7*b*, excepting 5*a* and 7*a*. The passage to, from, and through Pond 5*a*, and other points where the chub sucker occurs, was open until 1907 and the crucial collecting was done in 1908. It will be noted that the distribution of the chub sucker in Ponds *7a* and *7b* is entirely similar to the distribution of the perch in Ponds *5a* and *b* and *5c*. The chub sucker is not in the part of the long Pond 7 which has been recently connected with the outlet.

It will be noted also that the tadpole cat and the spotted bull head have not been taken in *5a*. The yellow bullhead has been taken from *7a* but once, but we may expect that it is a resident of the locality. One would expect to find it in Ponds *5b* and *5c*.

With the exception of Ponds *14a* and *14b*, there are no noteworthy peculiarities. The study was begun in *14a*. This was partially drained in the winter of 1908–9 and it was therefore necessary to turn attention to *14b*. This was probably partially drained previous to 1892 by the East Chicago ship canal, but again renewed through a dam made by the building of the Wabash railroad grade between the point of draining and the part of the pond studied about 1892. This probably accounts for the presence of only a single species in Pond *14b*. A single juvenile blue-spotted sunfish was found in Pond 56 which is directly connected with the Calumet River where it is common.

2. *Ecological Age of Ponds.*—The ecological age of the ponds is determined by an inspection of (*a*) the amount of bare bottom, (*b*) the amount and kind of vegetation, and (*c*) the amount of humus. It is a well-established fact that an entirely new pond (in the matter of recent separation from a lake, like Lake Michigan) has little vegetation and very little or no humus. Both vegetation and humus come only with age. Age-determination is so simple that no difficulty usually is experienced by one trained in plant ecology, in arranging a series of ponds in the order of their ecological age. In the matter of the kind of vegetation we have had the advice of Dr. H. C. Cowles.

Pond 1 is the youngest, because it has the kind of vegetation that grows in young ponds, more bare sand bottom, least humus, and least vegetation. For similar reasons Ponds *5b* and *5c* stand second in the matter of age. Because of human interference, which has kept the vegetation down in Pond *5c*, it is probably ecologically younger than *5b*. The outlet is probably intermediate between 5 and 7. Ponds *7a* and *7b* stand next, but without any difference as far as one can see.

From the standpoint of bare bottom and humus Pond 5*a* is very much advanced. Being located near the outlet, it has become filled with decaying vegetation and a floating bog, now destroyed, indicated an advanced stage comparable with 14. It differs from 14 in possessing qualitatively "younger" vegetation and some other characters of youth. The differences between 14*a* and 14*b* have already been discussed.

In Table II. the ponds or parts of ponds are arranged in order of ecological age and the distribution of fish is shown.

TABLE II.

DISTRIBUTION OF FISH: PONDS ARRANGED ACCORDING TO ECOLOGICAL AGE.

Common Name.	Scientific Name.	Ponds.							
		1	5*b* 5*c*	7*a* 7*b*	5*a*	14*a*	14*b*	56 58	52
Large-mouthed black bass	*Micropterus salmoides*	*							
Blue gill	*Lepomis pallidus*	*							
Blue-spotted sunfish	*Lepomis cyanellus*	*						*	
Pumpkin seed	*Eupomotis gibbosus*	*							
Warmouth bass	*Chænobryttus gulosus*	*							
Yellow perch	*Perca flavescens*	*	*						
Chub sucker	*Erimyzon sucetta*	*	*	*					
Spotted bullhead	*Ameiurus nebulosus*	*	*	*					
Pickerel	*Esox vermiculatus*	*	*	*	*	*		*	
Mud minnow	*Umbra limi*	*	*	*	*	*		*	
Golden shiner	*Abramis crysoleucas*		*	*	*	*			
Yellow bullhead	*Ameiurus natalis*			*					
Black bullhead	*Ameiurus melas*			*	*	*	*	*?	
Dogfish	*Amia calva* (*juvenile*)								*

When the ponds are arranged according to ecological age, there is a noticeable symmetry in the distribution of the fish, and we are at once interested in its meaning.

IV. INTERPRETATION AND DISCUSSION OF THE DATA.

We have noted from the tables and discussion of tables that the present distribution of the fish is correlated with (A) the age of the ponds (Table II.) and (B) the length of time that channels have been open (Table I.).

It is a well-known fact that ponds fill with plant detritus. With the filling of ponds, the conditions change progressively in a definite direction. In a pond which remains in the same general physiographic relations during a considerable period of time, there is a succession of conditions due to the accumulation of detritus just as there is a succession of conditions in a stream due to physiographic changes.

Since the channels of communication between the different ponds have been open until recently, the present arrangement of the species of fish is very probably a *behavior adjustment* to the changed and changing conditions just so far as barriers have permitted.

We see from the arrangement and mode of origin of the ponds that our oldest pond—number 14, Map I., and Fig. 1, was once in the same relation to the lake as Pond 1 is now. At such a time it was in a condition similar to that of the present Pond 1. Ponds 5 and 7 are intermediate in conditions between Ponds 1 and 14. We have the same general basis for the discussion of ecological succession as in the streams. The changes in these ponds have depended mainly upon physiographic stability within each pond, rather than upon physiographic changes, and have been due to the *action of the organisms present on their own environment.*[1] This is true because after a given pond is once separated from the lake (Fig. 1) the changes due to the organisms go on without regard to further separation and the lowering of the lake level. Evidently the level of the water has remained much the same in the ponds after their separation from the lake proper regardless of the lowering of the lake level (Fig. 1).

The method of deducing succession herein employed is similar to that used in the case of the streams. The easily observable fact that animals occupying similar conditions are ecologically similar (*i. e.*, similar in habits and some main features of their physiology of external relations) is used as a starting point, and the conclusions drawn are to the effect that when the older habitat was in the stage of a younger habitat, it was occupied by *fishes ecologically similar to those now in the younger habitat.* Whether they were the same or different species is often of little importance to ecological succession. With this simple explanation as a background, and with the use of Figure 1, we will

[1] The changes which are caused by the filling of the pond with plant material are physiographic, but the biological aspect is the more important, and we may discuss the changes as caused by biological forces.

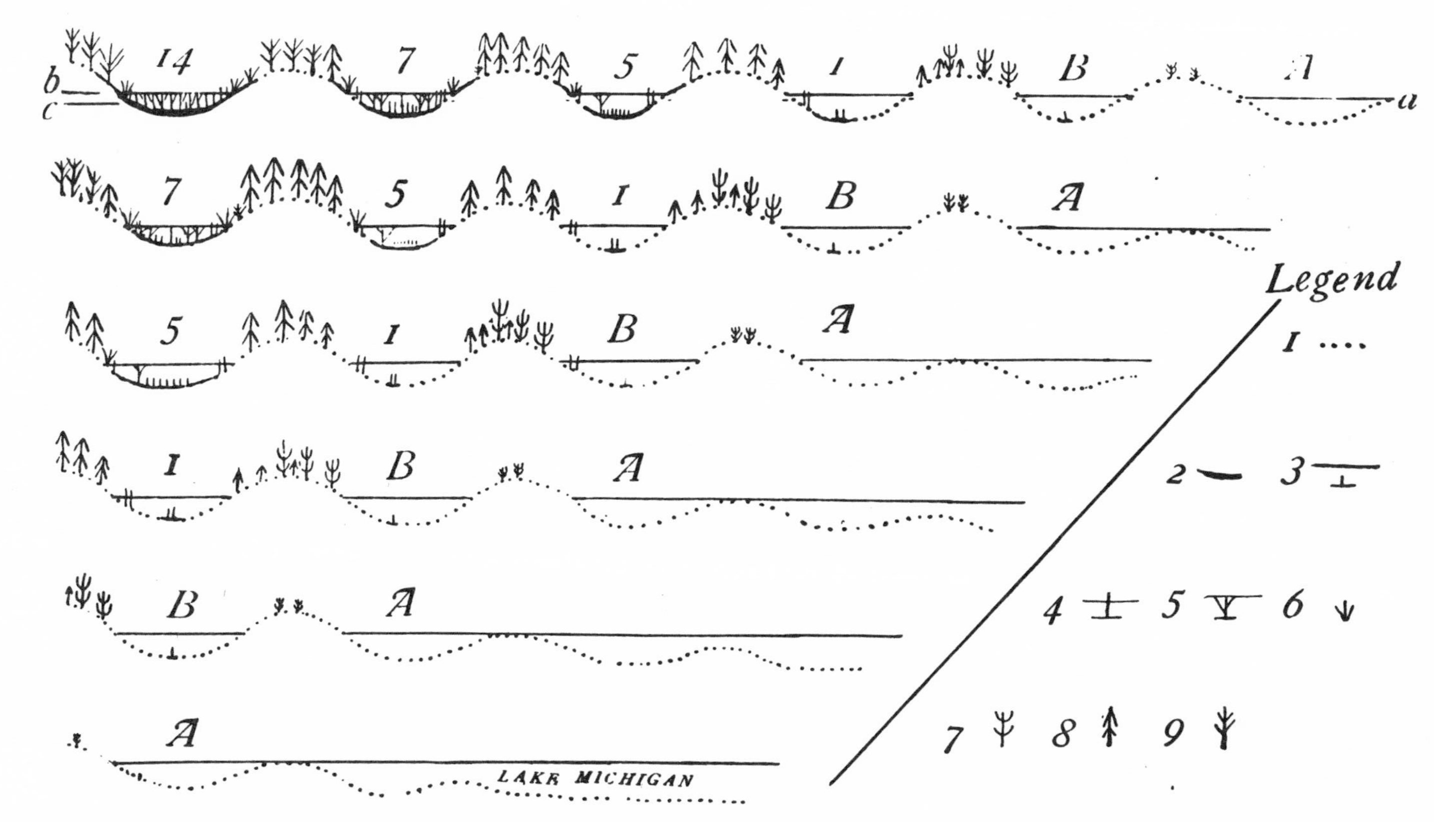

Fig. 1. Diagram showing the relation of the horizontal and vertical series of pond stages. The horizontal series at the top represents the present ponds with the intermediate ones omitted and hypothetical stages *A* and *B* added. The line *a–b* represents the ground water level; the line *a–c*, indicated by occasional dashes, the level of the lake. When read from bottom to top the entire figure represents the history of the area; showing the addition of ponds near the lake shore and the aging of the others.

The left-hand vertical series (A, B, 1, 5, 7, 14,) represents the history of the present Pond 14, and is used as the basis for the discussion of succession. The legend is as follows: (1) sandy bottom, (2) humus, (3) plants that do not reach the surface, (4) bulrushes, (5) aquatic plants which reach to the surface of the water, (6) shrubs, (7) cottonwoods, (8) pines, (9) oaks.

state, as well as practicable, the succession in these ponds with particular reference to fish.

1. *Statement of Ecological Succession.*—A statement of succession can be best made in the form of a history of Pond 14 (Fig. 1). From our knowledge of the origin of the ponds and their present topography, it is reasonably certain that there was a time when Pond 14 was more closely associated with the lake than Pond 1 is at present (Fig. 1, hypothetical stage B). At such a time Pond 14 contained less vegetation than we now find in Pond 1. For a knowledge of the ecological character of the fish which inhabit such ponds, we have collected from a pond of the same origin as those made the subject of the present study. This one has maintained a close connection with Lake Michigan and receives the waves of the lake during the spring and winter storms. It flows into the lake during every highwater period. This pond is at Beach Station, four miles north of Waukegan, Ill. Near its outer end it presents a clear bottom of sand and gravel, little vegetation and no humus. In this outer portion we have collected the pike (*Esox lucius*), which prefers clear, clean, cool water (Forbes and Richardson, '08); the red horse (*Moxostoma aureolum*), which dies quickly in the aquarium if the water is the least impure and succumbs to impure conditions in its native waters (Forbes and Richardson, '08); *Notropis cayuga*, common in Cayuga Lake and the lower course of its tributaries (Reed and Wright, '09). As compared with Illinois waters, these streams would be counted clear. We also found *Notropis cornutus*, which shows a marked preference for clear waters (Forbes and Richardson, '08).

When Pond 14 was in a very early stage (hypothetical stage B, Fig. 1) it must have been occupied by fishes which were ecologically similar to the red horse, the pike, the cayuga minnow, and the common shiner. This is a community of species which may be characterized as requiring clear, clean waters, clean bare bottom (especially during the breeding season), and little vegetation. Such fishes may be designated as *pioneer ecological types*. To this group might be added such fishes as the common perch, which is hardy and lives in a wide range of conditions.

Following the history of Pond 14 further, we note that as the vegetation grew, humus accumulated in the deeper parts, and forms ecologically similar to the black bass, the blue gill, the pumpkinseed, and the white crappie (*Pomoxis annularis*) which is widely distributed (Forbes and Richardson, '08), but which Meek and Hildebrand ('10) record only from streams and ponds which are ecologically young, must have made their appearance, if any such fishes were available. All but one of these species were taken from the pond at Beach but chiefly from the older parts. They were not clearly separated from the other species, but conditions were graded and no barriers present. We would expect the absence of some of the pioneer ecological types, such as the pike and the red horse, from such a community.

When the conditions in Pond 14 which made the habitat suitable for the last mentioned fish community had progressed a little further, *all the organisms present must have sufficiently affected the habitat to render the continued existence of some of the pioneer ecological types impossible, and at the same time prepared the way for new ecological types*, such as the blue spotted sunfish, the little pickerel, the chub sucker, the warmouth bass, the spotted bullhead and tucked away in the mud of the deeper portions a few fishes ecologically similar to mud minnows and tadpole cats.

This is the condition of our Pond 1 at present—a shallow pond with considerable accumulations of humus in the deeper parts, and much of its bottom covered with *Chara* and other plants. In this connection it should be noted that Meek and Hildebrand ('10) record the crappie and top minnow (1900) from Pond 1 (Map II.), but they are no longer found there.

It should be remembered that the parts of a pond which are so situated as to most easily allow the accumulation of plant detritus will be first to become unfavorable to pioneer fishes, while the parts which are swept clean by wave action remain in a good condition for a much longer period of time. Stages in pond development such as the present Pond 1 are the most complex of all, and are more like the larger lakes. Such stages are in a condition to support a greater diversity of ecological types than the later stages. While still retaining a part of its pioneer fishes, Pond 1 supports the forms which belong to other older conditions (mud minnow and tadpole cat).

Up to such a stage, Pond 14 must have become from the first more and more favorable to diversity of ecological types, and accordingly possessed at such a time its greatest number of species of fish. When Pond 14 was at a stage comparable to the present Pond 1, the fish community present and all the other organisms associated with it, so acted on their environment (just as they are acting on their environment in Pond 1 at present) as to make the habitat less favorable to the fish of the earlier pond stages, and more and more favorable for those dependent upon and tolerating dense vegetation, absence of bare bottom, and lower oxygen content. As a result of this action of the biota on its environment, the fishes of ecological constitution similar to that of the sunfishes and basses now present in Pond 1, disappeared either by emigration or death. In the absence of these, and in the more favorable conditions of competition and denser vegetation, fish, such as the golden shiner, were able to find a suitable habitat. At such a stage Pond 14 possessed a fish community of the ecological character of that now found in Pond 5*c*. Here the pioneer element is reduced to a single species, the perch, which is very hardy (Hankinson, '07).

The same process continued and caused the disappearance of the perch and like ecological types. A fish community ecologically like that now in Pond 5*b* (Table I., column 3) then existed. The absence of the perch in 5*b* and its presence in 5*c* may be explained, judging from the general habits of the perch (Forbes and Richardson, '08), by the fact that neither pond appears to be favorable for perch and they have been able to move out of Pond 5*b*, but not out of Pond 5*c*. An experimental comparison of the behavior of perch from Pond 5*c* and other perch habitats would have an important bearing on our problem.

The fish community of Pond 5*b* is made up of the chub sucker and the golden shiner, which are abundant, the spotted bullhead, the tadpole cat and the mud minnow. The spotted bullhead is the only one known to use bare bottom for nesting. There is only a little bare bottom in Pond 5*b*. The spotted bullhead usually builds its nest under cover (Eycleshymer, '01).

When such a fish community occupied Pond 14, the biota present gradually changed its own environmental conditions as the former stages had done. The nature of the changes are evident now when we can compare Ponds 5*c* and 5*b* with Ponds 7*a* and 7*b*. In ponds 7*a* and 7*b* we find water lilies and bladderwort in abundance. These are barely present in 5*b* and 5*c*. Ponds 7*a* and 7*b* have much less bare bottom than 5*b* and 5*c*.

FIG. 2. Shows Pond 1 at the extreme low water of the drought of 1908. In the spring the old boat is usually covered with water. In the foreground a large area of bare sand bottom is shown; to the right a few rushes and sedges. The absence of shrubs near the water's edge should be noted.

This is present in favorable situations very near the water's edge.

The fish community of 7*b* is made up of the same species as the community of 5*b*, with the addition of the black bullhead. 7*a* lacks the chub sucker, and here as has been noted the chub sucker bears the same relation to the two parts of Pond 7 as the perch does to Ponds 5*b* and 5*c*. It is evident that it moved out of the pond from which a channel of exit has been open.

The first step in the transformation of the fish community of Pond 14, which was ecologically similar to the present 7*b*, into the next later stage, was the loss of such ecological types as the chub sucker and the addition of such as the black bullhead. The latter is a well-known "mud-lover" which inhabits the oldest stages of pond succession. As the conditions which organisms produced increased in intensity in the directions indicated, the species ecologically similar to the tadpole cat and the spotted bullhead disappeared, and with them probably the golden shiner also. With the loss of these we have the fish community which was in Pond 14*a* before draining. It is illustrated by Pond 5*a* also, in which the plant and animal detritus has collected

FIG. 3. Showing Pond 14 at moderate low water. In contrast with 1, we see that it is choked with vegetation and the margin occupied by shrubs and bulrushes, etc.

because it is located at the outlet where the channel becomes shallower and narrower. Ecological types, such as the pickerel, the black bullhead, and the mud minnow make up this community in Pond 14. They usually continue until the pond becomes temporary when they are destroyed by drying.

2. *The Future of the Pond.*—Ecology is one of the few biological sciences in which prediction is possible. I shall venture to predict the fate, in a state of nature, of a pond like Pond 14*a* *was* at the

point where the study was first begun, before the draining took place.

Had Pond 14*a* not been disturbed by man, it would have continued to fill with humus. Its center would have come to be occupied by the cat tails, bulrushes, *Proserpinaca*, and *Equisetum*, all of which invade from the sides. When such a stage was reached it would be subject to almost complete drying in extreme droughts and the fish would be eliminated in the order of their ability to withstand draught, and in relation to accidents of distribution in parts of the pond drying to various degrees.

In September, 1908, we were able to obtain evidence on this point from some of the older ponds. A group of black bullheads, grass pickerel and mud minnows were found in the lowest part of one of the sections of pond 58. The pickerel were dead and badly decayed. The bullheads were dead and just beginning to decay, but the mudminnows, fully a bushel of them, were still alive although without water. This lot included some of the largest individuals collected.

When Pond 14 reached such a stage, its fish content would become very problematical, dependent upon the distribution of the different species with respect to deeper and shallower parts of the pond, and the length of the drought, as related to the resistance of the different species. Evidence for this is seen in Pond 14*b*, where *Ameiurus melas* is the only species present since a probable partial drying of the pond.

Indeed, we may carry the prediction one step farther. When such a pond becomes of this sort the plants that take hold of it are the grasses, sedges, etc., which fill the soil with roots, and form hummocks. If accessible to them, such ponds become the breeding place of the dog fish. During the drought of 1908 I found a school of half grown dog fish in such a pond, which answered the description of the favorite breeding place of the dog fish as given by Reighard ('02). Such a stage as this will be followed by the invasion of shrubs and the final destruction of the pond as an aquatic habitat.

We have in ponds a progressive change in the conditions and a progressive change in the ecological and physiological character of the fish communities, and a succession or evolution of fish communities through emigration, death, immigration, and the modification of *mores* by external stimuli.

V. Ecological Succession and Succession of Species.

The difference between ecological and geological succession were suggested in my preceding paper. We noted that ecological succession deals with the succession of *ecological types*. We noted also that geological succession is due, in the main, to the death of a given set of *species* and the evolution of new ones throughout geological time. While this is true of the broader aspects, in the more detailed cases and especially in dealing with recent post-glacial fossils, the palæontologist often encounters a vertical succession of fossils which have been left behind by the migrations of a succession of species over the locality of fossilization (Warming, '09, p. 362, Adams, '05 and '09, Sharff, 07, esp. Chap. IX. and citations). Palæontologists may also encounter vertical succession of fossils in situations where such succession as we have been describing has taken place.

1. *Vertical Succession of Fossils.*—Steenstrup ('41) found from the study of fossils in moors that one kind of vegetation succeeded another. Various other workers (Andersson, '97) have found similar arrangements. In the case before us the fossilization of species would give a vertical succession of fossils.

Turning again to the diagram, we note that the skeletal parts of any fish in the earliest stages indicated by hypothetical stage B might have been preserved as fossils. The accumulations of humus which lead up to stage 1 would have covered the fossils of stage B, the fish of stage B would be present, if at all, as fossils at the bottom of the pond.

Likewise, the accumulations of humus which led to stage 5 covered those skeletal parts of the fish of stage 1, and the fish of stage 1 should be found as fossils overlying those of stage B and underlying those of stage 5. Again, the accumulations of humus which led up to stage 7 covered fossils which were present and added at stage 5, and the fish preserved as fossils from stage 5 would lie above all those preserved as fossils from the younger stages and below those of stage 7. Fossils from stage 14 and the intermediate stages would lie at the top of the series and above those of stage 7.

However, no such arrangement of fossils has been found in these ponds. No attempt to ascertain whether or not they are present has been made.

2. *Relation to Climatic Changes.*—As has been indicated, succession is due to changes in conditions which make it impossible for a given group of organisms to continue to live at a certain locality. Accordingly, such a group gradually disappears and is gradually succeeded by a group which is adapted to the new conditions.

The changes in conditions referred to are climatic or physiographic changes, and others independent of physiography and climate. The last type of change is usually due to the action of the organisms themselves. These three forces may act separately or together. The latter is probably the more common. In case the changes are climatic, both the ecological succession and the succession of species over a locality might be partly due to migration of a succession of species. Narthorst ('70, fide Warming, '09, p. 362) found that the oldest fossils underlying moors are of arctic tundra plants. In post-glacial dispersal arctic-tundra plants and animals were succeeded by conifers and associated animals, and by deciduous trees and associated animals (Adams, '05).

3. *Relation of Horizontal Series and Vertical Succession.*—Cowles ('01) pointed out the fact that there is frequently a horizontal series of successional stages (habitats of different ages) associated with continuous physiographic processes, such as the deposition of sand along shores. The diagram (Fig. 1) illustrates both the horizontal and vertical series of conditions with which we have been dealing. Cowles ('01) further pointed out that the horizontal series of plant communities must bear a close ecological resemblance or be practically identical with the vertical series of past plant communities which have succeeded one another over a given locality in the older part of the horizontal series. He pointed out further that the horizontal series may be taken as an index of what the vertical series has been. He would except the first stages of plant communities which occupied a locality at the close of the ice age and which as has been stated, are arctic plants. These arctic plants, however, must have affected

the soil or pond bottom in much the same way as the pioneer plants of the present climate would affect them. The differences resulting from the change of climate are then, those of detail rather than of principle.

4. *The Relation of Ecological Succession to Species of Fish.*—If fish were found fossil in the bottom of Pond 14, in the order which we have indicated, one might conclude at once that it constituted a proof of ecological succession. This seems to have been the general impression of zoölogists who have heard the presentation of these data. The question has been asked, "Do you find fossil in 14 all the fish which you mention as occurring in succession there?" My answer to the effect that no such fact has been discovered seems to have been regarded as constituting a refutation of the entire statement of ecological succession.

If fish were found fossil in the order described above, and the species and order of species, the same as now found in the horizontal series of ponds, we would have some important data bearing on succession, migration, and other matters of interest to be discussed presently, but this would yield *no crucial evidence for or against ecological succession.* Ecological succession is based upon physiology, habits, behavior, mode of life, and the like, which I have proposed to call *mores* (opposed to the term form). Unless the mores of the morphological species found fossil were the same as the *mores* of the same morphological species at present, they would have no weight in the matter, and it would be impossible to ascertain mores from fossils if such fossils were found.

If the same one or more species were found fossil in each and all of the vertical stages of a pond like 14, the evidence would not refute the proposition of ecological succession because the physiological characters of the individuals of a given species living in the early stages could have been very different from those of individuals living in later stages, without the differences being shown in the preservable skeletal structure. Furthermore the modifiability of animal behavior seems well established. The same species of plant may remain in a number of different pond stages. Such plants show suitable functional responses manifested by different growth-forms in different stages.

The proof of ecological succession must rest on the results of experimental work on the animals of the different stages of such a horizontal series. *It matters not from the point of view of ecological succession what the forms are, so long as the mores are different in the different stages of the horizontal series, and the character of the mores is correlated with the environmental conditions.* When such differences and relations are found, ecological succession is essentially established. It is of course recognized that within rather uncertain limits the *mores* of a morphological species remain, in a general way, the same throughout its geographic range. This kind of reasoning must be pursued with caution. When applied in detail or to "common" species which are wide ranging, it is not at all trustworthy. If, for example, an examination of the deposits in the bottom of a pond, such as Pond 14, showed the presence of a fish which now habitually inhabits the *youngest stages* of pond succession, let it be a more northern, or the same species as those in the horizontal series under consideration, it would constitute some evidence for ecological succession which could be further checked by the study of the modifiability of the *mores* of that species at present. However, in no case can this kind of evidence be crucial.

Furthermore, the conditions such as were in primaeval Pond 14 could have been reached by a pond like Pond 1 in a few hundreds of years. In a pond with a soil more suitable than sand for the growth of aquatic plants, these changes could take place within the life time of the builder of such a pond (Knauthe, '07, p. 575).

VI. General Discussion.

We noted in the preceding paper that physiographic analysis is a method. Here we have employed another method which, while not physiographic, is similar in principle, since it deals with succession over a given locality. The cause of this succession is biological and the biological interest is proportionately greater.

We have here the same attempt to discover something of the physiological character of the organism as a whole, and to classify organisms on the basis of the conditions in which they are most nearly in physiological equilibrium. Again, the *mores* of the fish of pond stages in other parts of the world are probably similar if the ponds are similar.

However, a legitimate question at this point would be: "What is the value of all this reconstruction and these complicated conceptions to biological science and the analysis of the organism?" Indeed, this is a question which we have asked ourselves repeatedly while working through the necessary plant data and all the scattered unorganized literature with which the investigator in this line must deal. The question has always been answered in a manner satisfactory to us because whatever value the papers on the ponds of the south end of Lake Michigan and other similar papers may possess, is due to the conception of pond succession acquired from the plant ecologists. Had I not acquired this knowledge I would have done as others have done—worked this excellent area over without seeing anything in it. The problems that have arisen in connection with this work have given a fresh point of view and a motive for investigation which has repaid the effort.

In connection with problems and motive for their solution, it should be noted that this is one of the most important things which the plant ecologists regard as of value in the conception of succession. Cowles, who has done more to stimulate work on succession and the use of physiographic method than any other American ecologist, regards functional response of plants (determined by watching plants and experimenting on plants) as much more fundamental than succession. This appears to be generally true of plant ecologists. Warming, who has been a great leader in ecology, emphasizes the physiological side.

We have, then, among the workers and originators of the use of this principle, a condition with respect to its place in biology which is quite different from what one would expect from simply noting what has been the dominant thing in their work. The relation of the historical and genetic side of ecology to more fundamental physiological and ecological problems, is similar to that of the historical and phylogenetic side of evolution, to the problems of biology and the motives for investigation. When Darwin framed the idea of evolution into a logic which was not refutable by academic attack, the facts of biology took shape, arranged themselves into orderly relations to each other. This made possible methods of work which were new, opened up new problems, the attempt at the solution of which has made modern biology what it is.

If we assume the attitude that nothing can be done in the organization of natural history materials into a science, we are closing the question to investigation, much as the idea of special creation closed other lines of biological investigation. Still an occasional biologist appears to take this point of view.

Aside from the general questions which we have been discussing, there are many practical applications of pond successsion to the study of behavior and to the economic and quantitative side of biology. These we will discusss in a succeeding paper under the head of the causes of succession in ponds.

VII. Summary.

1. There is a series of ponds at the south end of Lake Michigan arranged in the order of their age; age is determined by the physiographic history; by the relative amount of humus and bare bottom, and by the quality and quantity of vegetation.

2. Species of fish are arranged in these ponds in an orderly fashion; the order is related to the age of the ponds.

3. The ponds of different ages represent stages in the history of older ponds.

4. The horizontal series of fish communities is ecologically representative of the succession of fish communities within the older ponds.

5. The method employed here is similar to physiographic analysis; the motives and possible results are similar but more strictly biological, because the causes of the succession are the organisms themselves.

Hull Zoölogical Laboratory,
University of Chicago,
April 19, 1911.

VIII. Acknowledgments and Bibliography.

1. *Acknowledgments.*—The writer is indebted to Dr. H. C. Cowles for assistance with the plant side of the pond problem; to Mr. W. C. Allee for assistance in mapping the ponds. The identification of the fish is by Dr. S. E. Meek and Mr. S. F.

Hildebrand; some of the collecting was done in coöperation with Dr. Meek and Mr. Hildebrand. Without the assistance of these gentlemen this paper could not have been written. The writer is also indebted to Mr. Ellis L. Michael for the reading of the manuscript.

Adams, C. C.

'05 Post-glacial Dispersal of North American Biota. Biol. Bull., Vol. IX., p. 53.

'08 Ecological Succession of Birds. The Auk, Vol. XXV., pp. 109–153 Bibliography.

'09 Isle Royale. Biological Survey of Michigan (Lansing). Climatic and Geological Succession, p. 45, p. 31. Succession of Birds, p. 134. Succession of Beetles, p. 160. Succession of Mammals, p. 390. Bibliography.

Alden, W. C.

'01 Geological Atlas of the United States. No. 81, Chicago Folio U. S. Geol. Surv. Maps.

Andersson, G.

'97 Die Geschichte der Vegetation Schwedens. Engler's Botanical Jahrbücher. B. 22, p. 433.

Atwood, W. W., and **Goldthwait, W. J.**

'08 The Physiography of the Evanston-Waukegan Region. Bull. 7, Illinois Geological Survey.

Blatchley, W. S.

'97 Geology of Lake and Porter Counties, Indiana. 22d Ann. Rep. Ind. Dept. Geol. and N. Resources, Map.

Chamberlain, T. C., and **Salisbury, R. D.**

'07 Geology. Vol. III. Henry Holt, New York.

Clements, F. E.

'05 Research Methods in Plant Ecology. Lincoln, Nebr.

Cowles, H. C.

'01 The Plant Societies of the Vicinity of Chicago. Bull. 2, Geog. Soc. of Chicago. Also Bot. Gaz., Vol. 31, pp. 73–108 and 145–182.

Dachnowski, A.

'08 The Toxic Properties of Bog Water and Bog Soil. Bot. Gaz., Vol. 46, p. 130.

Gilbert, G. K.

'85 Topographic features of Lake Shores. 5th Ann. Rep. Dir. U. S. Geol. Surv., pp. 69–123.

'98 Recent Earth Movements in the Great Lakes Region. Ann. Rep. U. S. Geol. Surv., 1896–7, Part II., pp. 601–647.

Goldthwait, J. W.

'07 Abandoned Shorelines of Eastern Wisconsin. Wisc. Geol. and Nat. Hist. Surv. Bull., XVII., Scientific Ser. 5.

Eycleshymer, A. C.

'01 Observations on the Breeding Habits of Ameiurus nebulosus. Am. Nat. Nat., Vol. XXXV., pp. 911–18.

Forbes, S. A., and Richardson, R. E.

'08 The Fishes of Illinois. Nat. Hist. Surv. of Illinois (State Laboratory). Icthyology. Vol. III.

Hankinson, T. L.

'97 Walnut Lake. Biol. Surv. of Michigan (Lansing). Rept. Geol. Surv., 1907, pp. 161-271.

Knauthe, K.

'07 Das Süsswasser. Neudamm.

Leverett, F.

'97 Pleistocene Features and Deposits of the Chicago Area. Chicago Acad. Sci. N. H. Surv. Bull., II.

Meek, S. E. and Hildebrand, S. F.

'10 Synoptic Lists of the Fishes Known to Occur within Fifty Miles of Chicago. Field Mus. Nat. Hist., Pub. 142, Zool. Ser., Vol. VII., No. 9.

Reed, H. D. and Wright, A. H.

'09 The Vertebrates of the Cayuga Lake Basin, N. Y. Proc. Am. Phil. Soc. XLVIII., No. 193, 1909, pp. 370-459.

Reighard, J.

'03 The Natural History of Amia Calva. Mark Anniversary Volume, pp. 57-109.

Salisbury, R. D., and Alden, W. C.

'01 Geography of the Chicago Area. Bull. I., Geog. Soc. of Chicago.

Shaler, N. S.

'92 The Origin and Nature of Soils. 12th Rep., Dir. U. S. Geol. Surv., 1889-1891, p. 317.

Shantz, H. L.

'06 A Study of the Vegetation of the Mesa Region East of Pikes Peak. The Bouteloua Formation. Bot. Gaz., Vol. 42, p. 179.

Sharff, R. F.

'07 European Animals; Their Geological History and Geographic Distribution.

Shelford, V. E.

'07 Preliminary Note on the Distribution of the Tiger Beetles (Cicindela) and its Relation to Plant Succession. Biol. Bull., Vol. XIV., pp. 9-14.

'10 Ecological Succession of Fish and its Bearing on Fish Culture. Ill. St. Acad., Trans., Vol. II., pp. 108-110.

Steenstrup, J. J. S.

'41. Geognostik-geologisk undersögelse af Skovmoserne Vednesdam og Lillemose i det nordlige Sjaeland ledsaget af sammenlingende Bemaerkninger, heutede fra Danmarks Skov-, Kjaer- og Lyngmoser i Almindelighed. Danske. Vid Selsk. Afhandl., IX., 1842 (fide Warming).

Warming, E.

'09 (Ecology of Plants. An Introduction to the study of Plant Communities. Oxford. Translation by Percy Groom.

6

Reprinted from *Ecology* **41**(1):34–49 (1960)

ORGANIC PRODUCTION AND TURNOVER IN OLD FIELD SUCCESSION

Eugene P. Odum

University of Georgia, Athens

The establishment of the AEC Savannah River Plant (hereafter designated as the SRP) in Aiken and Barnwell counties, South Carolina, has provided an unusual opportunity for studies on secondary succession. During 1951-52 the approximately 6,000 human inhabitants moved out of an area of 315 square miles to make way for the establishment of the SRP; the entire area was then closed to the public. At that time about 67,000 acres (one-third of the land area) were in croplands and pastures, cotton and corn being the chief crops. From 1952, the first growing season following the cessation of agriculture, to the present time a faculty-graduate student team from the University of Georgia has been engaged in a series of ecological studies on fields of various soil types and locations within the SRP reservation. We have been especially interested in the pattern of metabolism of the "old-field ecosystem" as a whole, the energy flow characteristics of major component populations, and in the role which these components play in the fate of radioactive isotopes experimentally introduced into the ecosystem.

Secondary succession on abandoned cropland has been described for a number of localities in southeastern United States (Oosting 1942, Kurz 1944, Bronck and Penfound 1945, Quarterman 1957). The most complete description is that of Oosting who tabulated the density and frequency of plants in a large series of fields and forests of various ages on the Piedmont of North Carolina. Our studies on secondary succession involve an entirely different approach from that of Oosting and other investigators cited above in that the functional rather than the descriptive aspect is being emphasized. Essentially, we are attempting to measure the energy flow through the different communities which develop on abandoned agricultural land, and thereby to investigate the fundamental relationships between energy and succession. In this paper we are concerned primarily with that part of the energy flow which constitutes the net primary production (i.e. apparent photosynthesis or total photosynthesis minus plant respiration), and secondarily with the overall heterotrophic utilization of the annual net production of the producers. That portion of the energy flow of the community which constitutes the respiration of the producers as well as the details of energy flow through consumer levels will be considered in other papers. The time period of primary consideration in this paper is the first five to seven years of succession, which essentially covers the period when annuals and biennials are dominant.

* This research is a part of the University of Georgia Ecological Studies on the AEC Savannah River Plant supported by a grant from the U. S. Atomic Energy Commission (Contract At(07-2)-10).

Methods

The method of measuring net primary production is illustrated in Table I. At approximately one month intervals during the growing season all plants were removed from a series of one square meter quadrats. The generally sandy soils of the area and the shallow rooted growth form of early seral species made it easy to pull up plants with little loss of root material. Each species was placed in a separate paper bag and taken to the laboratory where the number and dry weight of each species was determined. In a number of cases soil was sifted after the plants had been pulled up to obtain an estimate of the rootlets not included in the harvest, but the amount of material obtained in this way was not large enough, nor variable enough, to warrant laborious sifting on a routine basis. The total weights of each species were tabulated as shown in Table I, thus providing an estimate of the standing crop at any one time as well as a basis for calculating the net production. The maximum weight of each species was listed in the final column and these values summed to provide the estimate of the annual net production. For annuals the entire weight of the plants (shoots and roots) was included. For perennials only the new growth was included since part of the weight of perennial species represents production from a previous year. Almost all of the important species in the old fields during the first few years were annuals, thus simplifying the procedure.

Sampling problems were minimal because the generally level topography and the sandy soils made homogeneous by years of "row crop" agriculture resulted in a relatively uniform post-agricultural vegetation. Since the objective of the

TABLE I. Seasonal change in standing crop of green plants in a 3-year abandoned field and calculation of total productivity uncorrected for animal consumption. (Field 3-412, 1954). Figures in grams dry biomass per square meter

Species	April	May	June	July	Aug.	Sept.	Net Production
Rumex........	19.0	0.7	1.6	1.4			19.0
Linaria........	1.2	0.1					1.2
Oenothera......	3.0	3.4	1.2				3.4
Specularis......	3.2	3.3					3.3
Festuca........	0.5	0.3	0.2		0.2	0.1	0.5
Gnaphalium ob..	4.8	20.3	13.4	33.9	28.2	0.3	33.9
Lepidium......		1.3	2.8	2.0	1.5		2.8
Cyperus.......	1.4	2.8	4.2	16.8	18.1	19.2	19.2
Leptilon.......			4.3	11.1	16.0	20.4	20.4
Heterotheca.....	0.6	3.0	8.8	11.9	32.0	50.0	50.0
Haplopappus...	0.4	3.8	18.5	27.7	85.0	110.4	110.4
Miscellaneous species.....	2.5	0.3	1.2	1.3	2.1	2.1	5.9
Totals.........	36.6	39.3	56.2	106 1	183.1	202.5	270.0

study was to measure the rate of change in a given area rather than to describe in great detail the standing crop at any one time, harvest quadrats were spaced 25 feet apart along a transect. Successive samples were taken immediately adjacent to quadrats harvested during the previous month so that the same combinations of locations were sampled at each period. Species-area curves (Oosting 1956, Goodall 1952) indicated that 10 quadarts provided an adequate sample during the first 2 or 3 years of natural succession; with the increase in species diversity in later years it was necessary to increase the number to 20 or more in some cases.

It is important to emphasize that harvests should be made at frequent intervals if a reasonable estimate of production is to be obtained. Note (Table I) that the annual net production is always greater than the standing crop at any one time since not all species in the community reach their maximum growth at the same time (as might, for example, a cultivated crop). As will be emphasized later, standing crop becomes less and less a measure of productivity as succession proceeds.

The estimates of net primary production are to be considered minimal since growth which may have died and become detached from the plant or eaten by animals within a sample interval would not be included. Our studies indicate that such losses within a month interval are small in so far as the old-field type of ecosystem is concerned. Losses to the litter as estimated by changes in litter weights proved to be quite small for the time interval considered, while immediate consumption by herbivores was likewise small. Most consumption was delayed until the organic production was in the form of seeds and detritus. Large mammalian grazers were absent from the areas studied (wild pigs and deer are a factor in some parts of SRP area but not in the areas selected for study). The two common rodents (Peromyscus and Mus) eat some vegetation but are primarily seed and insect eaters. Rabbits and cotton rats (Sigmodon), important herbivores in later stages of succession, were of little importance in the early stages of succession. All the birds are either seed or insect consumers. Insect herbivores were the chief consumers of living plants but population size was small in undisturbed old-fields (Cross 1956). When our energy flow analyses of consumer populations have been completed we expect to be able to estimate the total consumption with some degree of precision. At present we can say that in the early stages of succession consumption of living plants within any one 30-day period is certainly less than 10% and probably less than 5% of the net production (for that period). Consequently, it is feasible to use a "short-term harvest method" for estimation of productivity in a system characterized by delayed heterotrophic utilization of the net primary production. In summary, what we are discussing in this paper is "net community production," that is, organic production in excess of that consumed by the community over short intervals of time. In the early stages of old-field succession such net community production closely approximates true net primary production.

At the time that harvests of living plants were made all of the litter materials, including dead plant material on the surface of the ground as well as the "standing dead," were removed from the quadrat, placed in a bag, and the dry weight later determined. In other words the term "litter" as used in this paper refers to all of the dead plant material above the surface of the soil. During the first several years of succession it was possible to separate out major dominant species from the litter and thus to estimate the rate of decomposition of individual species.

During the autumn of 1951, the cultivated crop or other vegetation in about 600 fields widely scattered over the SRP area was recorded on overlays attached to sectional areal photographs. Each field was given a number and prefixed with another number referring to the map section (for example, 9-111 is field 111 located on map section 9). A set of the maps and overlays are on file both at SRP and at the University of Georgia. These reference fields were visited periodically during the years and were useful in establishing the general pattern of succession in relation to broad soil types. In those fields selected for quantitative study, soil samples from the top 6 in.

and from the 18-24 in. level (A_2 or B horizon) were taken. Determinations of nutrient and organic matter levels, pH, exchange capacity, and texture were made by the Soils Laboratory of the University of Georgia. Additional mechanical analyses employing the Boyoucous hydrometer method were made in our Ecological Laboratory when it became apparent that the silt-clay fraction was a key to the ecology of the fields. The Boyoucous method works well for coastal plain soils which are easily dispersed with hexametaphosphate (Calgon). Excellent data on rainfall and temperatures were available from the 6 weather stations which are being maintained in different parts of the SRP reservation.

Two large fields, 9-111 and 3-412, were selected for an especially intensive analysis. Net production in these two fields, which represent two of the common soil types of the area, have been made each year since the beginning of the study in 1952. Likewise, studies of major animal populations in these fields have been made by various graduate student members of the research team.

Shortly after the establishment of the SRP an extensive program of pine tree planting was begun by the U. S. Forest Service. Pines, of course, are a part of natural succession and would eventually invade the fields. However, because of the very large size of many fields, the raw soil conditions and the frequent absence of a natural seed source, natural reforestation would be slow; hence artificial reforestation was economically desirable. By 1957 most of the original 600 reference fields had been planted, but the special study fields have been reserved and will not be artifically planted. There has been very little invasion of woody plants in these large fields during the first 7 years.

Acknowledgments

Two "generations" of graduate students have helped in the productivity studies while at the same time carrying out their own independent research. Edward J. Kuenzler, William H. Cross, Leslie B. Davenport, and Robert Pearson contributed much time and energy taking the monthly samples during the first 3 years of the study while Clyde E. Connell, Larry D. Caldwell, Wallace Tarpley, John B. Gentry, and William K. Willard were active during the next several years. Dr. Robert A. Norris took a special interest in the work during 1956-58 when he was Resident Ecologist on SRP. He revisited many of the original reference fields and did some special mapping during the period when the diversity of species was increasing. Dr. Wilbur Duncan, Department of Botany of the University of Georgia, has checked numerous identifications for us. We are, also, indebted to John P. Hatcher, Project Forester for SRP, and his staff for helping out in numerous ways. Finally, Karl Herde, Chief, Radiation Protection Branch of SRP, not only cheerfully made the necessary arrangements for us to work in complete freedom within the area but he, also, took an interest in all the studies and often accompanied us into the field.

Physiography and Soils of the SRP Area

The SRP area is located on the Upper Coastal Plain physiographic province in Aiken, Barnwell and (to a small extent) Allendale counties, South Carolina. The area borders the Savannah River for about 27 miles; it is about 150 river miles from the ocean, and 14 airline miles downstream from Augusta, Georgia, on the edge of the Piedmont Plateau. As shown in Figure 1 two distinct physiographic subregions are represented, namely,

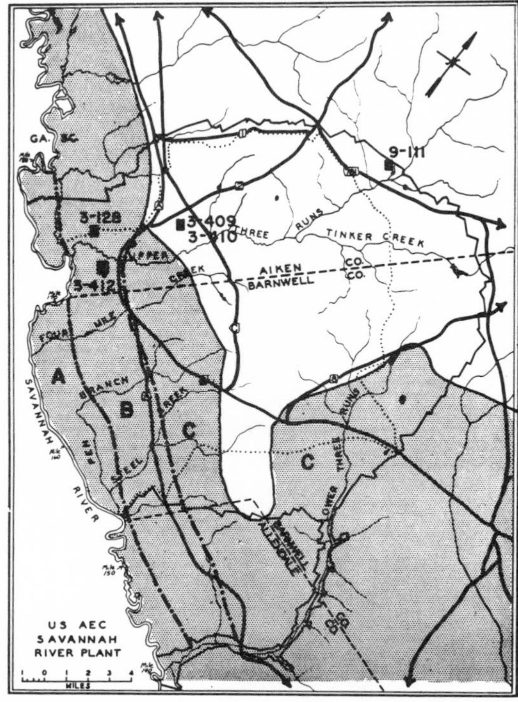

Fig. 1. The AEC Savannah River Plant reservation showing principal streams, highways (A, B, C, 1, 2, etc.), physiographic subregions and location of special study fields (9—111, 3—409, 3—410, 3—128 and 3—412). The shaded area is the Coastal Terraces subregion and the unshaded area the Aiken Plateau or "sandhills" subregion. Three terraces are recognizable as follows: A—Wicomico; B—Sunderland; C—Brandywine.

TABLE II. Characteristics of major soil types of SRP old-fields

Soil Type	Number of Fields		Nutrients-pounds Per Acre						% Organic	Texture %		
			pH	NO_3	P_2O_5	K_2O	CaO	MgO		Sand	Silt	Clay
Lakeland Sand (Deep phase)	4	A_p	5.7	5	37	27	288	8	1.07	92	5	3
		A_2	5.3	5	5	16	200	5	.15	91	5	4
Lakeland Sand	8	A_p	5.7	9	77	43	451	3	1.10	91	5	4
		A_2	5.3	6	5	7	108	2	0.12	86	6	8
Kalmia Sand-Loamy-Sand	7	A_p	5.8	5	55	42	413	5	0.98	89	6	5
		A_2	5.6	2	6	16	450	3	0.11	87	7	6
Norfolk Loamy-Sand	2	A_p	5.8	10	75	60	500	10	1.45	88	7	5
		B_1	5.4	5	20	10	250	10	0.32	78	5	17
Ruston Loamy-Sand	3	A_p	5.7	5	27	36	683	23	1.00	91	5	4
		B_1	5.2	5	5	54	350	10	0.20	78	8	14
Cahaba Loamy-Sand	3	A_p	6.6		197	202	2100	52	2.04	80	9	11
		B_1	6.3		132	104	1500	50	0.55	75	7	18
Myatt Sandy-loam	1	A_p	5.4	60	15	5	850	T	4.10	60	13	27
		B_1	6.5	10	20	5	850	T	1.14	42	12	46
Izagora Sandy-loam	2	A_p	5.1	5	12	35	300	5	2.97	72	23	5
		B_1	—	—	—	—	—	—	—	52	16	32

Cation Exchange Capacity (in millequivelents cations):—
Lakeland Sand (1 field) A_p=3.05; A_2=1.25.
Norfolk Loamy-Sand (1 field) A_p=3.75; B_1=2.20.
Cahaba Loamy-Sand (2 fields) A_p=4.10; B_1= 2.75.
Izagora Sandy-Loam (1 field) A_p=8.49; B_1=2.70.

the pleistocene Coastal Terraces (below 270 feet elevation) and the Aiken Plateau (above the 270 foot contour). At least 3 terraces are recognizable within the former subregion as shown in Figure 1. The lowest terrace (Wicomico) is the present very broad Savannah River floodplain which is covered with dense swamp forest. The higher terraces have a level to gently rolling topography and were intensively cultivated where drainage and soils were favorable. The Aiken Plateau is relatively hilly and deeply dissected by small streams. There are extensive areas of "sand hill" vegetation of scrub oak and long-leaf pine forests along the ridges and many of the farms in this region were small and marginal in agricultural productivity. The geology and physiography of the South Carolina Coastal Plain has been well described by Cooke (1936).

As is characteristic of the "geologically young" Coastal Plain in general, soil profiles on the SRP area are "immature," and a large number of different soil types have been delimited. Small differences in elevation, drainage and parent material may produce striking differences in color, texture, organic content and other characteristics, so that distinct soil types may exist side by side in an area which might at first appear to be uniform. Sixty-three soil types have been described from the Edisto Soil Conservation District which includes the SRP according to a list furnished by P. W. Montgomery, State Soil Scientist of South Carolina. However, only a few of these are important agricultural soils and hence of primary interest in the study of old-field succession. Well-drained to moderately well-drained soils which were most widely used for the standard crops of the area (cotton, corn, peanuts, peas, watermelons, etc.) fall largely into 5 series as follows: the Lakeland and Ruston series of the Aiken Plateau and the Kalmia, Norfolk and Cahaba series of the Terraces. Poorly drained soils of the Myatt and Izagora series were also cultivated to a limited extent. Fields representative of all 7 of these series have been included in our studies. Characteristics of the major soil types in the 30 special study fields are summarized in Table II, while complete profile descriptions of the two major study areas and also of two other fields of contrasting soil types are shown in Table III.

In most of the soils cultivated prior to the establishment of SRP the topsoil or A_p horizon (i.e. "plowed" layer) is 6-18 in. thick, ranges in color from light to dark gray, and in texture from 60% to 95% sand. The subsoil, which technically is the A_2 horizon in very sandy types and the B horizon in heavier soils, is usually sharply marked off and brightly colored yellow or red in well drained situations. The Myatt and Izagora soils

TABLE III. Profile descriptions of soils of four SRP fields

Field Number & Soil Type	Horizon	Depth lower edge (inches)	Color	Structure	Texture % Sand-silt-clay
9-111 Lakeland Sand	A_p	8	grayish-brown	weak medium crumb	89— 5—6
	A_2	38	yellowish-brown	weak medium crumb	86— 8— 6
	B_1	(very narrow and weakly developed)			
	B_2	70	yellowish-red	weak medium sub-angular blocky	70— 4—26
	C	—	red mottled with yellow		
3-412 Cahaba Loamy-Sand	A_p	12	dark gray brown	weak medium crumb	85— 7— 8
	B_1	15	dark red-brown	weak medium crumb	80— 5—15
	B_2	60	yellowish-red	weak medium crumb	81— 4—15
	C	—	reddish-yellow	weak medium crumb with some quartz gravel	83— 3—14
3-409 Lakeland Sand (deep phase)	A_p	6	light olive gray	weak medium crumb	94— 3— 3
	A_2	24	pale yellow	weak medium crumb	92— 3— 5
	A_3	60	yellow	weak medium granular	85— 7— 8
	B_1	Not developed			
	B_2	66	yellow	weak medium granular	70—12—18
	C	—	red mottled with yellow	sandy clay	53—10—37
3-410 Ruston Loamy-Sand	A_p	12	brownish-gray	weak crumb	89— 6— 5
	B_1	42	yellowish-red	weak sub-angular	79— 8—13
	B_2	84	red	sub-angular, blocky	61— 3—36
	C	—	red	sub-angular, blocky	65— 2—33

* Field descriptions made by P. W. Montgomery assisted by P. Wakefield. Textural analyses made in the University of Georgia Ecological Laboratory by E. W. Kuenzler.

have a white to bluish clayey B horizon. The agricultural soils fall naturally into three groups: (1) the "yellow sands," light sandy soils with a yellow subsoil, and including the Lakeland, Kalmia and Norfolk types, (2) the "red sands," somewhat heavier soils with orange or red subsoils and including the Ruston and Cahaba types, and (3) the poorly-drained soils with wet, heavy subsoils and including the Myatt, Izagora and other types. All of these soils have a low pH and a relatively low content of organic matter.

The texture of the soil, and especially the texture of the subsoil, has proved to be a key factor in the ecology of the communities which have developed on the former agricultural land. The amount of clay and silt in the subsoil largely controls the water and nutrient holding capacities of the soil as a whole and thereby the availability of water and nutrients to the plants. The series of soil types as listed in Table II actually forms a gradient from the lightest and driest soils to the wettest and heaviest soils that were used in ordinary "row crop" agriculture. The nutrient and organic matter levels listed in Table II are based on analyses made during the first year of the study. As shown in Table IV there has been a tendency in the lighter soils for phosphorus, calcium and organic matter to leach downward. Thus, 3 years after the end of cultivation these materials had decreased in the topsoil and increased in the subsoil, on the average.

The General Pattern of Succession

A check-list of the most important plants of the early stages of old-field succession is found in the Appendix. In this paper dominance (i.e. importance in the community) is considered in terms of the net production of dry matter and not in terms of the number or height of individuals. A dominant, for our purposes, is arbitrarily defined as any species producing one gram or more of dry matter per square meter per growing season irrespective of the time in the year when it was produced. Species producing less than one gram were placed in a miscellaneous category and not listed separately in the tables that follow. Although many genera were represented by several species in the fields, usually only one species in a genus was important enough productivity-wise at any one time or place to be a dominant by the above definition. Consequently in the discussions and tables that follow, the genus name suffices to designate the species in many cases.

The changes which occurred in field 9-111 as shown in Table V illustrate the general pattern of succession in relation to net production on the well-drained soils. Woody forbs were major dominants during the first 5 years or so, with horseweed (*Leptilon canadense*), yellow aster (*Haplopappus divaricatus*) and camphorweed (*Heterotheca subaxillaris*) usually appearing in that order. The growth form of these three forbs is illustrated in the photograph of Figure 2. Leptilon was al-

TABLE IV. Soil nutrient levels in December 1951 (last year of cultivation) and in December 1954 (after three years of natural succession). (Figures are in pounds per acre; T = trace)

Field No.	Crop in 1951	Soil Horizon	P_2O_5 1951	P_2O_5 1954	K_2O 1951	K_2O 1954	CaO 1951	CaO 1954	% Organic 1951	% Organic 1954
5-146 Kalmia loamy-sand	Cotton	A_p	40	45	40	44	700	200	0.92	0.69
		A_2	T	18	T	22	100	600	0.08	0.19
2-105 Lakeland loamy-sand	Cotton	A_p	150	70	65	36	850	300	1.18	0.80
		A_2	T	9	50	20	100	200	0.69	0.22
9-111 Lakeland sand	Corn	A_p	70	20	T	28	550	200	1.31	1.00
		A_2	T	10	40	24	300	200	0.17	0.22
9-112 Lakeland sand	Cotton	A_p	180	18	50	18	300	300	1.07	0.80
		A_2	T	3	T	24	50	200	0.26	0.19
1-392 Izagora loamy-sand	Peanuts	A_p	T	17	T	30	100	100	0.88	0.80
		A_2	10	10	T	14	100	400	0.17	0.19
5-150 Kalmia sand	Corn	A_p	65	63	25	16	1100	100	1.18	0.58
		A_2	T	19	T	12	200	50	0.10	0.43
Averages		A_p	84	39	30	29	600	200	1.09	0.79
		A_2	2	12	15	19	142	275	0.24	0.24
Totals both horizons			86	51	45	48	742	475	1.33	1.03

TABLE V. Net Primary Productivity (gms dry wt/m²/year) of Field 9-111

Species	Years Abandoned 1 (1952)	2 (1953)	3 (1954)	5 (1956)
Linaria	18.3			
Leptilon	352.6	39 6	1.1	0.3
Gnaphalium purpureum	1.6	2.2	7.8	11.0
G. obtusifolium	41.6		6.3	45.5
Digitaria	74.9			6.1
Rumex		3.9	4.4	29.6
Lechea		111.2	94.9	18.3
Hypericum		19.5	59.8	33.2
Oenothara		0 8	8.1	1.5
Cyperus		0.4		0.7
Haplopappus		149 0	24 2	72.2
Specularia			1 3	
Lepidium			3.2	.4
Andropogon			1.4	16.7
Festuca			15 8	.3
Eragrostis			44.52	
Diodea				14.1
Heterotheca				14.4
Triplasis				21.4
Leptoloma				3.4
Miscellaneous	4.9	5.1	21.1	16.1
Totals	494.0	331.7	282.9	308.6
Number of dominant species	5	6	13	15

FIG. 2. The growth form of three major forb dominants. From left to right: *Heterotheca subaxillaris* (individual from previous growing season), *Haplopappus divaricatus* (held by hand of observer), *Leptilon canadense* (partly dwarfed 2nd year individual), and *Heterotheca* (individual from current season).

most always by far the main producer during the first growing season. By late September of the first year these tall weeds often reached 6 feet or more in height and formed dense, uniform stands. Crabgrass (Digitaria) frequently formed an understory and usually was the second most important species productivity-wise. During the second and subsequent years Leptilon was invariably present in the study fields but always in a stunted form (two feet high or less). Although the number of individuals per m² was often greater in these years the contribution of the species to total net production declined rapidly as shown in Table V. Yellow aster (*Haplopappus divaricatus*) replaced horseweed as the major dominant during the second year in most cases. Tall, uniform stands of this species were quite spectacular when the yellow flowers were at their peak in September and October. Figure 3a shows a typical stand of Haplopappus on a fertile area (Field 3-412) as it appeared in the winter of the second year of abandonment when the stalks were bare of leaves. The yellow aster continued to be an important dominant during the third and fourth years, and in some fields was still important after 7 years. As shown in Table V the number of dominants increased with time; after the second year there was never a clear-cut dominance by one or two species. Field 9-111 was atypical in that Lechea was prominant; in most other fields of the Lakeland soil series, or other well-drained types this plant was only a minor component. Also, Heterotheca was less important in field 9-111 than in most other fields. In field 3-412, for example, it became a major dominant by the third year (Table I) and continued important in subsequent years. It is important to note that the Coastal Plain species of Heterotheca (i.e., *H. subaxillaris*) is quite different in growth form and seral position as compared with the Piedmont species, *Heterotheca latifolia,* which has recently been described as replacing horseweed as a first year dominant in abandoned croplands in some areas (Keever 1955).

During the spring of the first growing season blue toadflax (Linaria) often formed a conspicuous blue carpet for a brief period, but these small plants made only a small contribution to total annual production. Rumex and *Gnaphalium purpureum* along with many other species were conspicuous during the spring and early summer of later years. From the third to the fifth or sixth year the large *Gnaphalium obtusifolium* often was a midsummer dominant and a major contributor to the annual production. In fact, the general pattern was one of change from a single crop of plants maturing in the fall, as was characteristic during the first two years, to three distinct seasonal crops maturing in spring, mid-summer and fall, as occurred in subsequent years. In Table I species forming these 3 crops in field 3-412 during the third year are grouped together. As will be discussed later the general pattern of succession resulted in greatly increased species diversity, a change in the seasonal distribution of production, a decline in the size of the terminal or fall standing crop, but not much change in total annual production (see also Fig. 3).

Broomsedge (Andropogon), a bunch grass destined eventually to become the major dominant of the perennial herbaceous stage which follows the annual "weed" stage and which precedes the invasion of shrubs and trees, appeared in many fields by the third year. However, Andropogon did not become important until about the fifth or sixth year when scattered clumps became noticeable (Figure 3b). By the eighth year broomsedge formed a closed stand only in certain small fields on heavier

FIG. 3. Field 3—412 as it appeared at the end of the second growing season (1953) and at the end of the fifth season (1956). Vegetation during the second year (Fig. 3a, upper photograph) consisted of an almost pure stand of Haplopappus, many individuals of which were almost as tall as the corn which had been previously grown in the field (1951). By the fifth year (Fig. 3b, lower photograph) the number of species was much greater and the height of the forbs much less, and there were no clear-cut dominance by any one or two species. The perennial Eupatorium (dark clumps) and Andropogon (light colored clumps in the background) were beginning to appear as widely scattered clumps in the matrix of annuals and biennials.

soil types, or locally in the large fields. In contrast to the one-species-per-genus of the forb dominants, the two common species of the genus Andropogon (*A. virginicus* and *A. ternarius*) appeared in equal abundance or in various ratios while one or more of three rarer species (*A. scoparius, glomeratus,* and *Elliottii*) sometimes showed local dominance. The factors affecting broomsedge invasion and the functional aspects of the broomsedge community will be considered in more detail in later papers.

As would be expected soil type greatly influenced the pattern of succession. Certain well-marked deviations from the above described pattern were associated with the more extreme soil types. The contrast during the first two years between the well-drained and poorly-drained types is shown in Table VI. While the annual productivity of these two particular fields was similar (the relation of soil type to productivity will be considered in a subsequent section), the distribution of productivity by species was different. In the poorly-drained field, the number of dominants was greater. Leptilon was much less important, and Haplopappus and Heterotheca were important in the first and second years respectively in contrast to a later dominance in the well-drained soil types.

TABLE VI. Comparison of species composition and net production (gms. dry matter/m^2/year) during the first two years of secondary succession on the extreme soil types used for row-crops (cotton and corn) prior to 1952

Field Number Soil Type % Silt-Clay Topsoil Subsoil	9-111 Lakeland Sand (well drained upland type) 10 13		3-128 Izagora Sandy-loam (poorly drained lowland type) 31 58	
Species	First Year (1952)	Second Year (1953)	First Year (1952)	Second Year (1953)
Linaria	18		9	
Leptilon	353	40	83	9
Gnaphalium purpureum	2	2	5	3
G. obtusifolium	4		5	25
Digitaria	75		150	
Rumex		4	5	10
Lechea		111		
Hypericum		20	7	36
Oenothera		1		1
Cyperus		1	53	50
Haplopappus		149	86	71
Cynodon			9	66
Diodia			9	5
Polypremum				21
Heterotheca				75
Others	3	1	5	2
Total Net Production	494	331	425	374
Number dominant species	5	6	12	12

The effect of soil type on species composition is shown more broadly in Table VII which summarizes the average terminal standing crop during the second and third years in a series of 19 fields arranged in a gradient from the dry sands to the wet silt-clay types. In general, the number of dominants and the speed of succession increases with increasing silt-clay percentage and/or nearness of the water table to the surface. Thus, Cyperus, perennial grasses such as Cynodon and Andropogon, and the woody perennials Eupatorium, Bignonia and Rubus invade the heavier or less well-drained soils more rapidly than they do the well-drained light soils. At the other extreme, in deep sands, the annual forb stages (i.e. Leptilon, Haplopappus, Heterotheca) tend gradually to give way to a stand of sand-hill grasses of the genera Aristida, Leptoloma, Panicum and others. Diodia was also usually more prominent in the lighter soils.

To summarize, the vegetation, especially the autumn aspect, appeared much the same during the first year regardless of the soil type, except that Haplopappus shared dominance with Leptilon in poorly drained sitautions. During the second and third years species composition began to show marked differences on different soil types, and the divergence increased during the next several years. Thus, by the third or fourth year one could identify the soil type and estimate approximately the silt-clay percentage of the subsoil from the composition of the vegetation. Part of the divergence in species composition resulted from a difference in the rate of succession (many species appearing sooner on the heavier soils as compared with the lighter ones), and part was the result of the virtual restriction of certain species to specific soil series (*Aristida tuberculosa* on deep sands or Polypremum on wet soils, for example). Thus, relative rates of succession can serve as ecological indicators of soil differences as well as specific species of plants which have a high "fidelity" for a particular condition.

The general pattern of upland old-field succession on SRP resembled that described by Oosting (1942) for the lower edge of the Piedmont Plateau in that the general sequence of life forms was similar (annual forbs to perennial grasses), and in that Leptilon was a major dominant during the first year. Succession on SRP differed from that of the Piedmont in the following respects: (1) The yellow-flowered Haplopappus and Heterotheca replaced the white aster (*Aster pilosus*) as second and third year dominants. (2) Ragweed (Ambrosia), often an important early seral dominant on the Piedmont, was generally absent on the uplands of SRP although of importance in some lowland fields (Table

TABLE VII. Effect of soil type and drainage on the terminal standing crop during second and third years of abandonment (1953-1954) as indicated by the grams of average dry biomass per square meter

Soil type:	Lakeland deep phase	Lakeland	Ruston	Cahaba	Kalmia	Myatt and Izagora
Physiographic subregion:	Plateau	Plateau	Plateau	Terrace	Terrace	Both
Drainage:	Extreme	Excellent	Excellent	Good	Moderate	Poor
% silt-clay in subsoil:	9	14	22	20	12	44
Number of fields:	3	3	3	3	4	3
A. OVERSTORY FORBS						
Haplopappus	38.6	96.8	30.6	247.4	102.9	39.4
Heterotheca	21.7	16.8	69.8	21.6	6.2	22.5
Leptilon		21.9	0.2	13.8	0.8	3.6
Lechea		21.7				
Gnaphalium ob.		3.1		0.5	13.8	
Eupatorium			40.0			
Ambrosia					9.0	
Bignonia					17.7	13.3
Rubus					28.0	30.5
Erigeron						23.5
B. UNDERSTORY FORBS AND SEDGES						
Hypericum	1.4	44.7	1.0	0.7	9.6	14.4
Diodia	38.1	5.1	3.9		8.6	1.6
Cyperus	0.2	0.3		6.6	55.1	3.1
Lespedeza	1.3	1.6		0.5	2.7	13.5
Polypremum					2.1	14.6
Cassia					0.4	1.0
C. GRASSES						
Aristida	21.7					
Leptoloma	14.2					
Cynodon	10.0	3.8	148.1			127.8
Digitaria				1.2	5.5	0.5
Andropogon		0.5	4.0	0.4		5.5
D. OTHERS	4.1	5.6	4.5	4.0	3.4	5.4
Totals	151.3	221.9	302.1	296.7	265.8	320.2
Number major Dominants	8	9	7	5	12	14

VII). (3) The rate of succession was slower and the annual and biennial forb stages lasted longer, especially on the light soils. For example, Oosting found that Andropogon rapidly invaded the clay Piedmont soils and established major dominance in 3 or 4 years in contrast to 6-8 years required for similar dominance on SRP. (4) The invasion of pines was very much slower, partly the result of the sandy soils and partly the result of the much greater average size of SRP fields and consequent greater distance from a seed source. The pattern of succession on some of the lowland SRP fields most nearly resembled Oosting's Piedmont fields, which is not surprising since the clay content of some of these fields approaches that of the Upland Piedmont, and since the climate of the lower Piedmont is not greatly different from that of the adjacent Upper Coastal Plain.

NET PRODUCTION AND SUCCESSION

Total annual net production of dry matter in the two intensive study fields for the first 7 years of succession is shown in Figure 4. Rainfall for each year and for each growing season is also shown. Temperatures during the seven years did not deviate more than 3° F from the average 10 year record of the SRP stations. As far as we could determine, temperature was not a major factor in bringing about annual differences in production rates during the period of study.

The trends in the two fields were quite similar, with production averaging greater in field 3-412 where soils were somewhat more fertile (Tables II and IV). In both fields production was much the greatest during the first year, declined in the second year, and remained approximately the same in subsequent years. The low values for 1954 seem clearly the result of low rainfall; in fact, 1954 was the only year which rainfall was appreciably below normal. The difference in production between the first (1952) and the second year (1953) can not be ascribed to rainfall since it was almost exactly the same for the two years. Production was also greater during the first year in 5 out of 6 other fields in which net production was measured during the first two years after the cessation of agriculture. (See

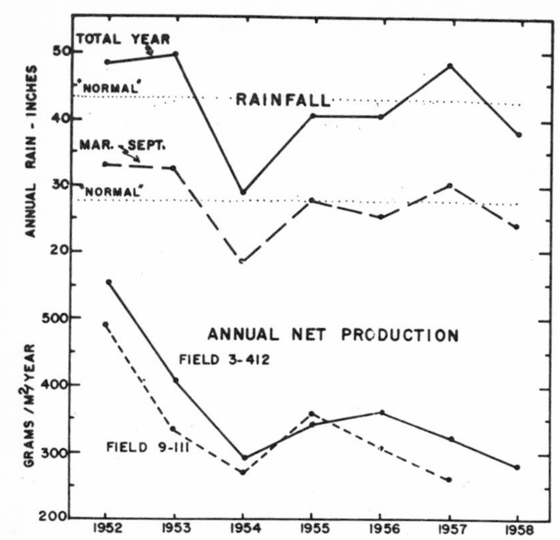

FIG. 4. Total annual net production of dry matter and rainfall during the first seven years of succession on the two intensive study fields.

Table VI for data on field 3-128). The difference was greater on light soils than on heavy soils. In the absence of any other logical explanation we attribute the high first year values to residual fertilizer remaining from the previous crop year, with phosphorus probably being the key factor. The only direct evidence for this was the fact that phosphorus (and nutrients in general) did decline in the topsoil where almost all of the roots of the annuals are located (Table III). The situation the first year resembled that of a temporary "bloom" such as often occurs in cultures where requisites have accumulated prior to the start of the culture.

To summarize, net production during the first year was generally of the order of magnitude of 500 g of dry matter per square meter per year and during subsequent years about 300 g, except when reduced by low rainfall or extreme soil type. In other words, after a one year "bloom," a steady-state in productivity was achieved even though species composition was constantly changing as was described in the previous section.

Net Production and Soil

The relationship between annual net production and the silt-clay content of the subsoil is shown in Figure 5. Each point on the graph represents a different field. In order to provide a more valid comparison in which other variables are reduced, values for the "bloom" year of 1952 and the "drought" year of 1954 are not included. Therefore, the points in Figure 5 are values for

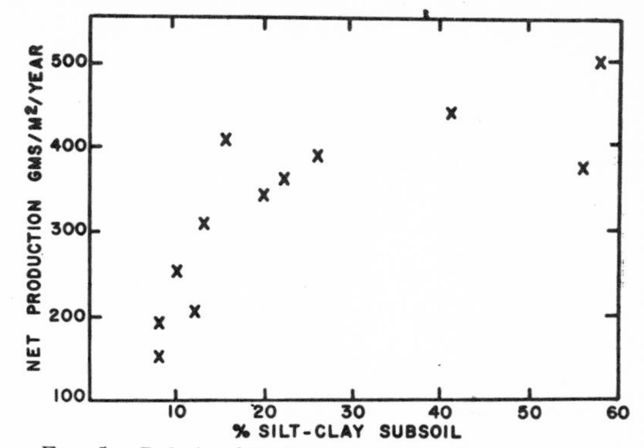

FIG. 5. Relationship between net production and the silt-clay fraction of the subsoil (A_2 or B_1 horizon). Each point on the graph represents annual production of a different field during one of the three years (1953, 1955, and 1956) when production was stabilized and rainfall was normal.

the years 1953, 1955 and 1956 when, as shown in Figure 4, rainfall was essentially normal and the stage of succession had little effect on total organic production. It is evident from Figure 5 that the amount of silt and clay, which functions as water and nutrient holding material, becomes critical below 20% and especially below 15%. Above 20% the amount of silt and clay had less effect on total annual production even though (as already described) species composition was influenced to a considerable degree by high silt-clay content. Apparently, there are somewhat different species combinations in the flora which are adapted to grow at different silt-clay levels. In the intermediate range of conditions (20 to 40%) these different combinations have approximately the same ability to produce organic matter, but in the low range of silt-clay the production of the adapted community may be reduced.

The Standing Crop of Living and Dead Material

In Figure 6 seasonal and succession changes in the "standing crop" of living and dead material are shown for Field 9-111. As would be expected, the amount of living plant material was minimum in winter and maximum at the end of the growing season in late September or early October, while the amount of dead material not yet incorporated into the soil (i.e. total litter) was minimum towards the end of the growing season and maximum after the killing frosts of late October or early November. The curves for both living and dead material tended to flatten out during succession since seasonal differences in both the rate of production and rate of decomposition became

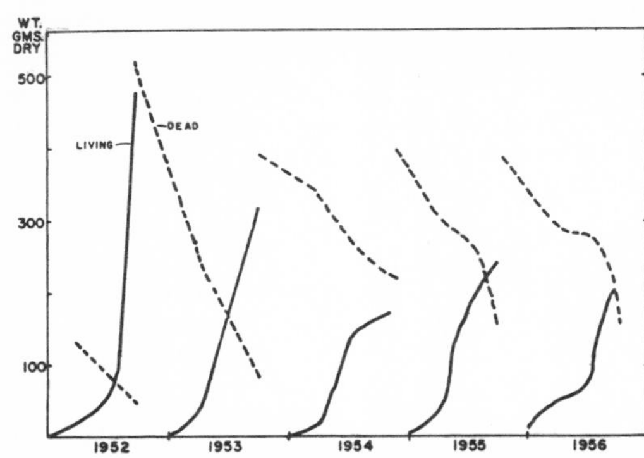

FIG. 6. Seasonal and successional changes in the standing crop of living and dead material on field 9—111 during the first five years after abandonment of agriculture. Solid line is a smoothed curve showing increase in dry matter during the growing season, while broken line is a similar curve showing decrease in dead material.

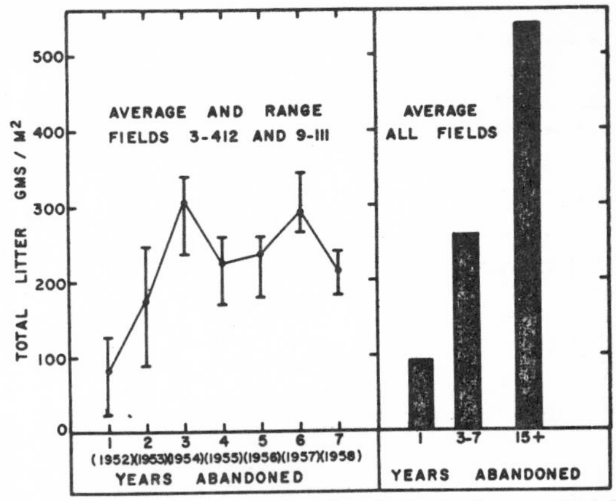

FIG. 7. Levels of litter (total dead plant material above ground) in fields of different ages. In the left hand diagram the average quantity (based on monthly samples) for the year is shown and also the lowest levels (usually occurring in September) and the highest levels (usually in November). In the bar graph on the right average levels are shown for one-year fields, 3-7 year forb fields and broomsedge fields.

less marked with the passage of time. An equilibrium between annual production and annual decomposition was essentially established after only 3 years.

Changes in total litter for both fields 9-111 and 3-412 are diagrammed in Figure 7 along with average values for all fields studied. After the harvesting of the last cultivated crops in the autumn of 1951 the small amount of litter which remained decomposed rapidly during the ensuing growing season of 1952. Since little new dead material was added until the end of the season the amount of litter reached a low point of less than 50 g of dry material per meter square in the early fall of 1952. During the next two years the amount of dead material increased and stabilized at around 250 to 350 g/m^2. It was evident that as long as annual and biennial plants remained dominant no further change in the seasonal pattern or average amount of litter would be expected. However, looking ahead to the broomsedge stage it would appear that the litter again increases and probably reaches another equilibrium as perennial grasses become dominant. As shown in Figure 7, the average amount of litter in two broomsedge fields estimated to be at least 15 years old (the exact age was unknown since these fields had already been abandoned for a number of years at the time of the establishment of the SRP) was of the order of 500 g/m^2, or about twice the average amount present in forb fields.

Although the litter increased during the first three years of succession, the amount of organic matter in the soil itself apparently did not increase (Table III). In fact, mineralization and leaching apparently removed more organic matter from the topsoil than was replaced from decomposition of the litter during these first few years. Ultimately, however, the organic content of the soils is destined to increase since the organic content of forest soils of the region is greater than that of the abandoned fields. Whether this increase occurs during the grass stage or during the forest stages of succession remains to be seen. As Olson (1958) has shown, the buildup of organic matter in sandy soils can be a very slow process.

Since stems and leaves of some of the major species could be identified in the litter samples it was possible to follow the decomposition of the above-ground portion of an annual crop by determining the per cent remaining at successive sample periods. In Figure 8 the disappearance of the fall-maturing, first-year Leptilon, the summer-maturing Gnaphalium, and the spring annuals are shown. The large, woody stalks of the first year Leptilon decayed slowly during the winter, more rapidly during the next summer until only the thick basal portions of the stems remained in the ground litter. Segments of the basal stem 2-6 in. long were very resistant to decay and resembled miniature "logs" lying on the surface of the soil. After 6 years these segments were still recognizable in the litter as were occasional sections of corn stalks remaining from the 1951 crop. In contrast, the Gnaphalium stalks were completely decomposed within one year, and the small spring annuals were decomposed by the end of the season in which they grew. Since average size of individual plant decreased and stems became less woody after the first year, the rate of decomposition of the annual crop increased until a temporary equilibrium between production and decomposition became established.

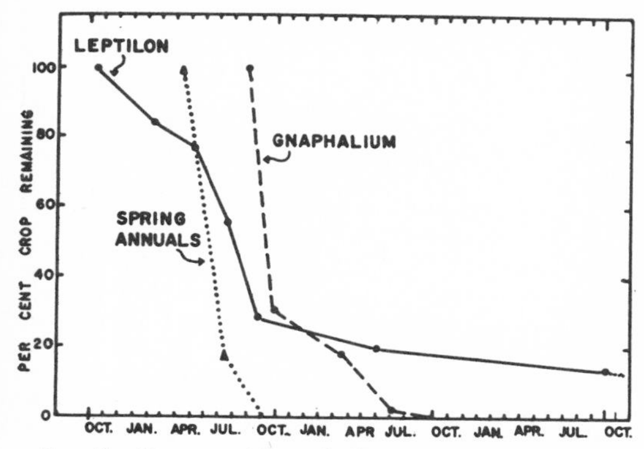

FIG. 8. Decomposition of the above-ground portions of annual crops of three types: (1) large, woody forbs as represented by first-year Leptilon, (2) summer annuals as represented by *Gnaphalium obtusifolium,* and (3) spring annuals including Rumex, Oenothera, *Gnaphalium purpureum,* Specularia and others. The per cent of the dry weight of identifiable materials in the litter at successive intervals after maturing and death of the crop is shown.

THE SEASONAL PATTERN OF NET PRODUCTION

In Table VIII and in Figure 9 net production data are given in terms of average daily rates for two-month intervals during the major growing season. As shown in Figure 8 production was low (about 1 gm./m^2/day) during spring and early summer of the first year and relatively high (over 5 gms./m^2/day) during the period from mid-July to mid-September. During the next several years of forb dominance, when total annual production was stabilized as previously described, the spring and early summer rates were higher but the late summer rate much lower so that the seasonal curve became much flatter (compared with the first year). Production was even more uniformly spread over the season in the two broomsedge fields studied since the spring rate was as high as or higher than that of late summer. The high spring production probably results from the fact that perennial plants were able to respond sooner than annual plants to the first favorable conditions of moisture and temperature in the spring. In fact, broomsedge often begins growth in February. It is also evident from the data that the first year "bloom" in production was the result of the high rate during the last two months of the season. Finally, as spring production increases and later summer production decreases with the progress of succession mid-summer values tend to become lower than either. High respiratory losses during the hot period may contribute to this pattern.

TABLE VIII. Seasonal and successional changes in net production rate (grams dry matter per square meter per day)

	Annual Forb Fields (Average 2 fields, 9-111 & 3-412)					Perennial Grass fields (Av. 2 fields 3-417 & 2-416)
Years abandoned.....	1	2	3	5	6	15
Year measured.......	(1952)	(1953)	(1954)	(1956)	(1957)	(1954)
Spring (March 15-May 15)..	1.0	1.4	1.1	1.2	1.6	2.2
Early Summer (May 15-July 15)........	0.7	1.6	1.8	1.4	1.4	1.5
Late Summer (July 15-Sept. 15).......	5.4	2.5	2.0	3.3	1.9	1.8
Average—6 months major growing season..........	2.7	1.8	1.6	1.8	1.6	1.9

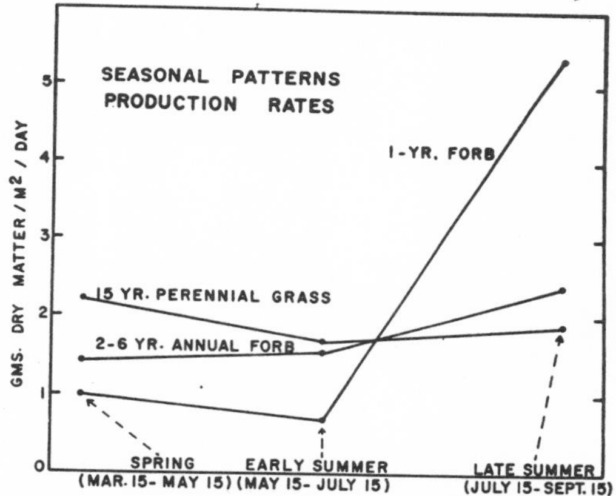

FIG. 9. Production rates for two-month intervals during the six-month major growing season for one-year fields, 2-7 year forb fields, and broomsedge fields (15+ years) illustrating the changing pattern of the seasonal distribution of production.

It is instructive to compare the old-field net production with that of other ecosystems. Comparable data on various natural and cultivated ecosystems have been recently summarized by Odum (1959: page 73, Table 7). The most fertile natural communities such as marsh grass or bottomland ragweed, as well as intensive agriculture such as sugar cane, average 10-20 g dry matter per square meter per day for periods of 6 months or more. For a 6 month growing season young forest plantations in England average 6, Nebraska tall grass prairie 3, Wyoming short grass grassland 0.5 and Nevada desert about 0.2 g/m²/day. Corn and wheat as intensively cultivated in northern Europe averages 4-8 g dry matter (including estimated straw, stubble and roots as well as the harvested grain), but the world average for both crops is 2.3. Thus, the productivity of SRP old-fields during the first year of abandonment was about the same as average corn or wheat, or natural prairie, but somewhat less than this in subsequent years (see Table VIII).

Discussion

Figure 10 summarizes the entire study and brings out certain points which seem worthy of emphasis. Six attributes which depict both the structure and the function of the community during the early stages of succession are shown. The bar graphs at the top of the Figure show the year to year changes in the relative importance of each of the six species which contributed the most to annual net production. The pattern of gradual and continuous change in species composition, as shown here, is a familiar one that has been described for other localities. The curve showing the increase in the number of species contributing at least 1 gm./m²/year to total organic production (i.e., our arbitrary definition of a "dominant") illustrates the increase in species diversity which occurred. The relationship between the terminal or maximum standing crop and the total annual production is shown in the next curve. Standing crop was an approximate measure of productivity during the first two years when production was concentrated into the last part of the season, but not in subsequent years when the standing crop at any one time was much smaller than the annual production. The rapid shift from an uneven to an even seasonal distribution of production is further shown by the curves of production rates. Finally, the stabilization of the average litter and the total annual production after about two years is shown in the bottom two curves in Figure 10.

If we consider that species composition and species diversity represent "structural" features of the community and that productivity is a "functional" attribute, then it is clear that structurally the community changed gradually and continuously but that functionally a temporary steady state was established during the period of study.

Previous studies of secondary succession on land have emphasized the changes in species composition and the structure of the standing crop; such changes are particularly striking in the early stages of succession. Typically, dominants replace one another from year to year, the relative abundance of any one species rarely is the same from one year to the next, and the number of species and the number of individuals often increases, at least in the early years. A pattern of this sort was quite evident in the SRP fields. Because changes in the taxonomic composition of the community with time are conspicuous, gradual and continuous, it has often been inferred that productivity also must change. In fact, it has often been assumed (without actual measurements) that productivity increases automatically with succession. For example, Oosting (1956) states the generally held view that succession progresses from "fractional utilization" to "maxi-

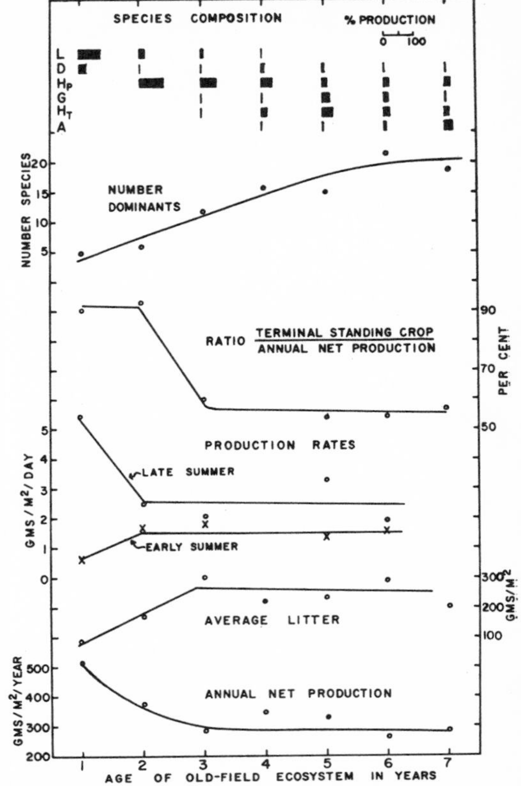

Fig. 10. Six features of the old-field ecosystem during the first seven years of natural succession comparing major structural attributes (species composition and diversity) and functional attributes (productivity). The relative importance in terms of the per cent of total annual production of six major taxa are shown in the top bar diagram. The six taxa are: L-Leptilon; D-Digitaria; H_p-Haplopappus; G-*Gnaphalium obtusifolium*; H_t-Heterotheca; A-Andropogon. The number of dominants includes all species contributing more than 1 g/m²/year to annual production. The relative stability of productivity features contrasts with marked changes in species composition and increase in species diversity.

mum utilization" and that a "peak in productivity" is reached at the climax. As is clearly shown in the present study, productivity does not necessarily change with a change in taxonomic composition, nor necessarily increase with succession. Nor does an increase in the number of species in the community necessarily increase total productivity, although increased diversity may well increase the stability of the community and its ability to adapt to changes in climate or to exploit various microenvironments in the habitat.

The reasons for the rapid change in species during the early stages of succession are not yet clear. Short life cycles, competition with later invaders and root anti-biotics are probably all involved in the succession of species. Keever (1950) has found some evidence that roots of Leptilon produce substances which inhibit the growth of second year individuals. Certainly, Leptilon, and to a lesser extent the other major forbs, decline in vigor after the first year of occupation of a site even though numerous seeds sprout.

While it is planned to continue the present study into the perennial grass and pine stages, the trends so far observed suggest a tentative hypothesis regarding the relationship of energy and succession. From the functional standpoint secondary succession seems to involve a series of temporary steady-states each associated with a major life form. For example, in so far as organic production and turnover is concerned, the years of forb dominance proved to be quite similar despite the almost complete turnover of species. It is probable that a second temporary steady-state equilibrium will be reached when broomsedge and associated perennials become fully established. Preliminary measurements indicate that the seasonal distribution of production is different, the level of organic matter higher, but the annual net production not much greater in broomsedge as compared with the forb stages. Invasion of pines will bring about other changes and another equilibrium which will tend to persist despite marked change in appearance of the community which will occur as numerous small trees give way to fewer large ones. The pine stage will likely be the final or climax stage on the lighter soils, especially if fire remains a regular part of the environment. On heavier or poorly drained soils succession may be expected to continue to a hardwood stage.

As we have seen, the forb stage on SRP fields began with a "bloom" in productivity. It seems likely that at least a temporary increase in productivity may be expected in the early stages of broomsedge and pine establishment. Such a "bloom" could result from the exploitation by the new life form of water and nutrients not available to the previous life form. Thus, the deeper roots of perennial grasses and trees probably will be able to obtain nutrients which have leached beyond the reach of the annuals. Once the accumulated requisites have been brought into the ecosystem, however, the community will adjust to the rate at which the total nutrient pool can be recycled and thus a steady-state in energy flow established. Assuming that climate and solar radiation remain the same, an established equilibrium will presumably persist until additional limiting materials are either brought into or lost from the system.

Summary

Energy flow through communities which develop on abandoned farmland is being studied on the AEC Savannah River Plant reservation. The present report is concerned with net primary production during the first seven years of secondary succession.

Net production was measured by what may be termed the "short-term harvest method" which involved harvesting at monthly intervals all organic materials from meter quadrats along a transect. Since heterotrophic utilization was mostly delayed in time during the early stages of succession the measured "net community production" approximates true net primary production.

The general pattern of succession was observed on 600 reference fields, productivity and soil measurements were made on 30 selected study fields representing the eight major agricultural soil series, and intensive studies were carried out on two of the study fields which represented two of the major soils and the two physiographic subregions of the SRP reservation.

Annual and biennial forbs produced most of the organic matter during the first five years with *Leptilon canadense, Haplopappus divaricatus,* and *Heterotheca subaxillaris* appearing as major dominants in that order on well-drained soils. The perennial grass, Andropogon ("broomsedge"), did not begin to contribute appreciably to production until the 5th to 7th year.

The vegetation was similar in general appearance on most soil types during the first year, but species composition showed increasing divergence on different soils in subsequent years. The silt-clay content of the subsoil (A_2 or B_1 horizons) was a key factor influencing species composition, the number of dominants (arbitrarily designated as species producing more than 1 gm. dry matter/m^2/year), and the rate of succession of species. In general, species composition changed more rapidly, broomsedge (Andropogon) and pines invaded more rapidly, and the number of dominants was greater on the heavier soils as compared with the light, sandy ones.

Organic matter and major mineral nutrients did not increase in the soil during the first three years; in fact, these materials decreased in the topsoil on sandy soils indicating that leaching exceeded the rate of decomposition and nutrient regeneration in the topsoil.

Net production during the first year was of the order of 500 grams of dry matter per square meter per year, and during subsequent years about 300 grams, except where reduced by low rainfall or extreme soil type. In other words, after a one year "bloom" (tentatively attributed to residual fertilizer or other requisites remaining from agricultural use), a steady-state in productivity was achieved even though species composition was changing from year to year.

Annual net production (in contrast to species composition) was not greatly different on different soil types as long as the water and nutrient holding silt-clay fraction was greater than 20%; below 15% there was a marked reduction in productivity.

A temporary equilibrium between annual net production and annual decomposition of dead plant material was established after three years when the average litter became stabilized at around 250-350 grams of dry matter per square meter. Preliminary measurements of fields already abandoned at the time of establishment of SRP indicated that the "standing crop" of dead matter may be expected to increase with the invasion of perennial grasses.

The seasonal pattern of production changed with succession. Production rate was low (about 1 $g/m^2/day$) during spring and summer of the first year and relatively high (about 5 $g/m^2/day$) during the period from mid-July to mid-September. During the subsequent years of forb dominance, when annual production became stabilized, spring and summer rates were higher but late summer rates much lower so that the seasonal curve became much flatter, and maximum or terminal standing crops became smaller.

The average production rate of 1-3 $g/m^2/day$ of SRP old-fields during the 6 month major growing season is about the same as world average wheat or corn, or natural prairie computed for the same time period.

If we consider that species composition and species diversity represent "structural" features of the community and that productivity is a "functional" attribute, then it is clear that structurally the "old-field" commmunity changed gradually and continuously, but that functionally a temporary steady-state was established during the period of forb dominance. The study clearly showed that productivity does not necessarily change with a change in species, nor necessarily increase with succession as has often been assumed.

The trends so far observed suggest a tentative hypothesis regarding the relationship of energy to succession. From the functional standpoint succession may involve a series of steady-states each associated with a major life form rather than a continuous change associated with species change as is usually postulated. It may be further suggested that an increase in productivity, even if only a temporary "bloom," would be most likely to occur in transition from one life form to another since a new life form may be able to utilize accumulated limiting materials not available to the previous life form.

Appendix: Checklist of Major SRP Old-field Dominants

Forbs (Dicots)

Rumex Acetosella L.	Sheep-sorrell
Lepidium virginicum L.	Pepper "grass"
Rubus hispidus L.	Southern dewberry
Cassia (Chamaecrista) fasciculata Michx.	Partridge-pea
Lespedeza striata (Thunb)	Annual Lespedeza
Hypericum gentianoides L.	Pineweed
Lechea villosa Ell.	Pineweed
Oenothera laciniata Hill	Evening-primrose
Polypremum procumbens L.	—
Linaria canadensis (L.)	Blue Toadflax
Bignonia radicans L.	Trumpet-creeper
Diodia teres Walt	Buttonweed
Specularia perfoliata (L.)	Venus's Looking-glass
Ambrosia artemisiifolia L.	Ragweed
Eupatorium capillifolium (Lam.)	Dog-fennel
Eupatorium compositifolium Walt	Dog-fennel
Heterotheca subaxillaris (Lam.)	Camphorweed
Haplopappus (Isopappus) divaricatus (Nutt.)	Yellow Aster
Erigeron annuus (L.)	Daisy-fleabane
Leptilon (Erigeron) canadensis (L.) Britton	Horseweed
Gnaphalium obtusifolium L.	Rabbit-tobacco
Gnaphalium purpureum L.	Purple Cudweed
Grasses and Sedges	(Monocots)
Festuca sciurea Nutt.	Fescue-grass
Eragrostis spectabilis (Pursh)	Tumble-grass

Triplasis purpurea (Walt)	Sand-grass
Aristida tuberculosa Nutt.	Three-awn Grass
Aristida purpurascens Pois	Arrowfeather
Cynodon dactylon (L.)	Bermuda Grass
Digitaria sanguinalis (L.)	Crab-grass
Leptoloma cognatum (Schultes)	Fall Witchgrass
Panicum virgatum L.	Switchgrass
Cenchrus pauciflorus Benth.	Field Sand-spur
Andropogon virginicus L.	Broom-sedge
Andropogon ternarius Michx.	Broom-sedge
Sorghum halepense Pers.	Johnson-grass
Cyperus rotundus L.	Nut-grass

REFERENCES

Bonck, J., and W. T. Penfound. 1945. Plant succession on abandoned farm land in the vicinity of New Orleans, Louisiana. Am. Midland Nat. **33**: 520-529.

Cooke, W. W. 1936. Geology of the Coastal Plain of South Carolina. Geo. Serv. Bull No. **867**.

Cross, W. H. 1956. The arthropod component of old field ecosystems; herb stratum population with special emphasis on the Orthoptera. Ph.D. Dissertation, Univ. of Georgia.

Goodall, D. W. 1952. Quantitative aspects of plant distribution. Biol. Rev. **27**: 194-245.

Keever, Catherine. 1950. Causes of succession on old fields of the Piedmont, North Carolina. Ecological Monog. **20**: 229-250.

———. 1955. *Heterotheca latifolia,* a new and aggressive exotic in Piedmont old-field succession. Ecology **36**: 732-739.

Kurz, Herman. 1944. Secondary forest succession in the Tallahassee Red Hills. Proc. Florida Acad. Sci. **7**: 59-100.

Odum, E. P. 1959. Fundamentals of Ecology (2nd Ed.). Philadelphia: W. B. Saunders. 546 pp.

Olson, J. S. 1958.. Rates of succession and soil changes on southern Lake Michigan sand dunes. Bot. Gaz. **119**: 125-170.

Oosting, H. J. 1942. An ecological analysis of the plant communities of Piedmont, North Carolina. Am. Midland Nat. **28**: 1-126.

———. 1956. A Study of Plant Communities (2nd Ed.). San Francisco: W. H. Freeman.

Quarterman, Elsie. 1957. Early plant succession on abandoned cropland in the central basin of Tennessee. Ecology **38**: 300-309.

Part III

MECHANISMS

Editor's Comments on Papers 7 Through 10

In the previous part the authors have interpreted their data to show that a rather orderly progression of communities may occur at a site. What mechanisms might cause this sequence? Why doesn't one group of plants and animals invade the disturbed site and persist indefinitely? In this chapter we will consider some of the mechanisms that could cause the sequential replacement of communities. Since this is an active area of research the papers are more current, and it is more difficult to identify key papers which mark a particular benchmark in the explanation of causality.

In the largest view, there are controls on succession due to the environment, the biota, and to the interaction of environment and biota. Environmental influences such as substrate effects, climate, water, fire and sedimentation are especially obvious to the student of sand dunes, glaciers, old fields and ponds. The papers of Cowles (Paper 2), Crafton and Wells (Paper 4), and Shelford (Paper 5) provide examples of these influences. Here we will consider other mechanisms influencing succession, especially mechanisms which involve the biota and involve interaction of plants to plants and animals to plants.

One of the first problems in succession is invasion of the site by the biota. In secondary succession seeds, root stocks, tubers, microorganisms and soil animals may persist on the site and initiate succession once a disturbance is removed. Henry J. Oosting [1903–1968], Professor of Botany at Duke University, and his student, Mary Humphreys, were among the first to study the population of seeds as related to succession (Paper 7). Their approach was to collect samples of soils from a spectrum of old-field successional communities from 1 to 112 years of age and from the mature oak-hickory forest characteristic of the North Carolina Piedmont to observe the germination of these seeds in soils in a greenhouse. In this early study the results were ambiguous. Some species retain seed viability over relatively long periods of time (as much as 50 to 75 years), others apparently do not. However, viable seed did occur in soil of all age classes, and seed germination was indicative of successional stage.

Other investigators have directly counted the numbers of seeds in the soil by flotation of seeds in K_2CO_3 or by sieving. The number of seeds or seed germinations per square meter of soil surface range from a few hundred to over 50,000. Oosting and Humphrey's estimates of 3,000 to 4,600 seeds in the *Andropogon* (Broomsedge) community is rather typical of these estimates.

Even if seeds, roots, and some animals and microorganisms persist in the soil of the disturbed area, usually there is considerable invasion of the site by other elements of the biota. Obviously two features are important—the size of the disturbed area and the mobility of the biota. As the size of the disturbed area becomes larger and larger, or more isolated from undisturbed areas, the more difficult it is to repopulate the area with plants and animals. Golley (1965), examining data of Odum and of Oosting, hypothesized that if the disturbed area is large enough, succession may be arrested at some intermediate stage. Isolated areas, such as islands, may never return to a community like that present before disturbance. Indeed, this phenomenon is one of the major underlying reasons for the concern of Gomez-Pompa et al. (1972) for the destruction of tropical forest. Terborgh (1975) has shown that tropical birds have relatively limited powers of dispersal and thus are strongly dependent on the size of the area. Terborgh uses this knowledge to suggest the size and pattern of natural reserves for tropical birds. If this is true for other tropical plants and animals, and there is further evidence that it is true, then Gomez-Pompa's concern is completely justified. In this instance, return to the original endpoint, as observed for the communities in the previous chapter, will not occur and the tropical forest as we know it will disappear as a resource.

Thus, mobility of plants and animals can be crucial in regulating the structure and function of the sere. Typically, the initial members of the flora and fauna are those which move easily under the influence of physical factors such as wind and water. As succession progresses, the interaction between plants and animals becomes more and more significant. Transport of disseminules by animals may be essential in these stages and if the animals are lost, for example, through over-hunting, the community may be different from that present before disturbance. Daniel H. Janzen, Professor of Zoology at the University of Pennsylvania, has been one of the most active students of these relationships between animals and seeds. His studies have been concentrated in tropical environments and, although not directly focused on succession, nevertheless provide insight into an important mechanism. Janzen (1969) suggests that seed predation exerts a strong pressure on plant evolution with the result that plants may respond to seed predation by increasing the number of seeds to satiate the predator, shifting size of seed to outside the feeding range of the predator, or developing biochemical toxins. Of course, the predator also responds by changing its seed hunting or handling capacity, with the results that Janzen finds many species of plants have a very limited set of animal seed predators associated with them. In another paper Janzen (1970) shows how seed predation can influence the pattern of vegetation by the consumption of seeds where they are abundant with consequent elimination of all seedlings of that species in specific locations.

The point of all this is that the species composition of the flora in succession is influenced by the presence of buried seed, seed viability over time, the effect of a variety of environmental factors on seed germination, dispersal of seed into the disturbed area by wind, water and animals, and the predation on seed and seedlings by the fauna. The observed floral composition in successional communities described in part two is a resultant of all these factors working together over time. And, of course, the type of disturbance to the community, whether fire, plowing, type of crop, time of last cultivation or the like, all influence the presence and viability of seeds and the presence of animals and thereby also control the species composition in the pioneer stages of succession.

There is another phenomena which is probably significant in altering the relationships between species and the species composition in succession. This is the chemical interaction through substances given off by a plant which inhibit other plants. These chemical interactions are termed allelopathy (Muller, 1966).

Probably the most notable and widely quoted extension of this concept to succession was the study of Catherine Keever, another student of H. J. Oosting. Keever (Paper 8) examined the dominants of three different stages of old-field succession; horseweed (*Leptilon canadense*), aster (*Aster pilosus*) and broomsedge (*Andropogon virginicus*). The reader will note that Keever's study covers a number of characteristics of these species besides the impact of chemical decomposition products on their growth. In this paper she concluded that the inhibition of horseweed by chemical substances derived from horseweed decomposition may explain the reason why horseweed does not persist as a dominant. However, at that time she was unable to show an inhibiting effect of one species on another.

In contrast, Elroy L. Rice of the University of Oklahoma, in a series of papers (1964, 1965a, b, c, 1969) has shown that chemical materials derived from plants may inhibit or stimulate growth of other species. These substances also may inhibit growth of nitrogen fixing and nitrifying bacteria, and Rice discusses the possible significance of this finding in Paper 9. Old fields in Oklahoma characteristically are low in available nitrogen and many of the common plant species of the pioneer weed stage of succession have low nitrogen requirements. Their ability to compete may be associated with the inhibition of nitrogen fixation.

It will be obvious to the reader that there are numerous features of the biota which can influence succession, but it should also be clear that a simple mechanistic explanation of succession is not possible. Biotic factors, dispersal, environmental influences, a variety of subtle animal-plant relationships, and plant inhibitors acting on other plants and on microorganisms may all play a role. Truly there is a rich array of possible mechanisms to explain succession.

We conclude this section with a paper that describes the effect of a single species, pin cherry (*Prunus pennsylvanica*), on the nutrient cycling in successional ecosystems (Paper 10). The study comes from an important series of investigations at Hubbard Brook Experimental Forest in New Hampshire under the direction of F. H. Bormann, Yale University and G. Likens, Cornell University. This paper is especially pertinent to the discussion of mechanisms because the authors, P. L. Marks and F. H. Bormann, suggest that the pin cherry plays a role in regulating nutrient conditions of the system. That is, successional species are "integral components of the larger ecosystem" even though they are absent from the steady state community. This paper, together with that of Odum (Paper 6), suggests that successional communities are not independent but rather are normal sequences adapted to the region. This dynamic view of succession will be considered in the next two parts.

REFERENCES

Golley, F. B. 1965. Structure and Function of an Old-Field Broomsedge Community. *Ecol. Monogr.* **35:**113–131.

Gomez-Pompa, A., C. Vazquez-Yanes, and S. Guevara. 1972. The Tropical Rain Forest: a Nonrenewable Resource. *Science* **177:**762–765.

Janzen, D. H. 1969. Seed-eaters versus Seed Size, Number, Toxicity, and Dispersal. *Evolution* **23:**1–27.

———. 1970. Herbivores and the Number of Tree Species in Tropical Forests. *Am. Nat.* **104:**501–528.

Muller, C. H. 1966. The Role of Chemical Inhibition (Allelopathy) in Vegetation Composition. *Torrey Bot. Club Bull.* **93:**332–351.

Rice, E. L. 1964. Inhibition of Nitrogen-Fixing and Nitrifying Bacteria by Seed Plants. I. *Ecology* **45:**824–837.

———. 1965a. Inhibition of Nitrogen-Fixing and Nitrifying Bacteria by Seed Plants. II. Characterization and Identification of Inhibitors. *Physiol. Plant.* **18:**225–268.

———. 1965b. Inhibition of Nitrogen-Fixing and Nitrifying Bacteria by Seed Plants. III. Comparison of Three Species of Euphorbia. *Oklahoma Acad. Sci. Proc.* **45:**43–44.

———. 1965c. Inhibition of Nitrogen-Fixing and Nitrifying Bacteria by Seed Plants. IV. The Inhibitors Produced by *Ambrosia elatior* L. and *Ambrosia psilostachya* DC. *Southwest. Nat.* **10:**248–255.

———. 1969. Inhibition of Nitrogen-Fixing and Nitrifying Bacteria by Seed Plants. VI. Inhibitors from *Euphorbia supina. Physiol. Plant.* **22:**1175–1183.

Terborgh, J. 1975. Faunal Equilibrium and the Design of Wildlife Preserves. In *Tropical Ecological Systems,* ed. F. B. Golley and E. Medina, pp. 369–380. New York: Springer Verlag.

7

Reprinted from *Torrey Bot. Club Bull.* **67**(4):253–273 (1940)

Buried Viable Seeds in a Successional Series of Old Field and Forest Soils

Henry J. Oosting and Mary E. Humphreys

(with two figures)

Introduction

That seeds may germinate after prolonged burial under natural conditions has long been known. Salter (1857) reported a remarkable flora that appeared on mud upturned in the deepening of Poole Harbor, England. Similar instances were discussed by Becquerel (1907) in his review of odd floras observed on soil from excavations such as canals, wells, and race tracks. Shull (1914) described an exposed pond-bed which produced abundant vegetation.

Undoubtedly these reports and other similar observations led to the experimental burial of seeds by Duvel (1902, 1905) and Beal (1905). Duvel believed he had demonstrated that viability increases with depth of burial although Goss (1924), after study of the data of the same experiment, maintained that depth of burial affected vitality of the seeds but little. In 1930, Darlington (1931) reported that the seeds of four species buried by Beal in 1879 were still viable, indicating that some seeds may remain viable for 50 years when buried under proper conditions. The history of our knowledge of life-span of seeds has been adequately reviewed by Crocker (1938).

During the past 25 years there have been several investigations of naturally buried seeds in relation to above-ground vegetation. Brenchley and Adam (1915) estimated the seed content of arable land, using natural germination in the field as an indication of the buried population. Later, taking soil samples from the field and observing germinations in the greenhouse, Brenchley (1918) found that the seed content of different soil levels varied with the past history of the land. Warington (1924) showed that the type of manure used on land affected its weed flora and hence its buried seed population. More recently Brenchley and Warington (1933) concluded that crop, soil type, and method of cultivation may influence the species and abundance of buried weed seeds.

Systematic studies of buried seeds in the forest floor have not been made since the report of Peter (1893), who found that the seeds of woody species occurred in soils of old forest while soils of new forest contained seeds of species common to cultivated land.

This suggested the present study of a series of sites whose past history is known, and which, vegetationally, bear a successional relationship to each other. By sampling soils from stands of a complete successional series (pioneers to climax) something should be learned of the length of time that naturally buried seeds may lie in the soil and remain viable. At the same time, there might be correlations between these buried seeds and the ages of the stands, and there might even be some clues to the mechanics of plant succession within the series.

Stands of all ages are readily available in the vicinity of Durham, N. C., for it is common practice to abandon fields when they cease to produce profitable crops. After abandonment, a weed population appears which shows a definite succession. The dominance of *Leptilon canadense* (L.) Britton denotes a field that has been lying fallow one year, while asters (*A. dumosus* L. and *A. ericoides* L.) and ragweed (*Ambrosia artemisiifolia* L.) indicate that the field has not been under cultivation for two years. At least three years without cultivation may be inferred from the dominance of *Andropogon* (chiefly *A. virginicus* L.) in an old field. When seed trees are available the fields are soon invaded by pines (*P. taeda* L. or *P. echinata* Mill.) which frequently become dominant within ten years. About a century later many of the pines will have been replaced by oaks and hickories, which indicate the ultimate climax forest.

Since this successional series invariably develops on old fields in this section, the seeds of the dominant species at least must be widely and evenly distributed. Some of these seeds probably do not germinate and they must lie buried in the forest floor. It seems plausible, therefore, that some seeds produced in one community would be present in the soil when subsequent communities occupy the site. By sampling the soil of representative communities of a successional series and subjecting these samples to uniform conditions favorable for germination, the resulting seedlings should be evidence of buried viable seeds.

METHODS

In the month of November, 1937, soil samples were obtained from twenty sites located in the Durham Division of the Duke Forest. Ten ages, determined by the time since abandonment, were included in the series. Each age was represented by two sites. The ages are typical of old field succession in the Piedmont, ranging from a field cultivated in 1937 through pine dominance to an oak-hickory forest. One pine stand (*P. echinata*) in each age class had been studied by Billings (1938). All stands were on Granville sandy loam or closely related soil types and none showed evidence of cutting or other serious disturbance. The following ages were sampled:

Dominant	*Years abandoned*
1. Cultivated in 1937	0
2. *Leptilon*	1
3. *Aster-Ambrosia*	2
4. *Andropogon*	5
	Age of dominants[1]
5. Pine	15
6. Pine	33
7. Pine	58
8. Pine	85
9. Pine	112
10. Oak-hickory	200 plus

Within each site and within each duplicate site, two rectangular sampling areas were chosen, as similar in slope, exposure and cover as could be found.[2] Each sampling area was 10 by 20 feet (paced). The areas were cross-marked with a cord at 2-foot intervals and the sampling points determined from Tippett's (1927) Random Sampling numbers. The soil sampler used was the outer case of one devised by Coile (1936) for obtaining undisturbed soil samples. This is a steel cylinder 5.3 inches deep and 3.7 inches in diameter, with a volume of about 57 cubic inches. To insure the same depth and volume of soil for each sample, the coarse leafy litter, down to the fermentation layer, was removed from the sampling points on the forest floor. Precautions were taken to prevent contamination by seeds from sources other than the samples. Twenty samples were taken from each sampling area and dumped together upon an oilcloth. The pile was thoroughly mixed and then quartered. From a randomly selected quarter, a volume of soil equal to four sample portions (228 cubic

[1] The fields dominated by pine were probably abandoned from five to ten years longer than the ages indicated for the trees. It is doubtful if the hardwood sites were ever cultivated, for the stands were very uneven-aged with many trees over 200 years old.

[2] We acknowledge with thanks the assistance of Professor F. X. Schumacher of the Duke School of Forestry who gave invaluable advice on sampling methods and statistical treatment of the data.

inches) was taken as a composite sample. Since sampling areas were duplicated in each site and two sites were sampled for each age class, the latter were each represented by four composite samples. These were bagged and stored at a temperature just above freezing until all samples were obtained.

All the soils were removed from the cold room on December 3 and each composite sample was placed in a wooden flat on the earthen floor of the greenhouse. To insure that no samples consciously received undue attention, the flats were filled and placed in the greenhouse in the order that the mixture of bags was removed from storage. The soils were kept moist by watering regularly. Since the samples, when spread in the flats, were only about an inch deep, some tended to dry out very rapidly. A thin mulch of powdered, sterilized sphagnum reduced evaporation. Two similarly mulched flats of sterilized soil were placed with the samples as a check on contamination by wind-borne seeds.

The first seedlings were removed on February 3, 1938, and as often thereafter as individuals matured to an identifiable condition they were removed. Competition was kept at a minimum by removing all but a few of the individuals of any species which was particularly abundant. Those remaining were permitted to flower and thus served to check identifications based on purely vegetative characters.

Several species, germinating early in the experiment, tended to flower at remarkably early stages in their development. These plants had scarcely made any vegetative growth and had the appearance of dwarfed alpine or arctic species. The phenomenon was probably a photoperiodic response to the short days of the winter and spring months. Later it was found that growth and consequently identification were facilitated by occasionally sprinkling the flats with a dilute solution of a standard commercial fertilizer.

RESULTS

Between February 3 and October 2, 1938, the germinations which resulted in the flats yielded 5,989 plants. All these plants were considered to be the product of seeds present in the soils when they were obtained from the several sites, for no seedlings appeared in the check flats. Germinations had apparently ceased when the last plants were removed in October. The numbers and distribution of plants for the duplicate sites of each age class are given in table 1. Unidentified seedlings, which numbered 125, are not included. Most of these did not mature sufficiently to develop recognizable characters.

The data are such that to discuss them it seemed desirable to know whether, for certain species, the germinations occurred in sufficient numbers, or were so distributed, that they could not reasonably be ascribed to chance. Accordingly the numbers of germinations for each of the 127 species were treated according to the method of analysis of variance of Fisher (Snedecor, 1937). When, in table 1 or in the discussion, a species is termed significant its germinations proved to have significant differences between age classes when subjected to the F test of Snedecor's table 10.2. Similarly it was determined which of the species within the wooded areas showed significant differences between pineland and hardwood. A further comparison was made within the areas that were not wooded. Here the germinations from the *Andropogon* sites were compared with the most recently abandoned fields. This division for comparison was chosen because the *Andropogon* field represents the best development of an herbaceous community before the dominance of upland fields by pine. Since forested areas represent something quite apart from land recently under cultivation, analyses were made to determine species significant in a comparison between the open and forested areas.

TABLE 1

List of Species and Numbers of Germinations in Soils from Duplicate Sites of Successive Age Classes

SPECIES APPEARING IN CULTIVATED FIELDS	AGES OF SITES										SIGNIFICANT FOR:			
	0	1	2	5	15	33	58	85	112	OH	Whole	Open vs. Forest	Early vs. Andropogon	Pine vs. Oak
Allium spp.	19	3	18	10	0	..	..	..	..	..	x	x	..	..
	5	4	33	14	1	..	..	..	..	..				
Alopecurus carolinianus Walt.	14	..	..	..	..	..	..	..	..	..	..	..	..	..
	0	..	..	..	..	..	..	..	..	..				
Andropogon spp.	1	1	0	26	3	..	2	1	1	7	x	x	x	x
	4	1	1	33	0	..	1	0	2	1				
Barbarea verna (Mill.) Asch.	2	..	2	..	..	..	..	..	..	..	..	x	..	..
	0	..	0	..	..	..	..	..	..	..				
Cerastium viscosum L.	0	1	1	1	..	0	..	..	..	..	..	..	x	..
	1	0	0	5	..	1	..	..	..	..				
Cerastium vulgatum L.	0	..	..	1	..	..	..	..	..	..	..	x	..	..
	5	..	..	2	..	..	..	..	..	..				
Cyperus compressus L.	84	39	0	22	2	0	..	..	..	..	..	x	x	..
	34	154	18	0	15	3	..	..	..	..				
Cyperus flavescens L.	88	45	0	9	40	0	..	0	1	..	..	x	..	..
	60	427	13	2	62	1	..	2	0	..				
Digitaria sanguinalis (L.) Scop.	92	189	20	61	..	..	..	0	..	..	x	x	x	..
	66	142	49	17	..	..	..	3	..	..				

TABLE 1—*Continued*

	Ages of Sites										Significant for:			
	0	1	2	5	15	33	58	85	112	OH	Whole	Open vs. Forest	Early vs. Andropogon	Pine vs. Oak
Erigeron and Aster spp.	1	8	1	1	3	0	2	0	1	2	..	x	..	..
	0	30	0	3	0	1	1	4	6	0				
Evonymus americanus L.	0	1	..	..	..	..	..	..	..	..	x	x	x	..
	1	2	..	..	..	..	..	..	..	..				
Fimbristylis autumnalis (L.) R. & S.	9	..	..	1	..	..	..	..	..	..	x	x	x	..
	6	..	..	0	..	..	..	..	..	..				
Fimbristylis laxa Vahl.	24	..	0	..	0	..	..	..	..	..	..	x	..	..
	6	..	1	..	1	..	..	..	..	..				
Gnaphalium purpureum L.	5	2	0	3	29	1	8	3	4	6	..	..	..	..
	0	34	1	6	8	4	0	3	3	7				
Houstonia spp.	0	..	1	..	..	..	2	148	31	31	..	x	..	..
	5	..	1	..	..	..	1	0	15	0				
Holosteum umbellatum L.	1	..	..	..	..	..	..	..	..	..	..	..	..	..
	0	..	..	..	..	..	..	..	..	..				
Juncus effusus L.	3	0	..	..	..	..	..	..	..	..	x	x	x	..
	1	1	..	..	..	..	..	..	..	..				
Krigia virginica (L.) Willd.	11	..	..	2	..	0	0	..	..	..	x	x	x	..
	15	..	..	3	..	1	1	..	..	..				
Linaria canadensis (L.) Dumont	8	10	0	0	19	3	3	1	..	..	x	x	x	x
	1	21	3	6	4	2	2	0	..	..				
Leptilon canadense (L.) Britton	3	7	1	0	..	..	0	..	0	3	..	x	..	x
	0	23	0	1	..	..	1	..	4	8				
Mollugo verticillata L.	4	1	2	3	0	1	0	1	..	..	..	x	..	x
	5	4	0	2	1	7	2	1	..	..				
Myosotis virginica (L.) BSP.	0	..	..	..	..	..	..	..	..	0	..	x	..	..
	3	..	..	..	..	..	..	..	..	1				
Oenothera laciniata Hill	1	2	2	0	5	..	..	..	..	..	..	x	..	..
	1	1	1	4	1	..	..	..	..	..				
Oxalis corniculata L.	4	0	2	4	..	..	5	1	9	5	..	x	..	x
	0	1	1	1	..	..	0	3	0	4				
Oxalis florida Salisb.	2	..	..	..	0	1	..	0	5	3	x	x	..	..
	0	..	..	..	2	0	..	2	5	0				
Oxalis stricta L.	14	1	5	2	0	0	4	0	2	0	..	x	..	..
	1	8	3	4	6	5	7	2	0	4				
Physalis virginiana Mill.	2	..	..	..	..	..	..	..	..	1	..	x	..	x
	0	..	..	..	..	..	..	..	..	0				

TABLE 1—*Continued*

	AGES OF SITES										SIGNIFICANT FOR:			
	0	1	2	5	15	33	58	85	112	OH	Whole	Open vs. Forest	Early vs. Andropogon	Pine vs. Oak
Plantago virginica L.	9	..	20	19	4	..	..	..	1	..	..	x	x	..
	1	..	0	67	3	..	..	..	0	..				
Polygonum Persicaria L.	2	..	..	0	1	..	..	..	..	..	..	x	..	..
	0	..	..	1	0	..	..	..	..	..				
Polypremum procumbens L.	1	8	0	56	385	572	1	135	7	..	x	x	..	..
	3	1	13	1	27	221	6	58	3	..				
Sagina decumbens (Ell.)	9	12	8	11	6	1	..	..	..	..	x	x	..	..
T. & G.	30	1	11	20	3	0	..	..	..	..				
Scleranthus annuus L.	0	..	..	0	1	..	..	..	..	..	..	x	..	..
	4	..	..	1	0	..	..	..	..	..				
Specularia perfoliata (L.)	2	2	10	4	10	..	1	1	..	..	..	x	x	..
A. DC.	3	6	0	14	1	..	0	0	..	..				
Stenophyllus capillaris (L.)	1	14	0	10	13	16	0	33	..	..	..	..	..	x
Britton	5	0	2	5	6	4	1	24	..	..				
Sisymbrium thalianum (L.)	0	23	9	0	..	..	..	1	..	..	x	x	x	..
J. Gay	9	68	2	8	..	..	..	0	..	..				
Veronica peregrina L.	60	0	8	0	3	2	..	0	2	0	..	x	..	..
	3	30	0	27	5	1	..	1	0	1				
SPECIES ADDED BY THE ONE-YEAR FIELDS														
Chenopodium ambrosioides L.	..	0	..	..	..	..	..	..	..	..	..	x	..	..
	..	2	..	..	..	..	..	..	..	..				
Cyperus sabulosus Mart. &	..	2	..	..	..	..	..	..	..	..	..	x	..	..
Schrad.	..	0	..	..	..	..	..	..	..	..				
Dactyloctenium aegyptium	..	0	..	..	..	..	..	..	..	..	..	..	..	..
(L.) Richter	..	1	..	..	..	..	..	..	..	..				
Danthonia spicata (L.) Beauv.	..	0	..	..	..	..	..	..	2	..	..	..	..	..
	..	2	..	..	..	..	..	..	0	..				
Draba verna L.	..	6	..	..	..	..	..	..	..	..	..	x	..	..
	..	54	..	..	..	..	..	..	..	..				
Eleusine indica (L.) Gaertn.	..	1	..	..	..	..	..	..	..	..	..	..	..	..
	..	0	..	..	..	..	..	..	..	..				
Hieracium Gronovii L.	..	0	1	1	0	..	..	..	..	0	..	x	..	
	..	1	1	1	2	..	..	..	..	1				
Oenothera biennis L.	..	0	..	..	..	..	..	..	..	..	..	..	..	..
	..	1	..	..	..	..	..	..	..	..				

TABLE 1—*Continued*

	Ages of sites										Significant for:			
	0	1	2	5	15	33	58	85	112	OH	Whole	Open vs. Forest	Early vs. Andropogon	Pine vs. Oak
Oxalis violacea L.	..	3	..	..	..	..	..	..	..	1	..	..	..	x
	..	0	..	..	..	..	..	..	..	3				
Solidago spp.	..	19	87	0	0	1	..	..	1	1	x	x	x	..
	..	14	78	5	12	1	..	..	12	0				
Trifolium arvense L.	..	0	..	7	..	..	..	..	..	..	..	x	..	..
	..	25	..	0	..	..	..	..	..	..				
Trifolium incarnatum L.	..	0	..	..	..	..	..	..	..	..	..	x	..	..
	..	3	..	..	..	..	..	..	..	..				
Xanthium cylindraceum Millsp. & Sherff	..	0	..	..	..	..	..	..	..	..	..	..	..	..
	..	1	..	..	..	..	..	..	..	..				
SPECIES ADDED BY THE TWO-YEAR FIELDS														
Amaranthus hybridus L.	..	..	1	..	..	..	..	..	..	..	..	..	..	..
	..	..	0	..	..	..	..	..	..	..				
Ambrosia artemisiifolia L.	..	..	1	0	1	1	..	..	..	..	..	x	..	..
	..	..	0	1	0	0	..	..	..	..				
Anthemis Cotula L.	..	..	0	0	1	0	..	..	0	0	..	x	..	x
	..	..	1	1	0	1	..	..	3	3				
Ascyrum hypericoides L.	..	..	0	2	1	..	0	..	1	..	x	x	x	..
	..	..	2	5	0	..	2	..	2	..				
Carara didyma (L.) Britton	..	..	0	..	..	..	..	..	..	..	..	..	..	..
	..	..	1	..	..	..	..	..	..	..				
Cyperus globulosus Aubl.	..	..	0	0	..	9	0	..	0	0	..	x	..	..
	..	..	1	3	..	0	2	..	2	1				
Diodia teres Walt.	..	..	1	..	..	..	..	..	..	..	..	x	..	..
	..	..	1	..	..	..	..	..	..	..				
Geranium carolinianum L.	..	..	2	..	..	..	..	..	..	..	..	x	..	..
	..	..	0	..	..	..	..	..	..	..				
Helenium tenuifolium Nutt.	..	..	2	..	..	..	..	..	..	..	..	x	..	..
	..	..	0	..	..	..	..	..	..	..				
Hypericum gentianoides (L.) BSP.	..	..	0	7	6	1	1	3	..	1	x	x	x	x
	..	..	3	15	7	3	0	0	..	0				
Hypericum mutilum L.	..	..	0	..	..	..	..	..	..	..	..	..	..	..
	..	..	1	..	..	..	..	..	..	..				
Lespedeza striata (Thunb.) H. & A.	..	..	8	86	..	..	..	..	..	..	x	x	x	..
	..	..	1	29	..	..	..	..	..	..				

TABLE 1—*Continued*

	AGES OF SITES										SIGNIFICANT FOR:			
	0	1	2	5	15	33	58	85	112	OH	Whole	Open vs. Forest	Early vs. Andropogon	Pine vs. Oak
Plantago aristata Michx.	..	..	0	1	..	..	..	..	..	..	..	x	..	..
	..	..	1	0	..	..	..	..	..	..				
Plantago heterophylla Nutt.	..	..	0	..	..	..	..	..	..	..	..	x	..	..
	..	..	8	..	..	..	..	..	..	..				
Pyrrhopappus carolinianus	..	..	0	..	..	..	..	..	..	..	..	..	..	..
(Walt.) DC.	..	..	1	..	..	..	..	..	..	..				
Solanum carolinense L.	..	..	0	..	..	..	1	..	0	..	..	..	..	..
	..	..	1	..	..	..	0	..	1	..				
Specularia biflora (R. & P.)	..	..	4	0	..	..	..	..	..	..	..	x	..	..
F. & M.	..	..	0	1	..	..	..	..	..	..				
Veronica arvensis L.	..	..	3	0	..	..	..	1	..	..	..	x	x	..
	..	..	0	18	..	..	..	0	..	..				
SPECIES ADDED BY THE FIVE-YEAR FIELDS														
Agrostis hyemalis (Walt.)	..	..	..	0	..	..	..	..	0	..	..	x	x	..
BSP.	..	..	..	2	..	..	..	..	2	..				
Aristida spp.	..	..	..	6	0	..	..	..	..	..	..	x	x	..
	..	..	..	1	2	..	..	..	..	..				
Cassia nictitans L.	..	..	..	4	..	..	..	..	..	..	x	x	x	..
	..	..	..	13	..	..	..	..	..	..				
Eragrostis pilosa (L.) Beauv.	..	..	..	0	..	..	..	..	..	..	..	..	x	..
	..	..	..	1	..	..	..	..	..	..				
Euphorbia Preslii Guss.	..	..	..	0	..	..	..	..	..	..	..	..	x	..
	..	..	..	1	..	..	..	..	..	..				
Festuca octoflora Walt.	..	..	..	0	0	..	..	..	0	..	..	x	..	..
	..	..	..	2	1	..	..	..	7	..				
Festuca ovina L.	..	..	..	7	..	..	..	..	..	..	x	x	x	..
	..	..	..	2	..	..	..	..	..	..				
Poa annua L.	..	..	..	1	..	..	..	..	..	..	..	..	x	..
	..	..	..	0	..	..	..	..	..	..				
Poa cuspidata Nutt.	..	..	..	15	..	..	..	..	..	..	..	x	x	..
	..	..	..	0	..	..	..	..	..	..				
SPECIES ADDED BY THE FIFTEEN-YEAR SITES														
Cyperus inflexus Muhl.	..	..	..	..	0	..	..	..	..	..	..	x	..	..
	..	..	..	..	1	..	..	..	..	..				

TABLE 1—*Continued*

	Ages of sites										Significant for:			
	0	1	2	5	15	33	58	85	112	OH	Whole	Open vs. Forest	Early vs. Andropogon	Pine vs. Oak
Eragrostis spectabilis	..	..	..	..	3	0	..	1	..	0	..	x	..	..
(Pursh.) Steud.	..	..	..	..	0	2	..	0	..	1				
Erechtites hieracifolia (L.)	..	..	..	..	1	..	..	..	1	1	x	x	..	x
Raf.	..	..	..	..	0	..	..	..	0	2				
Pagesia acuminata (Walt.)	..	..	..	..	0	..	1	..	..	..	..	x	..	x
Pennell	..	..	..	..	1	..	0	..	..	..				
Silene antirrhina L.	..	..	..	..	3	..	..	..	..	..	..	x	..	..
	..	..	..	..	0	..	..	..	..	..				
SPECIES ADDED TO THE THIRTY-THREE-YEAR SITES														
Chrysopsis graminifolia	..	..	..	..	..	0	..	..	..	..	..	x	..	..
(Michx.) Nutt.	..	..	..	..	..	2	..	..	..	..				
Eupatorium capillifolium	..	..	..	..	..	0	..	..	..	..	..	x	..	..
(Lam.) Small	..	..	..	..	..	1	..	..	..	..				
Panicum dichotomum	..	..	..	..	..	5	..	..	2	..	x	x	..	x
	..	..	..	..	..	3	..	..	0	..				
Panicum sphaerocarpon Ell.	..	..	..	..	..	11	0	0	0	..	..	x	..	x
	..	..	..	..	..	3	1	1	3	..				
Panicum villosissimum Nash.	..	..	..	..	..	0	0	3	4	1	..	x	..	..
	..	..	..	..	..	5	2	0	2	0				
Paspalum setaceum Michx.	..	..	..	..	..	1	..	..	..	..	..	x	..	..
	..	..	..	..	..	0	..	..	..	..				
Potentilla pumila Poir	..	..	..	..	..	0	2	..	..	0	..	x	..	x
	..	..	..	..	..	1	0	..	..	3				
Trichostema dichotomum L.	..	..	..	..	..	0	..	..	..	..	..	x	..	..
	..	..	..	..	..	1	..	..	..	..				
Viola sp.	..	..	..	..	..	0	..	..	..	..	..	x	..	..
	..	..	..	..	..	1	..	..	..	..				
SPECIES ADDED BY THE FIFTY-EIGHT-YEAR SITES														
Broussonetia papyrifera (L.)	..	..	..	..	..	..	1	..	..	..	..	x	..	..
Vent.	..	..	..	..	..	..	0	..	..	..				
Chrysopsis mariana (L.) Nutt.	..	..	..	..	..	..	0	..	..	..	..	x	..	..
	..	..	..	..	..	..	1	..	..	..				
Cyperus strigosus L.	..	..	..	..	..	..	5	..	..	5	..	x	..	x
	..	..	..	..	..	..	0	..	..	0				

TABLE 1—*Continued*

	AGES OF SITES										SIGNIFICANT FOR:			
	0	1	2	5	15	33	58	85	112	OH	Whole	Open vs. Forest	Early vs. Andropogon	Pine vs. Oak
Gnaphalium obtusifolium L.	..	..	..	..	..	..	0	..	..	..	..	x	..	..
	..	..	..	..	..	..	1	..	..	..				
Panicum anceps Michx.	..	..	..	..	..	..	0	..	..	..	..	x	..	..
	..	..	..	..	..	..	1	..	..	..				
Panicum philadelphicum Bernh.	..	..	..	..	..	..	1	..	..	..	..	x	..	..
	..	..	..	..	..	..	0	..	..	..				
Panicum xalapense HBK.	..	..	..	..	..	..	1	0	..	0	..	x	..	x
	..	..	..	..	..	..	0	1	..	3				
Rubus sp.	..	..	..	..	..	..	1	..	..	3	..	x	..	x
	..	..	..	..	..	..	0	..	..	0				
Sisyrinchium gramineum Curtis	..	..	..	..	..	..	1	..	..	..	..	x	..	..
	..	..	..	..	..	..	0	..	..	..				
Verbascum Thapsus L.	..	..	..	..	..	..	2	..	..	..	..	x	..	..
	..	..	..	..	..	..	0	..	..	..				
SPECIES ADDED BY THE EIGHTY-FIVE-YEAR SITES														
Diodia virginiana L.	..	..	..	..	..	..	..	1	..	..	..	x	..	..
	..	..	..	..	..	..	..	0	..	..				
Hypoxis hirsuta (L.) Coville	..	..	..	..	..	..	..	0	7	..	..	..	..	..
	..	..	..	..	..	..	..	10	0	..				
Juncus scirpoides Lam.	..	..	..	..	..	..	..	14	..	..	..	x	..	..
	..	..	..	..	..	..	..	0	..	..				
Oxydendrum arboreum (L.) DC.	..	..	..	..	..	..	..	1	..	..	..	x	..	..
	..	..	..	..	..	..	..	0	..	..				
Panicum commutatum Schultes	..	..	..	..	..	..	..	1	..	..	..	x	..	..
	..	..	..	..	..	..	..	0	..	..				
Sambucus canadensis L.	..	..	..	..	..	..	..	0	..	..	..	x	..	..
	..	..	..	..	..	..	..	2	..	..				
Ulmus alata Michx.	..	..	..	..	..	..	..	0	..	..	..	x	..	..
	..	..	..	..	..	..	..	1	..	..				
SPECIES ADDED BY THE 112-YEAR SITES														
Agrostis alba L.	..	..	..	..	..	..	..	..	1	..	..	x	..	..
	..	..	..	..	..	..	..	..	0	..				
Cyperus cylindricus (Ell.) Britton	..	..	..	..	..	..	..	..	0	..	..	x	..	..
	..	..	..	..	..	..	..	..	1	..				

TABLE 1—*Concluded*

	AGES OF SITES										SIGNIFICANT FOR:			
	0	1	2	5	15	33	58	85	112	OH	Whole	Open vs. Forest	Early vs. Andropogon	Pine vs. Oak
Liquidambar styraciflua L.	..	..	..	..	..	..	..	..	1	..	..	x	..	..
	..	..	..	..	..	..	..	..	0	..				
Panicum lanuginosum Ell.	..	..	..	..	..	..	..	..	0	..	..	x	..	..
	..	..	..	..	..	..	..	..	1	..				
Prunella vulgaris L.	..	..	..	..	..	..	..	..	0	1	..	x	..	x
	..	..	..	..	..	..	..	..	1	0				
Rhus copallina L.	..	..	..	..	..	..	..	..	1	..	..	x	..	..
	..	..	..	..	..	..	..	..	0	..				
Scleria pauciflora Muhl.	..	..	..	..	..	..	..	..	0	..	..	x	..	..
	..	..	..	..	..	..	..	..	1	..				
Trifolium procumbens L.	..	..	..	..	..	..	..	..	0	..	..	x	..	..
	..	..	..	..	..	..	..	..	1	..				
SPECIES ADDED BY THE OAK-HICKORY SITES														
Cercis canadensis L.	..	..	..	..	..	..	..	..	..	1	..	x	..	x
	..	..	..	..	..	..	..	..	..	0				
Eupatorium hyssopifolium L.	..	..	..	..	..	..	..	..	..	0	..	x	..	x
	..	..	..	..	..	..	..	..	..	1				
Galium pilosum Ait.	..	..	..	..	..	..	..	..	..	1	..	x	..	x
	..	..	..	..	..	..	..	..	..	0				
Hedeoma pulegioides (L.)	..	..	..	..	..	..	..	..	..	0	..	x	..	x
Pers.	..	..	..	..	..	..	..	..	..	1				
Lactuca sp.	..	..	..	..	..	..	..	..	..	1	..	x	..	x
	..	..	..	..	..	..	..	..	..	0				
Morus rubra L.	..	..	..	..	..	..	..	..	..	0	..	x	..	x
	..	..	..	..	..	..	..	..	..	1				
Panicum Ashei Pearson	..	..	..	..	..	..	..	..	..	0	..	x	..	x
	..	..	..	..	..	..	..	..	..	2				
Panicum Boscii Poir.	..	..	..	..	..	..	..	..	..	1	x	x	..	x
	..	..	..	..	..	..	..	..	..	3				
Scutellaria sp.	..	..	..	..	..	..	..	..	..	1	..	x	..	x
	..	..	..	..	..	..	..	..	..	0				
Sonchus asper (L.) Hill	..	..	..	..	..	..	..	..	..	0	..	x	..	x
	..	..	..	..	..	..	..	..	..	1				
Vaccinium sp.	..	..	..	..	..	..	..	..	..	0	..	x	..	x
	..	..	..	..	..	..	..	..	..	1				
Viburnum affine Bush.	..	..	..	..	..	..	..	..	..	0	..	x	..	x
	..	..	..	..	..	..	..	..	..	1				

DISCUSSION

1. *Origin of the germinating seeds*

The majority of the germinations must have been derived from naturally buried seeds, for there was an average of 3.8 seedlings per square inch of natural surface sampled. Soils of fields abandoned one year yielded 1,463 germinations from samples with a field surface area totaling only slightly over one square foot (171.8 square inches) and a volume of about half a cubic foot. This, combined with the fact that over 90 per cent of the total germinations from each of the fields abandoned one, two, or five years were of species found growing in successionally younger fields, seems highly indicative of buried viable seeds.

It might be supposed that germinations in the forest soils resulted from seeds which were transported to the areas sampled, and that seeds from plants of an old-field community may effect an entrance into the interior of a pine stand. However, it seems improbable that many of the seeds could sift through the litter and fermentation zone, because of the thickness of these two layers after the stand is more than 20 years old. Billings (1938) charted the changes in the soil profile from an *Andropogon* field through the successional series of pine stands used in this investigation. The "plowed" horizon decreases, because it is becoming slowly incorporated in the A_2 horizon beneath it and the new A_1 horizon forming above it. The seeds buried in the plowed horizon are therefore incorporated in the A_1 and A_2 as the profile changes; they are not actually compacted in a layer of soil of decreasing width. Above the mineral soil there is a gradual accumulation of litter with the accompanying formation of the fermentation layer. The litter layer increases from nothing in an old field to well over one and a half inches in the mature pine stands. The fermentation zone gradually reaches a depth of over half an inch in the mature stands. It is conceivable that, if the seeds were retained in this zone, conditions would be favorable for their germination in the damp mycelial mat of the fermentation layer soon after their entrance into the stand. After germination, survival of the seedlings would depend upon their ability to compete with the new environmental condition. Any seeds germinating in the forest obviously could not be present to account for germinations in the flats. Those seeds of the old-field species which do germinate and survive in pine stands are limited in number (Billings) and decrease rapidly with each succeeding age class. Most of them probably do not reach the seed production stage before they die. Hence they are not responsible for the viable seeds in the forest floor.

The germinations of open-field species are undoubtedly products of seeds that were incorporated in the soil during cultivation or that entered during the early years of abandonment. It is reasonable that these species would contribute most since their seeds have not had to penetrate a litter and fermentation layer in order to be mixed in the "plowed" horizon. Seeds of characteristic forest species could not have been produced in the old fields. Since they appeared only in the flats of forest soil, they must have the ability to effect an entrance into the soil and their seedlings must be so adapted that they survive where old-field species cannot.

It is doubtful that any germinations in the oak-hickory soils were produced by seeds which lay buried in the soil since their formation in some earlier successional community. All evidence indicates that the oak-hickory sites have never been cultivated. Studies now in progress indicate that several of these species, typical of the early stages in old-field succession, are practically eliminated as pine stands mature but that they may again be characteristically present in the mature hardwood forest. Their reappearance may be correlated with the opening of the stands as they become over-mature and possibly with the changed leaf litter.

2. Distribution of numbers of germinations

Table 1 gives the number of germinations for the species as distributed in the different age classes. Those species germinating in soils from the greatest number of age classes are, without exception, species which appear first (successionally) in the most recently abandoned field. Only *Gnaphalium purpureum* and *Polypremum procumbens* were represented in all age

Fig. 1. Total germinations and total species that appeared in the soils from each age class.

classes. The species of recently abandoned fields not only appeared in the greatest number of age classes but also made up the largest percentage of total species in each age class except oak-hickory. Every age class had "new species," meaning species which did not appear in soils from any lesser-aged stand.

Figure 1, in which all germinations are graphed, shows that the greatest number of germinations occurred in the soils of the 1-year fields while the greatest number of species appeared in soils from the 5-year fields. However, all that can be said of the total seed population should properly apply to the 21 species significant for the whole experiment. Their germinations are graphed in figure 2. The two peaks of germinations are here reversed, with the 33-year stand having by far the greatest number. This high value is the result of the excessive number of germinations of *Polypremum procumbens*, a significant species but, in other age classes, of much less importance. Of the significant species, the greatest number was again in the 5-year soils.

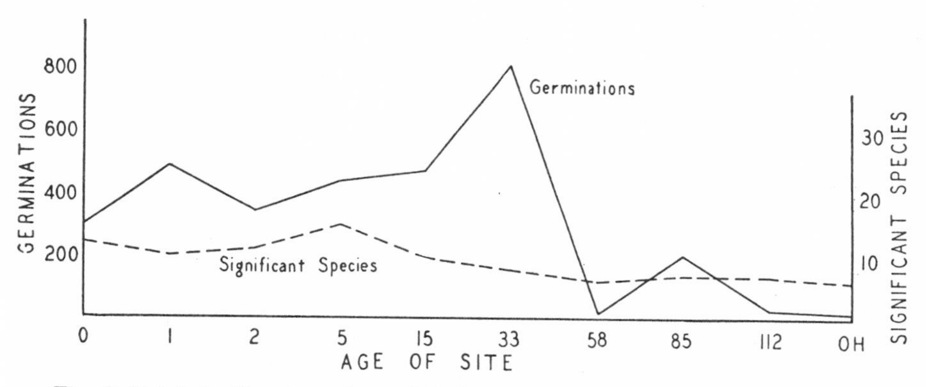

Fig. 2. Total significant species and their germinations for each age class.

Consideration of the distribution of significant species yields only a few generalizations. Those species producing sufficient germinations to be significant for the whole appeared first in open fields and younger pine stands. In spite of the numerous species and high germination counts for open fields, there were relatively few (25) species significant in the comparison between early fields and *Andropogon* dominance. Twelve of these species appeared in the cultivated field and eight did not come till *Andropogon* was dominant. The values serve to emphasize a correlation with the actual vegetation. Although many species are always present, only a few are dominant in size and numbers. The distinctness of the *Andropogon* field may be judged from its addition of nine species of which eight were significant in a comparison with earlier fields.

The comparison of open sites with wooded sites gives the most striking values for significant species. High percentages of the species added by each age of field were significant in this comparison, the cultivated field having 86 per cent. The consistent significance here is interpreted as an indication of the open-field nature of all these species. On the other hand, species appearing first in wooded stands were, without exception, all significant in the comparison. It cannot be entirely a matter of chance that those species which appear only in the soil from wooded stands should all then appear in significant numbers. Their seeds must not be transported elsewhere, or, as is more probable, must require the conditions of forest soil to retain viability.

The pine-hardwood comparison shows several early old-field species carrying over into pine stands in such numbers as to be significant. Those significant for pine were added in the young-to-middle-aged stands. The 85-year (mature) stand added no significant species and the over-mature (112-year) stand only one. Added species for the hardwood stand were, as before, all significant, these last added species being the portion of the viable seed population which is directly correlated with the dominants.

3. Relationships to Vegetation and Succession

The germinations of individual species become of special interest when considered in relation to their occurrence under natural conditions. All of the pine sites sampled in this study have been adequately described (Billings, 1938) on a phytosociological basis, and unpublished quadrat data (Oosting) are available for the oak-hickory stands and for herb stages of old-field succession. The general trend of old-field herbs begins with the abrupt appearance of large numbers of individuals which rarely maintain their importance for more than a year. The succeeding year finds the numbers materially reduced and thereafter there is a more or less gradual decline as trees become dominant. Many herbs of open old fields disappear entirely when trees appear, most are gone when the stand has attained middle age.

If the seeds of early old-field species are not especially viable, a natural break in the germination data should appear between the 5-year field and the 15-year pine. Also, if certain species germinate in soils from wooded stands in which, under natural conditions, Billings did not find them, these germinations should be excellent evidence of buried viable seeds, for the sampling is estimated at only 1 in 10,000.

The list of germinations includes 18 species of open-field herbs which appeared consistently for all age classes through 85 years or more. Of these species only three are recorded as growing naturally on the wooded sites and these only in the youngest stand. All are species which occur consistently in old fields, usually with high frequencies and densities. Obviously they have been eliminated from membership in the forest community, although some are able to maintain themselves a few years longer than others. The general tendency for numbers of germinations to decrease with increased age class is apparent and is a strong argument for the existence of buried viable seeds, the viability decreasing with age. Reasoning on this basis, more evidence is available. Germinations of several open-field species are constantly present for all sites through the lesser age classes of pine. It might be concluded that seeds of these species had all germinated, or, what is more likely, that they were less viable and consequently died in the soil.

Certain inconsistencies in the distribution of numbers for individual species cannot be explained. *Polypremum procumbens* is a typical open-field herb and is not recorded as growing in pine forest, but it produced 412 seedlings in the 15-year age class and the remarkable number of 793 in the 33-year class. No other species approached these figures in total germinations. In contrast, the species of *Houstonia*, usually abundant in field and forest alike, produced no germinations in the 15- or 33-year soils although numerous in the soils from older stands.

In general, the counts for species that are statistically significant and are also listed in the quadrat records of the forested stands show no correlation with the above-ground vegetation. Of these species, only ten appear in the vegetation lists and many other species on the lists did not produce germinations in soil from any age class.

For herbaceous stands the numbers of germinations for an age class may sometimes be correlated with dominance in the previous age class and again may be an indication of what species will be important in the succeeding class. If, regardless of significance, the herb field species with highest germinations are considered by age classes, certain relationships to the field vegetation become apparent. The dominant weed species on a cultivated field at the end of the growing season is *Digitaria sanguinalis.* After a year of abandonment *Digitaria* is even more abundant and *Leptilon canadense* forms an open over-story three feet tall or more. A field abandoned two years is usually characterized by *Aster ericoides* or *Ambrosia artemisiifolia* or both. Thereafter a mixture of species is possible with *Andropogon* rapidly gaining dominance so that by the fourth year it is usually the most important species.

The species with high germinations are all regularly found in the old fields, although some have only seasonal importance. The species of *Sagina, Krigia, Sisymbrium,* and *Draba* are small in stature and abundant only vernally. Others appear only locally or are dwarfed by the larger and more conspicuous dominants.

The high germinations for *Digitaria* and the trend in numbers correlate well with the distribution under field conditions. It is surprising that the field just cultivated yielded no *Leptilon* seedlings for, under normal conditions, it would surely have supported a goodly number of individuals the next year if left fallow. The number of *Leptilon* germinations for the 1-year field is related to the number of plants growing there and indicates their probable presence in the field the next year. Actually, 2-year fields have numerous *Leptilon* plants but they are invariably dwarfed and depauperate. Indications are that by the third year most of the seeds will have germinated, and probably the dwarfed plants produce few seeds, so that *Leptilon* is soon eliminated from the old-field flora. The germination evidence supports these observations.

Germinations from 2-year fields are remarkably unrelated to aboveground observations. Normally *Aster ericoides* and *Ambrosia* are the dominants, as they were on the fields sampled. Peculiarly, not more than one germination of *Ambrosia* was recorded for any age class. Difficulties with identifications made it necessary to lump all germination counts of *Aster* and *Erigeron*, but the combined value may be used as indicative of *Aster* importance. Evidently the *Aster* germinations from the 1-year field are indicators of the *Aster* dominance to come in the next year. Although *Aster* was a dominant on the 2-year field only one germination occurred in the samples. There must have been many more seeds present. Since the samples were collected in the fall and, except for the brief storage in the cold room, thereafter never exposed to winter temperatures, it is possible that the seeds require an after-ripening period. This deficiency may also have contributed to the lack of germinations of *Ambrosia* (*A. trifida* L. requires after-ripening; Davis, 1929) and possibly other species as well.

The 5-year field had a good stand of *Andropogon virginicus*, characteristic for the age class. The clumps were uniformly distributed and spaced from about 6 inches to 2 feet apart. The appearance was that of a pure stand with complete dominance. However, between the clumps were numerous undersized plants of several species, almost all of which had been present on younger fields. The germination lists indicate that little or no *Andropogon* would have appeared on the 2-year field during its third year of abandonment. This conforms with field conditions, for the major appearance of *Andropogon* is apt to be in the fourth year. Germinations in the 5-year samples indicate a potential increase in the *Andropogon* population although actually it had probably about reached its maximum.

Many species present in abundance in the early fields produced no germinations in the wooded stands. This may be an indication that the seeds of these species cannot retain their vitality for a long period of years. Evidence to support this idea is the fairly sharp decrease in the number of seeds germinating in soils from the 15-year pine as compared to the 5-year field. *Allium* drops from 24 to 5; *Digitaria* from 78 to 0; *Plantago virginica* from 86 to 7; *Cassia* from 17 to 0; *Lespedeza* from 115 to 0; and *Veronica peregrina* from 27 to 8. All of these species are statistically significant in the comparison between open and forested sites. To be sure, this radical decrease in germination may have been the result of germinations occurring naturally in the period between the field and the 15-year pine and not a loss of viability. These data are insufficient to prove either point.

Of the nine species added to the germination lists by the 5-year field seven were grasses. Numbers of germinations were not high and were rather in proportion to the importance of the species in this grass stage of old field succession. They represent an ephemeral condition which, with *Andropogon*, gradually disappears when trees become dominant. Billings (l. c.) found a few plants of *Andropogon* and *Eragrostis pilosa* in the 15-year pine stand, but *Andropogon*, *Aristida* and *Festuca octoflora* were the only species to produce germinations for this age class and they yielded only 3, 2 and 1, respectively.

Age classes beyond the establishment of pine show little correlation between natural vegetation and germinations. Numbers of herbs in pine forests are very small compared to those in open fields. With the relatively small soil samples used it is perhaps unwise to draw any conclusions. It is certain that, under natural conditions, field herbs decrease abruptly in total numbers and species with the development of pine. In a young stand of pine most of the old field herbs are eliminated and by middle age only scattered individuals remain, if any at all. At the same time new species of forest herbs have appeared, gradually increasing in abundance though never approaching open field numbers. The germination records show that viable seeds in the soil of these stands tend to include more and more species of wooded areas as the stands mature. However, the numbers never approach the concentrations of open fields.

When a forest stand becomes mature or over-mature, disturbances such as falling trees, etc., are not uncommon. If such a disturbance opens the crown cover and disturbs the accumulated litter there will immediately appear a crop of herbaceous species not at all related to the surrounding forest stand. These are the same species which, although characteristic of herb field succession, may appear sporadically in the forest quadrat records with very low densities and frequencies. They are likewise the same species which, producing tremendous numbers of germinations in open field soils, dropped off abruptly in soils where pine became dominant but showed scattered germinations for age classes up to 112 years. Both the irregular appearances in undisturbed stands and the mass appearance in disturbed areas may reasonably be interpreted as resulting from buried viable seeds, for the presence of these seeds is here demonstrated by germination records.

It is disappointing to find no seedlings of the woody dominants in the soil from the forested sites. Sixteen species of woody seedlings were recorded with five of the species occurring in the oak-hickory site.

SUMMARY AND CONCLUSIONS

1. It is known that buried seeds may remain viable for long periods and studies have been made of naturally buried seeds as related to the vegetation.

2. To determine possible relations between buried seeds and past and future vegetation, soil samples were taken from a series of abandoned fields of known age and vegetative composition: a field under cultivation that season; fields fallow for 1, 2, and 5 years; shortleaf pine stands of 15, 33, 58, 85 and 112 years of age; and an oak-hickory forest. All stands had previously been studied phytosociologically and it is known that the series is representative of old field succession in the area.

3. Exposed to greenhouse conditions for 37 weeks the samples produced 5,989 seedlings. The seedlings represented 127 species of which 16 were woody ones.

4. The highest total germinations were produced by soil from the field which had been abandoned but one year. The 5-year field produced the greatest number of species.

5. The germination of seeds of several species in soil from habitats in which the parent plants do not grow indicates the possibility that, under natural conditions, seeds may lie buried for long periods and retain their viability.

6. Probably some seeds do not retain their viability under natural conditions, for several species which were very numerous in the stands of one age class produced few or no germinations in soil samples from the next succeeding class.

7. Statistical analyses of the occurrences of each of the species were made to ascertain which were significant throughout the series; in comparisons between open and forest; between the 5-year field and the preceding fields; and between pine and hardwood. The distribution of significant species serves to emphasize a relationship between vegetation and buried seeds: the distinctness of *Andropogon* fields; a difference between field and forest and between pine and hardwood; and that forest seeds apparently require forest conditions to retain their viability.

8. The germinations demonstrate the presence of viable seeds in the soil of all age classes sampled, but age and origin of the seeds remain problematical. They show a succession of species, as do the plants above ground, and, in general, they are indicative of that same succession.

DEPARTMENT OF BOTANY
DUKE UNIVERSITY
DURHAM, NORTH CAROLINA

Literature Cited

Beal, W. J. 1905. The vitality of seeds. Bot. Gaz. **40:** 140–143.

Becquerel, P. 1907. Recherches sur la vie latente des graines. Ann. Sci. Nat. IX, **5:** 193–311.

Billings, W. D. 1938. The structure and development of old field short-leaf pine stands and certain associated physical properties of the soil. Ecol. Monogr. **8:** 437–499.

Brenchley, W. E. and Adam, H. 1915. Recolonization of cultivated land allowed to revert to natural conditions. Jour. Ecol. **3:** 193–216.

Brenchley, W. E. 1918. Buried weed seeds. Jour. Agr. Sci. **9:** 1–31.

Brenchley, W. E. and Warington, K. 1933. Influence of crop, soil and methods of cultivation upon the relative abundance of viable seeds. Jour. Ecol. **21:** 103–127.

Coile, T. S. 1936. Soil samplers. Soil Sci. **42:** 139–142.

Crocker, W. 1938. Life-span seeds. Bot. Rev. **4:** 235–274.

Darlington, H. T. 1931. The 50-year period for Dr. Beal's seed viability experiment. Am. Jour. Bot. **18:** 262–265.

Davis, W. E. 1930. Primary dormancy, after-ripening, and the development of secondary dormancy in embryos of *Ambrosia trifida*. Am. Jour. Bot. **17:** 58–76.

Duvel, J. W. T. 1902. Seeds buried in soil. Science N. S. **17:** 872–873.

————. 1905. The vitality of buried seeds. U. S. Bur. Plant Ind. Bull. **83:** 7–20.

Goss, W. L. 1924. The vitality of buried seeds. Jour. Agr. Res. **29:** 349–362.

Peter, A. 1893. Culturversuche mit "ruhenden" Samen. Nachr. Ges. Wiss. Gött. **17:** 673–691.

Salter, James. 1857. On the vitality of seeds after prolonged submersion in the sea. Jour. Linn. Soc. **1:** 140–142.

Shull, G. H. 1914. The longevity of submerged seeds. Plant World **17:** 329–337.

Snedecor, G. W. 1937. Statistical methods applied to experiments in agriculture and biology. 1–341. Ames, Iowa.

Tippett, L. H. C. 1927. Random sampling numbers. Tracts for computers, No. 15. London.

Warington, K. 1924. The influence of manuring on the weed flora of arable land. Jour. Ecol. **12:** 111–126.

8

Reprinted from *Ecol. Mono.* **20:**231–250 (1950) by permission of the publisher, Duke University Press, Durham, N. C.

CAUSES OF SUCCESSION ON OLD FIELDS OF THE PIEDMONT, NORTH CAROLINA

CATHERINE KEEVER

Department of Botany, Duke University Durham, N. C.

INTRODUCTION

REVIEW OF LITERATURE ON OLD FIELD SUCCESSION IN THE PIEDMONT OF NORTH CAROLINA

SEQUENCE OF SUCCESSION

Several studies of the succession of plants following abandonment of farm land in the Piedmont of North Carolina have been made during the last fifteen years. Much of the emphasis in these studies has been placed on the arborescent stages involving pine as the original arborescent invader and its gradual replacement by climax oaks and hickories. Only two studies have dealt with details of the herbaceous stages of succession which almost always precede pine. Crafton & Wells (1934) discuss the early stages of succession based on general observations, quadrat studies of several communities on different types of soil, and on invasion of plants into spaded squares. Oosting (1942) gives quantitative data based on quadrat counts in fifteen fields of the first three years of abandonment, as well as extensive data on the arborescent stages of succession. There is general agreement, with only minor variations, on the sequence of dominating species in the early herbaceous stages of succession.

Crabgrass (*Digitaria sanguinalis* (L.) Scop.) usually is dominant in fields in the late summer and fall following cultivation. During the first year of abandonment of a field horseweed (*Leptilon canadense* L.) and crabgrass are almost always dominant; but horseweed, being a plant four to six feet tall, is the conspicuous species. Occasionally ragweed (*Ambrosia elatior* L.) is present in large enough numbers to share dominance with horseweed and crabgrass. On badly eroded fields *Diodia teres* Walt. and *Aristida dichotoma* Michx. are often the dominant plants, at least locally. During the second year following abandonment, crabgrass and horseweed are still present, but the horseweed plants are only about six inches tall and are hardly noticeable among the four- to six-foot asters (*Aster pilosus* Willd.) that are usually dominant in such fields. On eroded second-year fields, aster may be entirely absent, and some combination of ragweed, Diodia and *Plantago aristata* Michx. may be dominant. The third year following abandonment broomsedge (usually *Andropogon virginicus* L.) assumes dominance which it maintains until replaced by pine, the seedlings of which may appear among the broomsedge as early as the third year. Pines may be taller than the broomsedge by the fifth year, and form closed stands in ten to fifteen years.

Apparently the type of soil, if it is not eroded, has little influence on the sequence or duration of these early stages of succession, and most abandoned fields in the Piedmont of North Carolina are dominated in turn by crabgrass, horseweed, aster, broomsedge, and then the first aborescent plant, pine.

CAUSES OF SUCCESSION

Some work has been done dealing with the causes of old field succession in the Piedmont of North Carolina. Coile (1940) concluded that no changes in soil characteristics appear to be related in a causal manner to loblolly pine succession; the invasion of pine into abandoned land is related to the coincidence of a good seed year and climatic conditions favorable to early development of seedlings. He says that the decline of pine dominance at the end of one generation is caused by the failure of pine seedlings to compete with established forest vegetation for soil moisture and nutrients because of differences in habit of root growth between pines and hardwoods. Oosting and Kramer (1946) found the available water at the margins of forests, where several species of pines become established, to be as low as within the stands where pine seedlings make little growth. They concluded that light is probably more significantly controlling than soil moisture in the establishment of pines under forest stands.

Crafton & Wells (1943) suggest that one of the keys to the causes of early stages of old field succession lies in the water relations. They consider why crabgrass precedes broomsedge in succession, and conclude, from greenhouse experiments and field observations, that seedlings of crabgrass are more drought resistant than are those of broomsedge; consequently broomsedge does not become dominant until tall weeds form a protective covering for its seedlings. They also suggest that the exclusion of crabgrass and other species from the tall weed communities is caused by their intolerance of low light.

OBJECTIVES OF PRESENT STUDY

THE PROBLEM IN GENERAL

Although these writers touch on causes of early old field succession there remain several questions which must be answered before the particular sequence can be explained adequately. If answers to these questions were available they might be applicable not only to early stages of secondary succession, but might also lead to a better understanding of the causes of succession in general. This study was initiated, therefore, with the broad objective of contributing to our knowledge of the causes of succession. The method was to attack the problem through old field succession as a particular example. Of the species involved, the major dominants of the first three years were given first consideration.

The problem resolved itself into several phases which are well represented by the following questions, then unanswered.

1. Why is horseweed, rather than some other plant, usually dominant in first year fields?
2. Why does horseweed fail to hold dominance more than one year?
3. Why is aster delayed in assuming dominance until the second year?
4. Why does aster fail to hold dominance after the second year?
5. Why is broomsedge delayed in assuming dominance until the third year?

It soon became apparent that progress would require consideration of certain puzzling questions involving minor variations from the normal sequence and duration of the successional stages. Here are the most pertinent ones:

1. Why are a few first-year fields clearly dominated by ragweed with almost complete absence of horseweed?
2. Why does aster occasionally share dominance with horseweed in first-year fields?
3. Why does second-year aster-dominance occasionally persist for an additional year or more?
4. What effect does erosion have on the general pattern of succession?

(The writer is indebted to Dr. H. J. Oosting for suggesting this study and for his advice and assistance.)

SPECIFIC MATTERS FOR STUDY

To answer the foregoing questions, a number of specific points concerning the life histories of the dominant plants and the environmental factors influencing the survival and growth of these species needed clarification. Although taxonomic manuals indicate the approximate time of flowering of horseweed, aster, and broomsedge, it was not known with certainty when the seeds are fully mature and whether the seemingly mature seeds are capable of immediate germination or require a period of dormancy. Differences in time of maturity and differences in time of germination of seeds of the dominant species might influence the sequence of dominance.

No species can become established and hold its place in a community unless the seedlings can survive and grow to maturity. Species vary as to optimum environmental conditions for survival and growth, and often the conditions that are favorable for germination and early growth are not the most favorable for the later growth of the species. Little was known concerning the survival and growth of the dominant species of old fields. It became necessary, therefore, to determine when and where the seeds germinate under field conditions. A major question to be answered concerned the dormancy of the seeds of the three dominants, and whether the failure of certain species to become established in old fields is caused by failure of seeds to germinate or by death of seedlings. Again it seemed necessary to learn what environmental conditions permit the seedlings to survive and grow and what conditions inhibit them.

Survival and growth of plants in natural communities are influenced by a number of factors working together in such a way that it is often difficult to determine what conditions separately or in combination cause the end results. Conditions of soil moisture in abandoned fields change with the increase of organic matter in the soil and the protection from evaporation by living plants and litter. A knowledge of the reaction of the particular species to variations in moisture conditions was needed to help answer the general problems of succession. A knowledge of the reaction of each species to the variations in light would help evaluate the effect of decreasing light in old fields on the trend of succession. Such information could best be gained by experimental methods in which all factors except the one being studied could be held as nearly constant as possible.

If horseweed produces some substance that inhibits growth of other horseweeds or stimulates the growth of asters, or if aster produces some substances that cause the decline of asters and stimulates the growth of broomsedge, this might prove to be one of the important causes of the sequence of succession. Such possibilities could best be checked by experiment.

The idea that one plant may produce some chemical compounds which are inhibitory or toxic to other plants was advanced by Pickering & Bedford (1914, 1917, 1919). Davis (1928) called attention to the injurious effects of black walnut on other species. Proebsting & Gilmore (1940) showed that neither exhaustion of plant nutrients or disease carried over from the last orchard could account for the failure of replanted peach orchards to make normal growth, but that peach roots added to virgin soil inhibited the growth of peach seedlings. Benedict (1941) showed that dried roots of bromegrass are inhibitory to bromegrass grown in sand cultures and suggested that an inhibitory substance in the roots may be responsible for the thinning of bromegrass stands after a few years. Bonner & Galston (1944) and Bonner (1946) found that water or nutrient solution in contact with roots of growing guayule plants accumulate substances which are toxic to the growth of guayule seedlings. Went (1942) in his study of annuals that grow near shrubs in the desert of the Southwest, suggested that the occurrence of one plant determines the presence of certain other plants through some unknown agent, presumably chemical. Went's observation that annuals grow near *Encelia farinosa* Gray led Gray & Bonner (1948) to consider the problem of whether or not growth inhibitors do arise from

Encelia plants, and they found that the leaves of Encelia, when placed on top of the sand culture in which tomatoes and other plants were growing, caused a striking inhibition of growth in these plants.

In most cultivated fields in this area harvest of a crop removes much of the aerial portion of the plants; consequently there is a minimum of organic matter in fields when they are abandoned. After a field is abandoned the plants that subsequently invade the area are left undisturbed to fall to the ground and eventually decay. This organic matter may influence the texture, water holding power, pH, and mineral content of the soil, and might contain some chemical substances that would be beneficial or harmful to plants growing in the soil containing it. Little was known concerning the effects of the dead plant parts of one species on the survival and growth of the same species or on other plants. The possibility that dead horseweed parts might inhibit the growth of other horseweed plants, either by producing some toxic substance or by causing an imbalance of minerals, needed to be considered. The effects of organic matter from aster or broomsedge upon other plants also needed to be known before the cause of old field succession could be determined.

Plants in a given community have mutual relationships, and all compete for water, light and available minerals in the soil. The effects of one species competing with individuals of the same species or of other species have been evaluated by a number of investigators. Brenchley (1919) planted mustard plants in pots with one, two, three, four, and five plants to a pot, and in comparing the final dry weights found the total weight of five plants in one pot to be about the same as that of the one plant in a pot. She found growth to be proportional to the amount of minerals in solution. Such knowledge of the effects of competition among dominant species in old field succession seemed necessary to answer some questions concerning the causes of old field succession.

GENERAL PROCEDURE

SEED MATURATION AND DORMANCY

A knowledge of as many details as possible of the life cycles of the three major dominants—horseweed, aster, and broomsedge—seemed necessary for an understanding of the causes of the particular sequence of early old field succession. Study of life cycles was begun with studies of seeds of the three species. Plants were observed in the field to determine time of seed maturity. Mature seeds were collected and tested for germination under a variety of conditions to find the elapsed time and environmental conditions affecting germination for each species.

GERMINATION, SURVIVAL AND GROWTH UNDER FIELD CONDITIONS

Further studies of life cycles, as well as of survival and growth of plants under field conditions, were undertaken by means of a series of permanent quadrats in fields of different ages. List counts of species were made at intervals during a year to determine the number of seeds germinating, the season and environmental conditions of germination, and the number of seedlings living and growing to maturity. A knowledge of the number and kinds of plants growing in each community was gained from these permanent quadrats and this gave information concerning competition among the three dominants and among the dominants and other species.

Observations of permanent quadrats indicated the desirability of obtaining information from a large number of fields that had slightly varied histories. Therefore surveys were made of thirty fields. These surveys involved not only examination of the present vegetation, but talking to the owners or tenants about the last crop, its time of last cultivation, and the amount of erosion in such fields.

In connection with the permanent quadrats and field surveys, a number of individuals of the three species were measured to help evaluate the amount of growth made by the plants at different ages and under different field conditions.

FACTORS INFLUENCING SURVIVAL AND GROWTH

Experiments were conducted in the greenhouse to determine the effects of variations of water and light on the survival and growth of the three dominants. Seedlings of each species were grown with three variations of light in combination with four variations of watering in such a way that the effect of each factor could be evaluated separately or in combination with the other.

Experiments were set up to see whether any one of the dominant plants produced some substance which, carried in soil water, would inhibit or stimulate its own growth or the growth of one of the other species. The three species were grown in separate pots, and all the water they received had percolated through the soil in which one of the species was growing. Size at various times and final dry weights were used to judge the effects of the different treatments.

Greenhouse experiments were carried out to check the effect of the dead plant parts of each of the old field dominants upon itself and upon each of the other species. Plants of each species were grown in soil containing chopped plant parts from each of the other species. This experiment gave information concerning the total effect of the specific organic matter on the growth of each species, but revealed little concerning the fundamental causes of the differences. A variation of this experiment using nutrient solution instead of distilled water was carried out to see whether the differences in growth between treatments was being caused by a mineral deficiency or by some other factor.

The effect of competition between broomsedge and aster was tested with a controlled field experiment in an attempt to solve the problem of why broomsedge replaces aster in old field succession. Young asters were transplanted to the vicinity of established broomsedge plants in the field, where other plants had been removed to eliminate their competition, and growth and survival of the asters in relation to distance from the broomsedge plant were studied.

METHODS, RESULTS, AND CONCLUSIONS

SEED MATURATION AND DORMANCY

METHODS. The three major dominants of old field succession, horseweed, aster, and broomsedge were observed in the field during two seasons to determine the time at which the seeds of each species matured. Mature seeds of the three species and also of ragweed were collected for testing germination under varying conditions. At weekly intervals during the months of November and December, 1947, seeds of each species were planted in two pots of soil, one pot being placed out of doors, and one in the germinating room of the greenhouse. If seeds of any species failed to germinate in the greenhouse, seeds were layered in damp sand in a cold room at a temperature of 10°C. and tested at intervals for germination. Seedlings grown in the greenhouse from the germinating seeds were compared from time to time with field specimens as an aid in recognizing seedlings in the quadrants being studied.

Three small plots in a first-year horseweed-dominated field were spaded on June 15, three others on July 15, and three on August 15. All plots were watched for seedlings of the species being studied to see when the seeds germinated in the field.

RESULTS. *Horseweed.* Horseweeds were observed blooming by the middle of July and some individuals were still blooming the last of October. The first mature seeds were collected early in August and germinated on damp filter paper about two weeks later. Horseweeds planted in soil at weekly intervals during November and December germinated outside through November, but did not germinate during the relatively cold month of December and many of the seedlings that had germinated during November died. Horseweed seeds that did not germinate outside when planted in December, germinated the following spring.

Horseweeds planted in the greenhouse germinated within two to five days regardless of the season. No seedlings of this species appeared in the spaded plots until the last week of August, at which time seedlings were found in abundance in the entire field.

Aster. A few scattered asters began to bloom in late September but most were in full bloom during late October and early November. The earliest mature seeds were collected the first of November from a few terminal heads of racemes. Aster did not germinate outside during November and December, but did show germination in the greenhouse seven to ten days after planting. Seeds planted outside germinated the following spring. No asters were seen at any time in the spaded plots.

Broomsedge. Soft immature seeds of broomsedge (*Andropogon virginicus* L.) were found the first of October, but firm seeds were not found until a month later. These seeds failed to germinate either in the greenhouse or outside during November or December Seeds were layered in the cold room on December 8, and the first samples of these layered seeds germinated in the greenhouse on January 10. No seedlings of broomsedge were seen in the spaded plots.

Ragweed. Ragweed seeds collected in October failed to germinate in the greenhouse or outside unless layered. Seeds that were layered in the cold room on November 13 first germinated in the greenhouse on February 13.

CONCLUSIONS. Horseweed seeds mature as early as the first of August and are capable of germination soon after maturity. They may continue to germinate at relatively low temperatures during the late fall.

Aster seeds are not mature before the first of November. They may germinate soon after maturity in a warm environment, but not in the normally cool outdoor conditions that prevail around Durham in November.

Broomsedge seeds are not mature before the first of November and will not germinate under any conditions without a period of cold dormancy.

Ragweed seeds require a period of cold dormancy before they will germinate.

GERMINATION, SURVIVAL AND GROWTH UNDER FIELD CONDITIONS

PERMANENT QUADRATS

METHODS. In the late winter of 1947-1948 a search was made near Durham, North Carolina, for fields in which to locate permanent quadrats for the study of life cycles, suvival, and growth. Fields that were not to be plowed that spring or summer were necessary and, since very little land was being abandoned deliberately because of the high price of farm products, such fields were difficult to locate. Such land as was being abandoned was usually badly eroded or otherwise unproductive. Nine fields were located, of which three had been last cultivated in 1945, three in 1946, and three in 1947.

Ten permanent quadrats one-fourth meter wide and one meter long were marked with stakes in each of the nine fields at intervals along a central line running the length of the field. In fields where furrows from the last cultivation were evident the arrangement was such that each plot included part of a ridge and a furrow. List counts of all species present were made in March, June, and September; the age of asters and broomsedge was estimated and each age group recorded separately. Counts of aster and broomsedge were made the following January in two fields of each age that had not been plowed by that time. When seedlings or young plants could not be recognized, samples were transplanted to pots in the greenhouse and later identified.

RESULTS. Species: Area curves made after each count indicated the one-fourth meter square quadrats were more than adequate for sampling at any season of the year.

Horseweed. Horseweed plants in the thirty quadrats in first-year fields showed an average density of

15.6 in March, 17.9 in June and 17.2 in September. In March these plants were rosettes ranging in diameter from one-half inch to six inches with many plants of the larger sizes. By June most of them had begun height growth, many being as much as one to two feet tall. In mid-July a large number were blooming, and although some were still only a few inches tall, the majority were between four and six feet tall. At the time of the last quadrat counts in late September, the plants that had bloomed first were nearly dead whereas some others were just beginning to bloom.

There were more than six times as many horseweeds per unit area in second-year fields as in first-year fields. The density in March was 97. In June the density had decreased to 78 and to 25 in September. In March most of the horseweed seedlings in second-year fields were less than a half inch in diameter, and none were seen that were more than one inch. Few finally grew to be more than a foot high, with the average being about six inches. These stunted plants began to bloom by the middle of July, and only a few remained alive until September.

Horseweed plants were found in only one third-year field. In this field there was an average of 123 seedlings per quadrat in March, 56 in June, and 18 in September. These plants in the third-year field were about the same size as those in the second-year fields.

Aster. There were only a few newly germinated asters (density 0.13) present in first-year fields in March, but by June the density had increased to 4.6, and in September it still averaged 4.6, but the decline in frequencies between June and September counts indicated that some of the seedlings had died and other seeds germinated.

The few asters beyond the seedling stage in the badly eroded second-year fields had a density of 0.07 in March and the same individuals lived through the season. Some aster seeds had germinated before March (density 4.9) in the second-year fields, and this number had increased by June to a density of 14.6. The survival of these plants could not be determined as the field containing the largest number of seedlings was plowed before September. The density of aster seedlings in the other two fields, however, was about the same in September as in June.

In the third-year fields aster plants with dead flowering stalks from the preceding year averaged 2.8 per plot. Most of these plants lived through the season and bloomed again in the fall (September density 2.4). Their height, however, was usually considerably less in their second year than in the first, as judged by the flower stalks still standing.

Some one-year old asters (density 8.0) were present in third-year fields. By June their density had dropped to 6.2, and the surviving plants were little larger than they were the year before.

The ground under the old asters in the third-year fields was almost completely covered with young aster seedlings, with a density in March of 321. Seedlings which survived to June (density 142) were scarcely larger than they were in March. September density had decreased to 45, including both young asters and seedlings of the current year which could not be distinguished from each other.

Broomsedge. There were no broomsedge seedlings in first-year fields in March, but in June their density was 2.2, and it increased to 4.6 by September. Of the two first-year fields checked for broomsedge in the following January, one showed the same number as in September and the other showed an increase from 4.4 in September to 5.8 in January. There was no way of knowing whether the same individuals were counted at the different times, or whether some died and other seeds germinated. Five one-year broomsedge plants were found in the ten quadrats of one of the first-year fields that had been abandoned late in June of the preceding year, and all five of these plants survived through the season.

No broomsedge seedlings were found in second-year fields in March. The June counts showed the low density of 0.8 and by September the value had increased to 1.4. Five one-year old plants that were found in the thirty quadrats in the second-year fields all lived throughout the period in which quadrat counts were made.

Broomsedge seedlings appeared earlier in third-year fields than in first- or second-year fields. The density in March was 1.0, and in June 19.8 and in September 61.7. In the two fields checked in January, the density in one remained the same as it was in September and in the other increased from 38.8 to 61.7. In third-year fields most of the broomsedge plants which were a year old survived, the density being 1.3 in March and 1.1. in September.

Ragweed. Young ragweed seedlings were found in all fields in March. In the first-year fields there was an increase in density from 1.4 in March to 3.5 in June, and nearly as high a count in September. The highest counts of ragweed were made in third-year fields, with densities of 6.3 in March, 3.5 in June, and 1.5 in September. Ragweeds in first-year fields, especially in one where the density of horseweeds and other plants was relatively low, usually grew to be branching plants three or four feet high. In second- and third-year fields the ragweeds were usually unbranched plants, less than a foot high except where they were growing in spaces unoccupied by many other plants.

Other Plants. There were 22 species of plants present in at least two of the three first-year fields studied. Aster, broomsedge, and pines, which later become dominants, are still in the seedling stages at this time. Horseweed, ragweed and crabgrass are he only ones of the 22 species that attain sufficient size or density to become dominants in the first year. A number of other plants that show relatively high densities and frequencies at some period of the year are small plants, no one of which alone covers enough area to be considered a dominant. Some of these species are *Sagina decumbens* (Ell.) Torr. and Gray, *Oenothera laciniata* Hill, *Plantago virginica* L., *Diodia teres* Walt., *Krigia virginica* (L.) Willd., *Polypremum procumbens* L., *Specularia perfoliata*

(L.) A. DC., *Draba verna* L., *Allium vineale* L., *Arabidopsis thaliana* (L.) Heynh.

Of the 25 species found in at least two of the three second-year fields, all but one, *Lespedeza striata* (Thunb.) H. & A., had been found in at least one of the first-year fields although sometimes with low densities. There were 23 species found in at least two of the third-year fields. Four of these, *Tecoma radicans* (L.) Juss., *Triodia flava* (L.) Smyth, *Solidago* spp., and *Erigeron racemosus* (Walt.) BSP. had not appeared in a field of a younger stage.

CONCLUSIONS. Horseweed is a winter annual, germinating in the fall, living as a rosette over winter, and blooming by mid-summer. Although the density is much lower in first-year fields than in second- or third-year fields, the percentage of survival and growth of horseweed plants is much greater in first-year fields than in the other two ages of fields.

Aster seeds germinate in the spring and early summer following their fall maturity and are present as small plants in first-year fields. These plants live over the winter, grow in height the second summer and bloom the second fall. The flower stalk dies, leaving a basal rosette of leaves that lives over the winter, and produces another flower stalk the following year. The density of aster seedlings under old aster plants in third-year fields is very high, but few seedlings survive and those that live make very little growth.

Some broomsedge seeds germinate during the spring following the fall in which they mature and others germinate during the summer and late fall. There are few broomsedge seedlings in the first- and second-year fields, but the survival of such seedlings is high. The survival of broomsedge plants beyond the seedling stage is almost 100 percent. A broomsedge plant blooms in its second fall and the clump continues to increase in circumference the third year and blooms again.

Ragweed is a summer annual; the seeds germinate in early spring and the plants bloom in late summer. The survival and growth of ragweed plants is greatest in fields where the density of other plants is least.

Horseweed, ragweed and crabgrass are the only plants present in first-year fields that have either density or size enough to become dominants.

FIELD SURVEYS

METHODS. Fifteen first-year fields near Durham and four first-year fields in Alexander County in the upper Piedmont of North Carolina, were surveyed in fall condition. The past cultivational history of each field was obtained from the owner or tenant. Ten quadrats, one-fourth by one meter in size were spaced at regular intervals in each of the fields and counts were made of asters and broomsedge by age classes. General observations as to the abundance of other species and amount of soil erosion were made.

Eleven second-year fields in fall condition were surveyed near Durham, histories were obtained, conditions of soil observed, and abundance of each species noted. Asters and broomsedge were counted in four of the eleven fields, using the same size and number of quadrats in each as in first-year fields. Several of these second-year fields had been observed during the search for permanent quadrat locations, but had been rejected because the owner was not sure they would not be plowed that season; consequently, in such fields the dominants of two seasons were observed.

RESULTS. *First-year fields.* Horseweed was clearly the dominant plant in ten of the nineteen first-year fields. All of these fields except one had been last cultivated in July or August of the preceding year. The last crop in seven of these horseweed-dominated fields had been corn; garden vegetables were the last crop in two, and sweet potatoes in the other one. This last field had been observed during the year following the October harvesting of potatoes. It appeared bare of all vegetation during the winter, and throughout the following summer possessed a dense cover of horseweed plants that were smaller and less mature than the horseweeds in neighboring fields. These retarded horseweed plants did not bloom until late September and October.

Three of the first-year fields were dominated by a mixture of horseweeds and asters. The last crop in two of these fields had been corn, and the last crop in the other one was garden vegetables. One of the corn fields and the garden had been last cultivated in June and the other corn field in July.

Six of the first-year fields were clearly dominated by ragweed with practically no horseweeds and only a few scattered blooming asters. Four of these had been plowed in the fall of the year before and planted in winter wheat, one had been planted in winter vetch following cotton the preceding fall, and the other was a cornfield that had been last cultivated in June and in which the soil was badly eroded.

Young asters were found in every first-year field studied, ranging in density from 0.1 in a large wheat-ragweed field that was a long distance from any visible seed source, to 28.6 in a garden plot that had a surrounding border of blooming asters. The average density of young asters in all nineteen fields was 5.0 with a frequency of 71.5 per cent. In general, the fields that showed greatest evidence of erosion had the lowest densities of young asters. In fields that had been recently eroded by torrential rains of November, 1948, many nearly dead young asters were observed with a large part of their root systems exposed where the soil had been washed away.

Broomsedge seedlings less than a year old were found in the quadrats of only nine of the nineteen fields, with an average density of 0.8. Only one older broomsedge plant was recorded in the entire 190 quadrats counted in first-year fields.

The other species present were in general the same ones found in the first-year fields studied with permanent quadrats and they occurred with a corresponding abundance.

Second-year fields. Of the eleven second-year fields all except two were dominated by asters. One of those was almost completely covered with *Plantago virginica* in the spring and with Diodia in the fall. The other had an almost complete cover of crabgrass in some areas and Diodia in others with the Dioda-dominated areas appearing much more eroded than the crabgrass-dominated parts. A few stunted horseweeds, scattered bunches of blooming broomsedge, and an occasional blooming aster were present.

The first-year histories of all but two of these second-year fields were determined by direct observation or by information obtained from the owners or tenants. Two were dominated by a mixture of horseweed and aster following a last crop of early abandoned corn. One was a mixture of horseweed and aster following a spring planting of lespedeza the preceding year. Three were ragweed-dominated following wheat, and one following sweet potatoes.

Blooming asters had an average density of 4.1. Blooming broomsedge had a density of 0.4. Broomsedge seedlings had a density of 0.15 in the quadrats of the four fields counted.

CONCLUSIONS. The time of year at which the last cultivation of a field takes place greatly influences the trend of succession on first-year fields. Fields last cultivated in early summer are usually dominated by a mixture of horseweeds and asters the following summer. Fields last cultivated in late summer after most of the aster seeds of the season have germinated are almost always dominated by horseweed. Fields last cultivated in late fall after most of the horseweed seeds have germinated are usually dominated by ragweed the following summer.

The density of young aster plants is greater in non-eroded fields than in eroded fields. Eroded spots in horseweed-dominated first-year fields are usually dominated by ragweed.

Most second-year fields are dominated by asters even though they are dominated the first year of abandonment by either a mixture of horseweeds and aster, by horseweed alone, or by ragweed. A few fields, usually badly eroded, are not aster-dominated the second year.

MEASUREMENT OF GROWTH

METHODS. Measurements of numerous individuals of aster and broomsedge were made in December, 1948, to determine sizes at different ages in different habitats. Five small sample plots were distributed at random in four fields that had been cultivated the preceding summer, and diameters of all horseweed plants falling within these areas were measured. Similar plots were located in four of the horseweed-dominated fields used for the first-year survey. The size of horseweed seedlings was compared with that of the seedlings found in zero-year fields.

All aster and broomsedge plants growing within the one-fourth by one meter quadrats in four of the first-year survey fields and four of the second-year survey fields were measured, with height measurements of aster and broomsedge seedlings, and both height and basal circumference of older broomsedge plants being taken. Blooming asters and all broomsedge plants beyond the seedling stage in the quadrats of two of the third-year permanent quadrat fields were measured in the same manner.

RESULTS. *Horseweed.* In the zero-year fields 106 horseweeds rosettes were measured. Only one of these plants was less than a half-inch in diameter; the remaining ones ranged in size from ½-4½ inches, averaging one inch in diameter. Of the 698 horseweed seedlings measured in the first-year fields 656 were one-fourth inch or less in diameter and none of the other 62 were more than one inch in diameter.

Aster. In first-year fields 55 young asters ranged in size from one to five inches in height and averaged 2.2 inches. The 194 young asters measured in second-year fields ranged in size from ½-4 inches and averaged 0.9 inches in height. Blooming asters (70) measured in second-year fields ranged in height from 12 to 57 inches and averaged 34 inches, but those (43) in third-year fields ranged in height from 3 to 67 inches and averaged 27 inches.

Broomsedge. Nineteen broomsedge seedlings measured in first-year fields were from 2 to 9 inches in height and averaged 5 inches. Six broomsedge seedlings found in the quadrats of the second-year survey fields were 3 to 7 inches high and averaged 5.3 inches. Nine broomsedge plants in second-year fields that were between one and two years old averaged 40 inches in height and 3 inches in basal circumference. Twenty-five broomsedge plants in third-year fields—one and a half year old plants being difficult to distinguish from older plants—averaged 39 inches in height and 10 inches in basal circumference.

CONCLUSIONS. Horseweed seedlings in second-year fields are only about one fourth as large as they are in first-year fields.

Young aster plants grow to be about two inches tall by fall in first-year fields, but in second-year fields where young asters are growing under mature asters, they are only about half as large.

Broomsedge seedlings grow as well in second-year fields as they do in first-year fields. Broomsedge plants reach their maximum height in second-year fields, but continue to increase in circumference the following year.

FACTORS INFLUENCING SURVIVAL AND GROWTH

WATER AND LIGHT

METHODS. Seedlings of horseweed, aster, and broomsedge were grown in the greenhouse during the summer of 1948 to see how they would respond to variations in water and light. Seeds of the three species were planted separately in pots of sandy loam, the asters being planted five days before horseweed and broomsedge so that the seedlings of this slowly germinating species would be approximately the same age as those of the other two species. The pots were placed in the greenhouse in such a way that one third received full sunlight, one third were shaded with cheesecloth over a rack in such a way as to receive one-half full sunlight, and the other third were shaded to reduce the light to one-fourth full sunlight.

All pots were watered daily until the seedlings were ten to fifteen days old. At that time the plants were thinned to forty to an eight-inch pot, or less in the case of asters where fewer than that number had germinated. Within each variation of light one pot of each species was watered every day, another pot of each species was watered every seven days, another every fourteen days, and a fourth pot of each species was watered at a twenty-one day interval. At the end of three weeks of differential watering, the surviving plants were counted. The differential watering was continued for another four weeks. The surviving plants were then removed from the soil, oven dried, and their average weight determined as a measure of the growth that had been made by the plants that had been subjected to different environmental conditions of light and water.

RESULTS. *Horseweed.* During the first three weeks of differential watering the mortality of horseweed seedlings was as follows:

Full sunlight
- 14 day watering 12%
- 21 day watering 75%

One-fourth full sunlight
- 21 day watering 50%

All the horseweed seedlings in the other variations of light and water were living at the end of the entire experiment. There was a further slight loss of plants during the second period of differential watering under the same treatments in which plants died during the first period as follows:

Full sunlight
- 14 day watering 03%
- (of plants surviving first watering period)
- 21 day watering 10%

One-fourth full sunlight
- 21 day watering 25%

Judging by the average dry weight of plants, the optimum conditions for horseweed growth were one-fourth full sunlight with a fourteen-day interval between waterings. In general the poorest growth was made in full sunlight. The best watering period for all variations of light was the two-week interval, and the poorest growth was made at the three-week watering interval.

Figure 1 shows the survival and growth of horseweed seedlings for the entire period of the experiment.

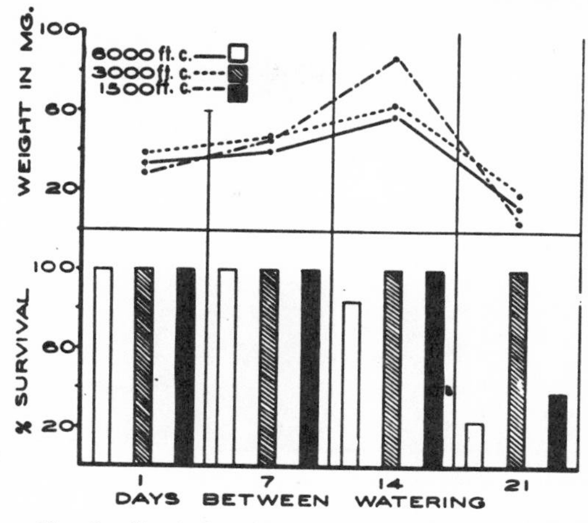

FIG. 1. Survival and growth of horseweed seedlings with variations in water and light. Differential watering started when plants were ten to fifteen days old. Growth expressed in final dry weight of surviving plants. Optimum conditions for growth are one-fourth full sunlight (1500 foot candles) with a fourteen day watering interval.

Aster. There were some deaths of asters in six of the twelve treatments as follows:

Full sunlight
- 7 day watering 50%
- 14 day watering 70%
- 21 day watering 70%

One-half full sunlight
- 14 day watering 53%
- 21 day watering 70%

One-fourth full sunlight
- 21 day watering 96%

All of the aster seedlings that survived the first period of differential watering were still living at the end of the second period.

Although the mortality among the asters was greatest in full sunlight, the plants that lived made the greatest growth in full sunlight. In every variation of watering the asters in full sunlight were definitely larger than those in the other two variations of light (Fig. 2).

Broomsedge. The deaths of broomsedge seedlings during the first period of differential watering were as follows:

Full sunlight
- 7 day watering 02%
- 14 day watering 50%
- 21 day watering 67%

One-half full sunlight
- 21 day watering 70%

One-fourth full sunlight
- 21 day watering 87%

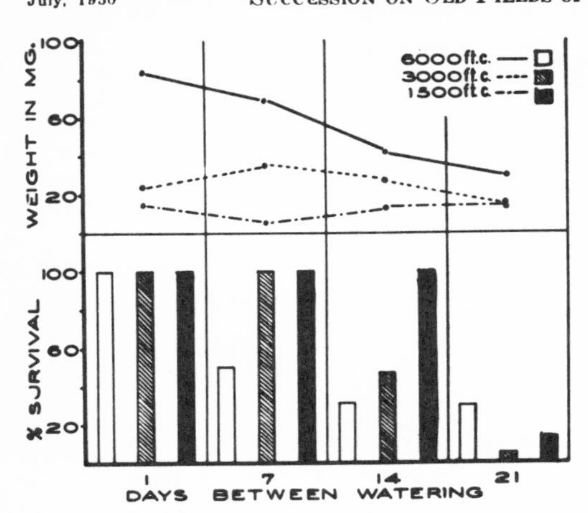

FIG. 2. Survival and growth of aster seedlings with variations in water and light. Differential watering started when plants were ten to fifteen days old. Growth expressed in final dry weight of surviving plants. Optimum conditions for growth of asters are abundant water and light.

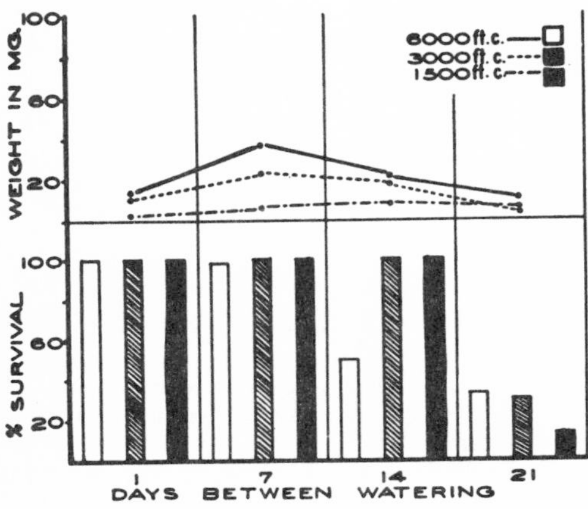

FIG. 3. Survival and growth of broomsedge seedlings with variations in light and water. Differential watering started when seedlings were ten to fifteen days old. Growth expressed in final dry weight of surviving plants. Optimum conditions for growth are full sunlight (6000 foot candles) with watering every seven days.

All broomsedge seedlings that survived the first period of differential watering lived until the end of the experiment (Fig. 3).

Seedlings in both conditions of reduced light soon showed a definite chlorosis, which became more pronounced with additional time, while those plants in full sunlight were dark green in color. Broomsedge seedlings showed their optimum growth in full sunlight with the seven day interval between watering. The seven-day watering interval produced the best growth in all three variations of light, with gradual decrease in growth as the interval between watering was lengthened. The plants grown in one-fourth full sunlight showed consistently less growth at every watering interval than the plants grown with more light.

CONCLUSIONS. Horseweed seedlings are drought-resistant, capable of enduring two to three weeks of drought without death of a large number of plants. They make their best growth in reduced light when watered every two weeks.

About half of the aster seedlings died in full sunlight when deprived of water for one week. Mortality is somewhat reduced with shading, but growth is also greatly reduced. Asters make their best growth with an abundance of both light and water.

Broomsedge seedlings can endure one to two weeks in full sunlight without severe loss of plants. Mortality is reduced with shading, but growth is also reduced. Broomsedge makes best growth in full sunlight when watered once a week.

SUBSTANCES PRODUCED BY LIVING PLANTS

METHODS. Experiments were carried out in the greenhouse during the late winter and early summer of 1948 to determine the effect of substances produced by living plants of each of the dominant species of old field succession upon the growth of others of the same species or upon individuals of the other two species. Clay pots were fitted with drain tubes and filled with unsterilized sandy loam soil from a field which had been cultivated the preceding year. Young plants of aster, horseweed, and broomsedge from fields were set one to a pot, except in the pots which were left without plants to be used as controls. When these donor plants were established and growing, receiver pots were set up; seeds of the three species being studied were planted separately in pots containing soil from the same source as that used in the donor pots.

The pots of seedlings of each species were arranged so that three pots of each species were watered with drip water from horseweed, three from aster, three from broomsedge and three from control pots containing soil in which no plants were growing. Each of the three sets of replications were placed at different exposures in a greenhouse room. The donor pots were watered with distilled water every three to five days when the receiver seedlings were young, and every five to seven days after the plants were well established. Receiver plants were given no water except that which percolated through the soil of the donor pots. None of the receiver plants showed any sign of wilting on this watering schedule.

Plants of other species were pulled out as they germinated and the receiver plants were gradually thinned, always leaving the largest plants, until only two plants remained in each pot.

The receiver plants were measured periodically, the diameter being used for the rosette stage of horseweed and height for the later stages of horseweed and all stages of aster and broomsedge. During the first week in August, 1948, when the receiver plants were five months old and some of the horseweed plants were beginning to bloom, they were removed from the soil, dried in an oven, and their weights compared. The data were analyzed to see if there were any significant differences in size at any stage that could be attributed to differences in treatment.

RESULTS. At the end of the experiment when the plants were five months old, most of the individuals which had been grown in the greenhouse much larger than plants of the same age that were growing in the fields under natural conditions. The horseweed plants ranged in size from 5 to 36 inches tall, the asters from 7 to 47 inches and the broomsedge from 6 to 34 inches.

Statistical analyses of the data for size measurements and final dry weights showed that there were more variations within each treatment than there were between treatments; consequently no differences in size at any time could be attributed to substances produced by living plants and passed to other plants in the soil water.

CONCLUSIONS. None of the old field dominants produced substances that are carried in soil water and thus inhibit or stimulate the growth of any one of the others.

DECOMPOSITION PRODUCTS

METHODS. Greenhouse experiments were carried out in the spring and summer of 1948 to determine the effect of decaying organic matter from each of the dominants of old field succession upon that species and upon the other two. Horseweed, aster, and broomsedge were grown in pots containing chopped roots or tops of one of the dominants mixed with sandy loam from a field that had been cultivated the preceding year. An attempt was made to estimate the amount of such plant parts that might normally be found in a given amount of old field soil at some stage in the succession. These estimates indicated that 115 grams of slightly damp roots or 80 grams of air-dry tops added to four thousand grams of soil in each pot would approximate the normal field conditions. Samples of each kind of plant part were ovendried to determine the amount of moisture in each. Three replications of each species were grown in soil containing one of the following kinds of organic matter: (1) horseweed roots, (2) horseweed tops, (3) aster roots, (4) aster tops, (5) broomsedge roots, (6) broomsedge tops. Three replications of each species to be used as controls were grown in soil containing no plant parts.

(1) Seeds were planted on February 26, 1948, and plants were watered with distilled water throughout the experiment. Other species were removed as they germinated and the desired species were gradually thinned, always leaving the largest plants until two plants remained in each pot. The plants were measured at intervals in the same way as in the preceding experiment. When the plants were five months old, they were removed from the pots, ovendried and their weights compared. The remains of the organic matter were sifted from the soil, washed, oven dried, and weighed. Final dry weight was compared with the original to determine the amount of decay that had taken place.

(2) If the plants in the preceding part (1) of the experiment should show a difference in growth in the soils containing different kinds of organic matter, it would not be known whether these differences were caused by some toxic substances, by mineral deficiencies, or by some other cause. In an effort to see the relation of mineral deficiencies to differential growth, should there be any, part of the experiment was repeated using a complete nutrient solution suggested by Hoagland and Arnon (1938) instead of distilled water so that there could be no mineral deficiencies. Horseweed plants were grown in sand, some of the pots containing horseweed roots, some containing horseweed tops and others containing no plant parts. Parts (1) and (2) of the experiment were carried out in the same general way except that in part (2) the plants were watered with nutrient solution instead of with distilled water.

(3) Horseweed plants were grown in unsterilized soil and watered with complete nutrient solutions with the same combinations of living plants and plant parts as used in the sand-nutrient solution experiment (2) to see if the soil organisms present in the soil but not in the sand would make a difference in growth.

RESULTS. *Horseweed.* Noticeable differences in the diameters of horseweed seedlings in the soil-distilled water part of the experiment were evident with two of the treatments as early as two weeks after the seeds germinated. The plants growing in pots containing horseweed roots were noticeably smaller than the controls and other treatments, and the seedlings growing in soil containing aster roots were consistently larger than any of the others.

When the plants were two months old, the six horseweed plants growing with horseweed roots averaged 4.3 cm. in diameter, the ones with aster roots 12.1 cm., and the ones in the controls 8.5 cm. Statistical analysis of the data for this stage showed that there was less than one chance in twenty that these variations in size could be attributed to conditions other than the differences in kind of organic matter in the soil. Figure 4 shows the relative size of plants with different treatments when they were two months old. Horseweeds growing in soil containing aster roots continued to be larger than the others during the entire five months the plants were growing, and statistical analysis showed this difference to be significant at the five percent level. As the plants increased in age there was so much variation within each treatment that the differences between treatments became insignificant. Horseweed

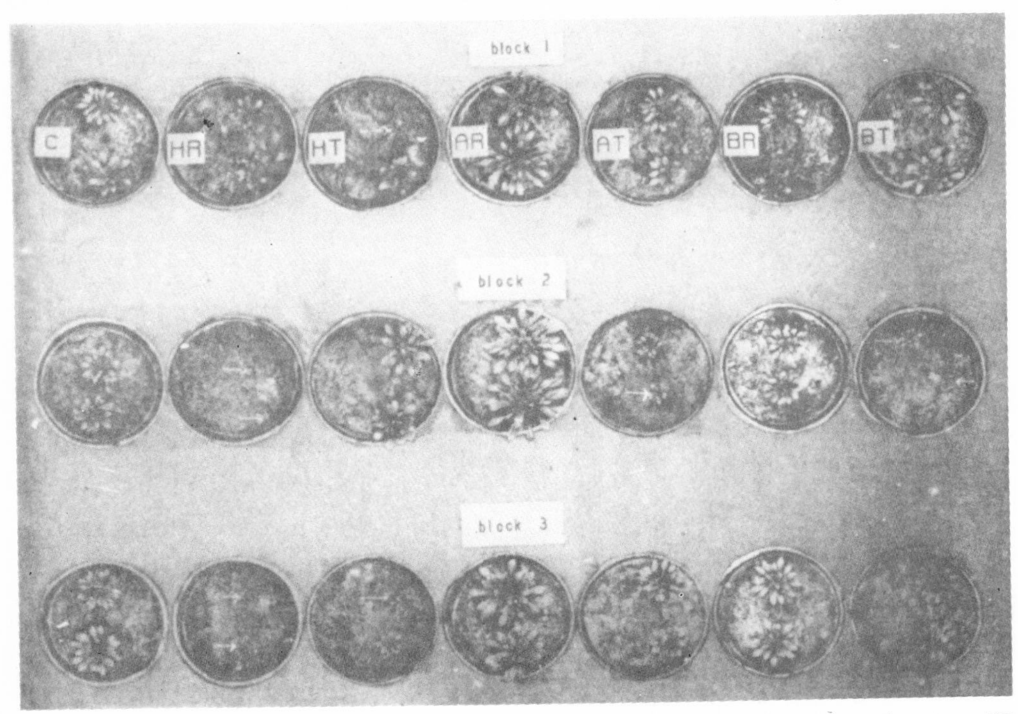

FIG. 4. Two-months-old horseweed plants growing in soil containing the following kinds of organic matter: HR- horseweed roots, HT- horseweed tops, AR- aster roots, AT- aster tops, BR- broomsedge roots, BT- broomsedge tops, C- Control (no organic matter added). Block 1, 2, and 3 are three replications of each treatment. Aster roots (AR) added to soil produced a significant increase in size, and horseweed roots (HR) resulted in reduced size of plants when compared with control. Other treatments produced no significant differences.

plants growing with horseweed roots remained smaller than the controls, but failed to show a significant difference in size after they were two months old.

In the sand-nutrient solution part of the experiment all plants grew luxuriantly, with little difference at any time in size of individuals. The six plants growing with horseweed roots at two months averaged 24.5 cm. in diameter, those with horseweed tops averaged 27.5 cm., and those in the controls 24.5. These differences are statistically insignificant at the 5 percent level.

In the soil-nutrient solution part of the experiment those plants growing in soil containing organic matter were consistently larger at every measurement than those growing in soil without additional organic matter. When the seedlings were six weeks old the diameters of plants growing with horesweed roots was 6 cm., those with horseweed tops 10.1 cm and those in the controls 4.1 cm. These differences are significant at the 5 percent level. When the plants were two and a half months old, the plants with horseweed roots averaged 23.3 cm., those with horseweed tops 23.3 cm., and in the control 21.6. The differences within the treatments were so small that the differences between the control and the other two treatments proved significant. The final dry weights showed that those plants grown with horseweed roots averaged (15.65 gms.) twice as much as the controls (7.75 gms.), and the differences proved to be significant. The dry weights of those plants grown in soil containing horseweed tops (8.88 gms.) were not significantly greater than the control.

Aster. Asters grown in the soil-distilled water part of the experiment showed a decided difference in size for different treatments when they were two months old. Figure 5 shows the asters at the age of two months when the stunting effect of horseweed roots and tops and broomsedge tops, and the stimulating effect of aster roots was quite apparent. Plants in soil containing aster roots were definitely the largest, averaging 14.1 cm. in height, and those in soil containing horseweed roots were significantly (at the 5 percent level) smaller than the controls, which averaged 11.0 cm. By the time the plants were three months old the variations within each treatment were so great that the differences between treatments were not significant except in the case of those plants growing with aster roots, which were almost twice

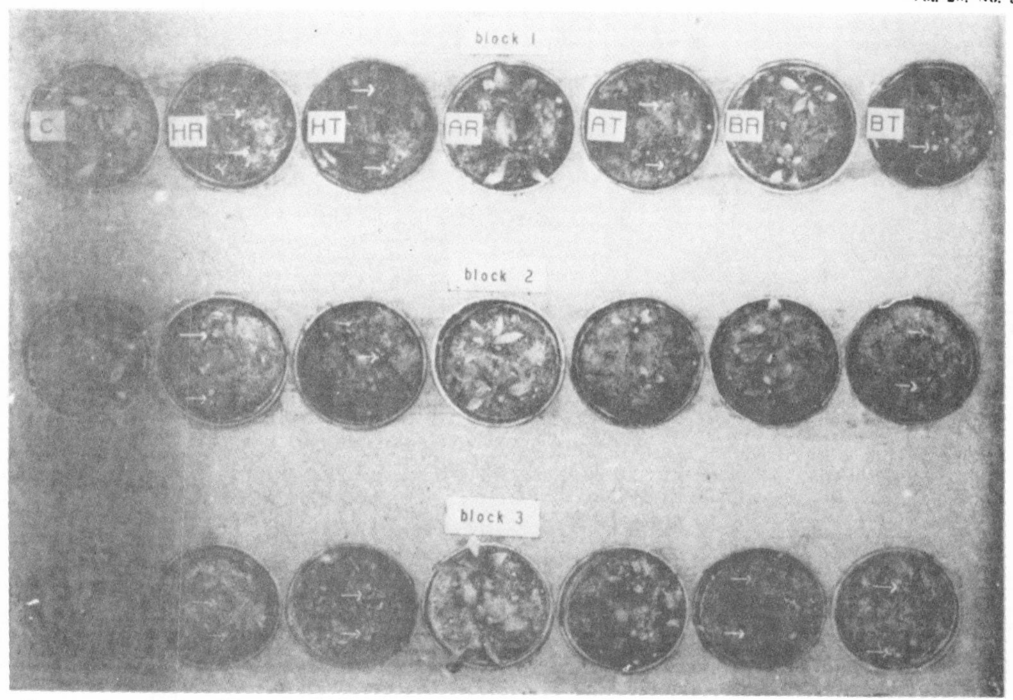

FIG. 5. Two-months-old aster plants growing in soil containing the following kinds of organic matter: HR- horseweed roots, HT horseweed tops, AR- aster roots, AT- aster tops, BR- broomsedge roots, BT- broomsedge tops, C- Control (soil with no organic matter added). Block 1, 2, and 3 are three replications of each treatment. Aster roots (AR) added to the soil produced a significant increase over the control and aster tops (AT), broomsedge tops (BT), horseweed tops (HT), and horseweed roots (HR) resulted in plants significantly smaller than the control.

as tall as any of the others. Although the height measurements in later stages failed to show significant differences, the final dry weights showed significant differences in four of the six treatments that were compared with the controls. At the end of the experiment the asters grown with aster roots were largest (average, 14.3 gms.); those in soil containing horseweed roots (3.58 gms.), horseweed tops (3.75 gms.), and broomsedge tops (3.38 gms.) were significantly smaller than the controls (10.75 gms.).

Broomsedge. Height measurement of broomsedge failed to be a fair evaluation of size, as it did not show the amount of tillering and circumference of clumps. There were no significant differences in height at any time among the plants grown with different treatments. Final dry weight, however, showed that those broomsedge plants grown in aster roots (av. wt. 8.86 gms.) were more than twice as heavy as those grown without organic matter (3.62 gms.). Plants grown in soil containing broomsedge roots (3.45 gms.) and broomsedge tops (3.27 gms.) were slightly smaller than the control and those with horseweed roots (5.34 gms.), horseweed tops (5.62 gms.), and aster tops (5.72 gms.) were slightly larger than the controls.

Decay and plant growth. The amounts of undecayed plant parts remaining in soil after plants had been growing in it for five months could not be determined accurately because it was difficult to separate small particles of organic matter from the soil. In spite of the possible experimental error, some information was obtained concerning the rate of decay of the different plant parts and its correlation with growth of the plants.

Ninety-eight percent of the aster roots placed in soil had decayed at the end of five months. All three species in soil containing aster roots made much greater growth than plants receiving any other treatment. Broomsedge roots showed the next highest amount of decay (71 percent) and no plants were stunted when grown with broomsedge roots. Broomsedge tops, with 67 percent decay, stunted horseweed in early stages and aster at all stages. Horseweed roots with 64 percent decay stunted horseweed in early stages and aster at all stages. Aster tops (32 percent decay) and horseweed tops (27 percent decay) significantly reduced the growth of asters.

When horseweeds were grown in sand and watered with nutrient solution, horseweed roots (30 percent decay) and horseweed tops (27 percent decay) did not affect the growth, and all plants grew equally well. When horseweed plants were grown in soil watered with nutrient solution, those plants growing with horseweed roots (98 percent decay) made much more growth than those growing with horseweed tops (65 percent decay). Plants growing in soil without added organic matter made less growth than those with the other two treatments.

CONCLUSIONS. The kind of decaying organic matter in the soil does influence growth of the dominant plants of old fields. Growth of plants seems to be related to the amount of decay of organic matter, and the amount of decay is not the same for different plant parts. Aster roots in soil decay more rapidly than other plant parts tested and produced the greatest growth in all three species. Horseweed roots in soil inhibit the growth of horseweed seedlings and of aster at all stages. Broomsedge tops, aster tops, and horseweed tops inhibit the growth of asters.

Plants grown in sand or soil containing organic matter to which nutrient solution has been added are not inhibited in growth.

COMPETITION BETWEEN ASTER AND BROOMSEDGE

METHODS. Field observations had shown that asters of all ages made little growth where broomsedge was increasing in number and size of clumps, and greenhouse experiments had indicated that living broomsedge produced no substances that inhibit the growth of asters. A controlled field experiment was set up to determine the effect of competition of broomsedge upon aster plants, and to determine if possible the basis of competition. It was hoped that an answer to the problem of why broomsedge replaces aster in old field succession might be partially solved by the results of such an experiment.

Three large broomsedge plants were selected in one of the third-year permanent quadrat fields. With each of these plants at the center, circles with one-meter radii were cleared by pulling up all the plants except the broomsedge and by removing the top inch of soil. On July 17, 1948, after the soil had been thoroughly wetted by a recent rain, young aster plants about an inch high that had been grown in the greenhouse were set at one-eighth meter intervals from the broomsedge on four radii of each circle. The plants that died during the first few days following resetting were replaced and by the end of a week all aster plants were established and living. On September 26, near the end of a five-week drought, the plants were checked for survival, the living plants were removed from the soil, and oven-dried. The average dry weights of all plants equidistant from a broomsedge plant were calculated.

The soil moisture conditions near the end of the five-week drought were determined in each of the experimental plots and also in the three third-year fields with permanent quadrats to see if water might be critical in competition. Soil samples were taken ⅛, ⅜, and ⅝ meters from the broomsedge plants at the center of each plot. The percentages of total water present, and the wilting percentage of each sample were determined, and from these data the water available to plants was calculated. Soil samples were taken beside quadrats one, five, and ten in each of th third-year permanent quadrat fields and available water calculated in like manner.

The distribution of broomsedge roots was determined by digging around the plants in the center of each plot to see how the extent of the root system corresponded with the survival of the asters.

RESULTS. The following table gives the percentage of plants dying and the average dry weight of the surviving plants for each distance from a broomsedge plant, and the available water present after drought at three of the positions.

Distance from broomsedge in meters	Percent of plants dying	Average dry weight of surviving plants in grams	Percent of available water in soil
⅛	33.3	0.0598	1.70
2/8	33.3	0.1140	
⅜	8.0	0.2000	3.52
4/8	0	0.3905	
⅝	8.8	0.4570	6.40
6/8	0	0.4200	
⅞	17.0	0.2060	

Most of the broomsedge roots extended to between two-eighth and three-eighth of a meter from the edge of each clump, with very few roots being found beyond this distance. More aster plants died in the regions where roots of broomsedge and aster were growing in the same soil than in other areas, and the surviving plants made less growth than elsewhere. The available water after drought was very low (1.70 percent) in soil occupied by broomsedge roots, but was greater outside the range of these roots (6.40 percent). Two of the twelve plants growing near the edge of the cleared plots died, and the surviving plants made less growth than ones located near the middle of the radii.

One of the third-year permanent quadrat fields which was clearly dominated by asters in 1947 and by broomsedge in 1948 showed the lowest available water of any of the third-year fields following a period of five weeks with very little rain. Soil samples from beside quadrats one, five, and ten showed the available water to be 1.83 percent, 2.58 percent, and 3.38 percent.

A second third-year field which was dominated by asters in 1947 showed a clear broomsedge dominance in 1948 on most of the field which was on a slight slope, but retained aster dominance in a small low area at the end of the field where the first quadrat was located. Soil samples showed the available water near the aster-dominated first quadrat to be 2.84 percent, but at quadrats five and ten which were up the slope in the broomsedge-dominated areas the values were 2.38 percent and 1.83 percent.

The other third-year field was aster-dominated in 1947 and although the number and size of broomsedge plants had increased slightly, the field was still aster-dominated in 1948. The available moisture was much higher in this field than in the other two, being 5.85 percent, 7.45 percent and 6.66 percent at the three stations tested.

CONCLUSIONS. The survival and growth of asters is poor when they are growing near thriving broomsedge plants. Competition for water may be one of the controlling factors in the replacement of aster dominance by broomsedge in old field succession.

CORRELATIONS AND GENERAL CONCLUSIONS

Causes of Old Field Succession

THE GENERAL TREND

The object of this study was to contribute to our knowledge of the causes of secondary succession by solving some of the problems concerning one particular example, namely, early old field succession in the Piedmont of North Carolina. To do this a number of general and specific questions needed to be answered. The foregoing experiments and observations were an attempt to answer as many of the specific questions as possible. The general questions usually could not be answered by one experiment or observations alone, but required evidence from several or all of the experiments and observations.

Although all of these questions have not been answered satisfactorily, several have been partially or nearly clarified, and the conclusions should add substantially to an understanding of succession in general.

WHY HORSEWEED IS USUALLY THE DOMINANT OF FIRST-YEAR FIELDS. The answer to the question of why horseweed rather than some other plant is usually the dominant of first-year fields lies largely in the facts of its life cycle, and to a much less extent in its response to the environmental conditions present in first-year fields. Corn, tobacco, cotton, garden vegetables, and other crops that are last cultivated in July or August occupy much of the farm land in the Piedmont of North Carolina. The fact that horseweed seeds are mature and ready to germinate at the time of year when most farm land in the Piedmont of North Carolina has been cultivated for the last time in a season gives it an advantage over other plants. The other old field species that possess genetic possibilities of being plants as large as horseweed are spring-germinating species, and many of them are slow-growing perennials that do not reach their maximum size the first year.

Horseweed is a composite which produces many small wind-borne seeds, so that an adequate supply of seeds is rarely lacking. Although the density of horseweed plants in first-year fields is usually high enough to insure a fair to good stand, the density is also low enough so that there is no severe competition among individuals of the same species. Under favorable conditions horseweed grows to be the largest plant of any species commonly found in first-year fields. Ragweed, also commonly present, sometimes approaches the size of horseweed, but being a spring-germinating species, gets started after horseweed is well established, and thus fails in competition with horseweed. Crabgrass is often present in large numbers, but even under the best environmental conditions rarely grows more than a foot high. Consequently it shares dominance with horseweed and does not replace it.

Horseweeds make their early growth in the fall, winter, and early spring when environmental conditions are favorable for the species. They are relatively drought-resistant plants, and a large number of them survive the fall droughts that normally occur in the Piedmont. They make their best growth in reduced light, and the short days with low light intensities during the fall and winter months are favorable for their growth. The ground cover of crabgrass plants that is often present in old fields at the time when horseweed seedlings are young is not harmful to them, but may be helpful in reducing light.

WHY HORSEWEED FAILS TO HOLD DOMINANCE AFTER THE FIRST YEAR. Greenhouse experiments showed that horseweed seedlings were definitely stunted when they were grown in soil containing decaying horseweed roots, even though competition had been removed. Field observations and measurements showed that horseweed seedlings growing in fields in which the dead horseweed plants of the preceding summer were still standing were much smaller than those in fields that had been cultivated the previous summer. Although the stunting of horseweed seedlings in second-year fields could be caused partially by the inhibiting effect of decaying horseweed roots in the soil, this could not have been the only cause, for horseweed plants growing with horseweed roots in the greenhouse where competition was not a factor, finally overcame the initial stunting effect. There were more than six times as many horseweed individuals per unit area in second-year fields as in first-year fields, and young asters were usually present in sufficient numbers to make competition a factor. The initial stunting effect of horseweed roots combined with the continued competition of numerous horseweed individuals and young asters could be enough to cause the loss of dominance by horseweed the second year. In eroded second-year fields where few young asters were present, the numerous horseweed seedlings were of the same small size as those in similar-aged fields containing many young asters, so it would seem that the inhibiting effect of horseweed roots and the competition among the large numbers of horseweed individuals could cause the small size of horseweeds without the added competition of asters.

The loss of dominance of horseweeds in second-year fields is not caused by the shading effect of asters or other plants. At the time the young horseweed plants are growing in second-year fields there are few other plants to diminish the amount of light reaching the seedlings; the bare stalks of the preceding horseweed stand obstruct little light and young asters are too small to interfere with light reaching the seedlings. Furthermore, horseweed seedlings grow best with the degree of reduction of light that is the normal condition for fall and winter.

Living aster plants produce no substance carried in soil water that inhibits the growth of horseweed plants; so living asters do not influence the decline of horseweed dominance in any way except in competing with them for water and minerals. As asters increase in size in the spring and summer of the second year, the already stunted horseweeds are at a still greater disadvantage, and never grow large enough to share dominance with asters in second-year fields.

WHY ASTER IS DELAYED IN ASSUMING DOMINANCE UNTIL THE SECOND YEAR. Much of the agricultural land in the Piedmont of North Carolina is cultivated for the last time in July or August. Aster seeds mature so late in the fall that the weather is normally too cool by that time for aster seeds to germinate. Consequently the germination is delayed until the following spring. Asters germinate in cultivated fields during the spring and early summer and are destroyed by the subsequent cultivation of such fields. No aster seedlings were found in the plots that were spaded on June 15 and later, and permanent quadrats showed an increase in aster seedlings between June and September in only one of the nine fields. This indicates that most of the aster seeds do germinate before the last of June, but that a few may be delayed in germination past that time. Oosting & Humphreys (1940) reported no buried viable seeds of aster found in soils around Durham that had previously been occupied by old field vegetation, so it is not likely that aster seeds brought to the surface by summer cultivations would germinate. Consequently the last cultivation of farm land in this area usually occurs after most of the asters have germinated and the young seedlings are thus destroyed. An occasional surviving aster was observed in a recently plowed field in the late summer of 1948, indicating that some asters do escape the plow. These few plants that are not destroyed by cultivation, or are derived from late germinating seeds, account for the occasional blooming aster seen in first-year horseweed-dominated fields.

Although aster seedlings show best survival with reduced light, they make best growth in full sunlight. Young asters growing in a horseweed-dominated field are somewhat shaded during the first few months of their life. This partially accounts for the fact that asters grown in the greenhouse with abundant water and no shading made much more growth than asters grown in the fields. Dead horseweed stalks cast but small shadows, so the asters are exposed to essentially full sunlight in the fall, winter, and following spring and summer after the previously dominant horseweeds die.

Decaying horseweed roots were found to have a greater stunting effect upon asters than they did upon horseweeds. Aster seedlings, however, are usually three to six months old before the horseweed plants, under which they grow, die and begin to decay. It cannot be said, however, that decaying horseweed roots do not inhibit the growth of asters to some extent, and that may be another reason why asters in the field grew more slowly than they did in the greenhouse.

Horseweeds, living or dead, do not stimulate the growth of asters. The living horseweeds may aid to some extent in the survival of aster seedlings by providing shade, but that same shade and the inhibiting effects of dead horseweed roots stunt the growth of asters. Therefore asters make some growth their first season in spite of the horseweeds, not because of them. Asters in the greenhouse bloomed their first season. It is possible that if no horseweeds or other plants were present in first-year fields to reduce the light and to compete for available soil moisture, asters might be dominant in a field in the fall of the first year of abandonment.

WHY ASTER FAILS TO HOLD DOMINANCE AFTER THE SECOND YEAR. In the water and light experiments in the greenhouse aster seedlings showed their best survival in reduced light, as long as water was available, but made their best growth in full sunlight with abundant water. Asters were definitely the least drought-resistant of any of the three species studied. With a two-week watering interval the survival of seedlings increased progressively as the light was reduced, but with a three-week watering interval the reverse was true; there was a greater survival in full sunlight and nearly all of the plants died in the reduced light.

In the spring of the third year there were a large number of seedlings present under the shade of the old asters, broomsedge and other vegetation. Most of these seedlings died before fall, and the surviving ones made very little growth.

The old asters that had bloomed the preceding year made considerable growth until about mid-summer, at which time many of them, especially the ones that were growing near large broomsedge plants, made little further growth and showed a rolling and yellowing of the leaves. Most of the old aster plants that had bloomed the preceding year lived and bloomed the second fall, but since they averaged only 27 inches high, they were not conspicuous in the fields where broomsedge plants averaged 39 inches.

In the field experiment where asters were placed in direct competition with large broomsedge plants, those asters whose roots occupied the same soil as that occupied by broomsedge roots showed greater mortality and less growth than those asters growing in soil not occupied by broomsedge roots. It is true that those plants close to the broomsedge were shaded a greater part of the day than those further away, but with all the other vegetation removed from the plots, all asters received full sunlight more than half of each day. The available soil moisture was less in the areas occupied by broomsedge roots than in soil not occupied by them. Young broomsedge plants are more drought-resistant than young asters, and older broomsedge plants in fields have a very high rate of survival. These facts, combined with the wide distribution of species of Andropogon in areas of North America where annual rainfall is low, suggests that broomsedge is relatively drought-resistant. When asters and broomsedge are competing for moisture in the same soil and when the available soil water is low, the broomsedge lives and continues to grow and the asters make little growth and finally die.

The soil moisture measurements in third-year fields confirm the idea that competition for water is one of the determining factors in the replacement of asters by broomsedge in old field succession. Those fields in which the available water is low during droughts are more clearly dominated by broomsedge than are those fields in which the available water is higher, and broomsedge dominance in such fields with higher moisture may be delayed a year until the broomsedge plants increase in size and number.

The reduced light in third-year fields is clearly one of the factors influencing the loss of dominance by asters in old fields. Few aster seedlings survive and grow to maturity in old fields after the vegetative cover of mature asters, broomsedge and other vegetation increases; the old asters fail to make sufficient growth in the reduced light to compete satisfactorily with the increasing broomsedge plants and finally die.

No evidence was found that living broomsedge or living asters produce any substances that inhibit the growth of asters. Both decaying horseweed roots and tops, which inhibit the growth of asters, are present in the soil at the time asters are losing dominance, but the actual amount of organic matter from the small horseweed plants that remain from second-year fields is too small to be an important factor in the elimination of asters.

WHY BROOMSEDGE IS DELAYED IN ASSUMING DOMINANCE UNTIL THE THIRD YEAR. There are few broomsedge seedlings of any age in old fields for the first two years after abandonment. Quadrat counts showed that there were few broomsedge seedlings in the first- and second-year fields at any time, and general observations even following rainy seasons, failed to show seedlings of this plant in numbers in any except third-year or older fields. Those few seedlings seen in the first- and second-year fields were usually found in open sunny portions of the quadrats rather than under the partial covering of crabgrass which was usually present.

No tests were made to see what percentage of broomsedge seeds germinate, but in all the greenhouse experiments involving planting of the three species, there was never any trouble in obtaining a full stand of broomsedge seedlings when layered seeds were used.

The water and light experiments showed that broomsedge seedlings are relatively drought-resistant. A large percentage of them are able to survive one to two weeks in full sunlight without watering with some increase in survival, but not in growth, with decreased lighting at the same watering intervals.

Broomsedge seeds are small and wind-borne, but are not so small as those of horseweed or aster, and may not be so easily transported. No tests were made to determine how far broomsedge seeds are commonly transported from their seed source. Although no old field in the Piedmont is far from a stand of broomsedge, it is possible that the seeds may not be carried far enough to produce a full stand. Oosting and Humphreys (1940) found three viable broomsedge seeds in 200 cubic inches of soil (maximum depth 5⅜ inches) taken from a zero-year field, one from a second-year field, and 37 from a fifth-year field. These data indicate the scarcity of viable seeds in old fields during the first two years of abandonment and the great number present after broomsedge becomes established. Those few plants that do appear in first-year fields do not produce mature seeds until the fall of the second year, and since the seeds require a period of cold dormancy they are not ready to germinate until the following spring. Consequently the spring of the third year is the first time an abundance of seeds is present in old fields.

Neither products of living horseweed or aster, nor products of their decay inhibit the growth of broomsedge. Broomsedge plants did grow somewhat better in soil containing organic matter from horseweed tops, horseweed roots, and aster-tops, and decidedly better in soil containing aster roots than in soil containing little organic matter.

Some environmental conditions in third-year fields are favorable for the growth of broomsedge plants and others are unfavorable. The abundance of organic matter found in soils of third-year fields seems to be beneficial to the growth of broomsedge plants, but the reduced light is unfavorable. Because broomsedge plants are more drought-resistant than asters, they are able to compete successfully with the asters and other vegetation present in third-year fields. The environmental conditions influence the survival and growth of broomsedge plants in fields once they are there, but the initiation of broomsedge dominance in old fields is probably retarded largely by the lack of seeds, and this condition is not remedied until the few first invaders are old enough to produce the seeds necessary for a full stand.

VARIATIONS FROM THE GENERAL TREND

An understanding of the causes of the general trend of succession in old fields of the Piedmont of North Carolina helps to explain variations from the general trend.

WHY SOME FIRST-YEAR FIELDS ARE RAGWEED-DOMINATED. Ragweed seeds require a period of cold dormancy and do not germinate until the spring following their maturation. In the fields studied the largest and most vigorous plants were found in the first-year fields and especially in those where the density of other plants was low. Although the spring density of ragweed plants was highest in third-year fields, the mortality of seedlings was also highest and the plants never grew to be large. Although no tests were made as to the light and water requirements of ragweed, field observations suggest that it does not compete successfully with other dominants.

Ragweed dominated all first-year fields in which wheat had been planted the preceding fall. Wheat is planted in this region in October and early November. The plowing of soil in preparation for wheat planting destroys the young horseweed seedlings that germinate in the early fall months. The plants from the horseweed seeds that germinate after wheat is planted are not able to compete successfully with the full stand of winter wheat, and remain small or die. Ragweed germinates the following spring and passes through its young seedling stage under the protection of the wheat. When the wheat is harvested in early June, it is removed as a competitor, full sunlight is available for the ragweed, horseweeds and most other plants are absent, and ragweed assumes complete dominance.

Young asters are present in the ragweed fields and succession proceeds normally to asters the following year. Young asters are of approximately the same size in these fields as in horseweed dominated ones, which indicates that any fast growing species which shades asters and competes with them for water retards their growth.

One field from which sweet potatoes were harvested in late fall was dominated by ragweed the following year and another such field was dominated by horseweed. The sweet potato field that was followed by horseweed was adjacent to a large field dominated by horseweed. The number of horseweed seeds available from the adjacent field was high and those seeds germinated in late fall or early spring, to produce a full crop of horseweed plants against which ragweed seedlings failed in competition.

Another ragweed-dominated field had been fall-plowed and planted in winter vetch with much the same results as fields with winter wheat.

Eroded edges and spots in fields otherwise dominated by horseweed are usually occupied by ragweeds. Horseweeds probably germinated in these areas the preceding fall, and although the flat rosettes of horseweed would have a tendency to keep the soil from being washed away from the roots, severe erosion could remove or cover up the entire plant. In areas which are swept clean of horseweed seedlings by winter rains, ragweeds germinate the following spring and grow to be fair sized plants without competition.

One extremely eroded field in which the corn crop of the preivous year had been last cultivated in June produced a pure stand of ragweed. Winter erosion had probably removed most of the horseweeds, asters, and other vegetation, and the spring-germinating ragweed grew without competition to be dominant in the field.

WHY ASTERS OCCASIONALLY SHARE DOMINANCE WITH HORSEWEEDS IN SOME FIELDS. Fields that are last cultivated in early summer before all of the asters have germinated usually have a mixture of horseweeds and asters as dominants the following year. The asters germinate in early summer, the horseweeds in the fall, and both species live together over the winter and share dominance the following year. Such fields appear to have horseweed as the only dominant when it blooms in July and August since it is then taller than the numerous asters, but aster appears to be dominant in the late fall when its blooms are conspicuous and the dead horseweed stalks are less noticeable.

WHY ASTER OCCASIONALLY PERSISTS IN DOMINANCE PAST THE SECOND YEAR. Asters usually give way to broomsedge in the third year because the seedlings and older asters fail in their competition for light and water with the broomsedge and other vegetation. The reduced light probably affects the seedlings as much as the competition for water, but with old asters water is a more critical factor than the reduction of light. In fields where there is no shortage of soil water because of type of soil, local drainage, or amount of precipitation, the old asters grow at least as tall as broomsedge and do not die as soon as they do in the fields where available water is low. In such fields, however, aster seedlings of the second or third generations rarely grow to maturity and the number and size of broomsedge plants increases within several years' time so that the old asters finally die, and the field is dominated by broomsedge.

THE EFFECT OF EROSION ON THE GENERAL TREND OF SUCCESSION. Erosion seems to affect aster dominance in second-year fields more than it does any other part of the normal sequence of succession. Young asters live for a full year as small upright plants less than three inches tall. Many nearly dead asters of this size were observed in eroded fields, following heavy rains, with the soil partially washed away from their root systems. Within a year's time few asters would escape one or more rains heavy enough to cause severe erosion in a field with some slope and no protecting terraces or cover crop.

Erosion causes some damage to horseweed, also, but to a smaller degree than to aster. Horseweed is a shorter-lived plant whose rosette form is somewhat soil binding. Horseweed passes through its younger stages at a time of year when precipitation is usually gradual rather than of the sudden storm type so common in the summer when asters are in their young stages.

Ragweed is common in eroded fields of both first and second years, because it seems to grow best where the competition of other plants has been removed in some way. Diodia is a small plant less than a foot high that often occupies eroded second-year fields. There were very few Diodia plants in any of the fields in the March counts, but it was present in fields in June, especially in eroded second-year fields. Some of these plants were present in third-year fields, but they rarely grew to be more than a few inches tall. This plant does not seem to be able to compete with other vegetation, and like ragweed is a short-lived summer annual that grows best when other vegetation is scant. For this reason it is commonly a dominant in eroded fields.

Causes of Secondary Succession in General

These partial answers to some of the questions concerning causes of secondary succession in old fields in the Piedmont of North Carolina should help in the understanding of causes of secondary succession in general. In his discussion of causes of succession Clements (1916) says that a plant or community reacts upon the environment in such a way that it may become less favorable to the organism responsible for the change and more favorable for other species. He says that a time ultimately comes when the reactions are more favorable to the occupants than to the invaders and that then the existing community becomes more or less permanent.

This study indicates that another important factor may enter into a particular sequence of dominant species in secondary succession. The peculiarities of the life cycles of these species, especially the time of year at which the seeds mature and germinate and the relation of the time of seed germination to the time at which secondary succession is initiated, often gives one species a decided advantage over another in becoming the invading dominant (Fig. 6). Weaver & Clements (1938) emphasize the importance of adaptations which faciliate seed dispersal as a major cause in determining which species will first become established. This may be the reason why broomsedge is delayed in assuming dominance, but in the case of horseweed and aster, which seem to be equally abundant in this area and equally capable of migration, the time of seed maturity is the chief factor that determines the sequence of dominance.

One species does not always influence the environment in such a way that it is made more favorable for the next species in the successional sequence. Aster could definitely grow better without the influences exerted by horseweed and would probably assume dominance a year earlier than it does if horseweed or some other large annual were not present.

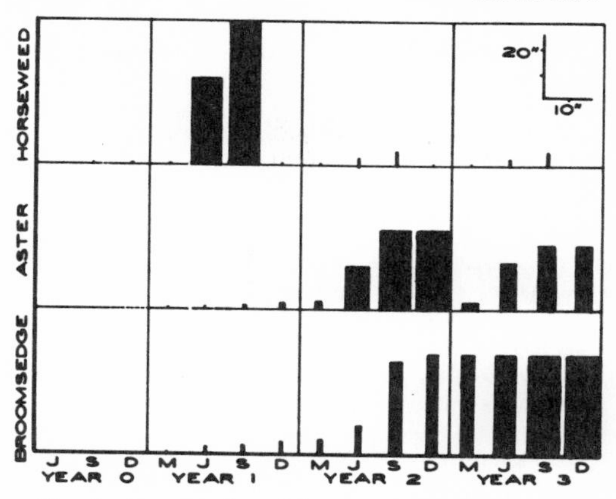

FIG. 6. Average size of horseweed, aster, and broomsedge plants in old fields for the first three years following abandonment. Height of column represents height of plant and base of column shows diameter at the widest point (all measurements in inches). M- March, J- June, S- September, D- December. Measurements for horseweed, an annual, are for different plants each year, but those for the perennials, aster and broomsedge, indicate sizes of the same plants. Young aster and broomsedge plants, though not in the graph, are present in second- and third-year fields. Because of the relative size, horseweed is the dominant of first-year fields; aster, of second-year fields, and broomsedge, of third-year fields.

There is some evidence that environmental conditions are more favorable for broomsedge after other plants have been growing in a field for several years so that the organic content of the soil increases, but this is not the chief reason why broomsedge is delayed in assuming dominance.

However, a species often makes its environment so unfavorable to itself that the species fails to survive. The influence of horseweed upon horseweed is the chief cause of the loss of dominance by the species. Aster influences its environment so that the survival and growth of seedlings of that species is so limited that the species can hold dominance only as long as the first generation of individuals can survive. The fact that broomsedge seedlings fail to make good growth in reduced light suggests that it, too, changes its environment so that conditions are unsuitable for reproduction.

Went (1942) suggests that the occurrence of one plant may influence the presence of certain other plants through some unknown agent, presumably chemical. If this were true it would be possible for chemical products of one species to stimulate or inhibit the growth of that plant or of other species and thus partially determine the sequence of succession of dominants in succession. This may be true in some situations, but no evidence was found to indicate that products of living dominants influence the general trend of succession in old fields of the Piedmont.

SUMMARY

1. The general trend of plant succession in abandoned fields of the Piedmont of North Carolina had been established before the beginning of this study. Crabgrass (*Digitaria sanguinalis* (L.) Scop.) is usually dominant in fields during the fall following their last cultivation for the season. Horseweed (*Leptilon canadense* L.) is usually dominant in first-year fields, but is sometimes replaced by ragweed (*Ambrosia elatior* L.). Aster (*Aster pilosus* Willd.) usually is the dominant of second-year fields, with some exceptions where ragweed or *Diodia teres* Walt. replaces aster. Broomsedge (usually *Andropogon virginicus* L.) assumes dominance the third year, and maintains this dominance until it is replaced by pines a few years later.

2. The purpose of this study was to contribute to the knowledge of causes of succession in general by investigating the causes of early old field succession in the Piedmont of North Carolina. For this purpose certain problems concerning the major dominants of the first three years of succession—horseweed, aster, and broomsedge—needed to be solved. These problems involved the reasons why each species assumes dominance at a particular time and why each loses dominance at a definite later time. Questions concerning variations from the general trend needed to be answered.

3. Answers to these questions were sought in the life cycles and responses to environmental factors of the three dominant species. Life cycles were studied by means of experimental tests of seed germination, permanent quadrats in fields of different ages, surveys of numerous fields and by measurements of size of plants growing in different habitats. The responses of each species to variations in water and light, to substances produced by living plants, and to decay products of organic matter in the soil were tested in greenhouse experiments.

4. The studies of life histories and environmental factors yielded several pertinent facts for each species as follows:

a. Horseweed seeds mature as early as August, and germinate with no dormant period in the late summer and fall. The plant lives over winter as a rosette, grows to maturity, blooms and dies by late summer of the following year. Horseweed is a drought-resistant species whose seedlings grow best in reduced light. Young horseweeds are stunted by decay products from horseweed roots and cannot compete satisfactorily with individuals of the same species or with other vegetation.

b. Aster seeds mature in the fall too late to germinate because of cool weather. They germinate the following spring and the seedlings grow to be about two to three inches tall the first year, live over winter, and bloom the following fall. The flower stalk dies and a basal rosette lives and produces another flower stalk the following year. Asters make their best growth with abundant water and light. Asters grown in the greenhouse with abundant water and light bloomed the first season. Aster plants were stunted by decay products of horseweed roots, horseweed tops, and broomsedge tops, when they were grown in the greenhouse in soil containing those plant parts. Aster plants appearing after the first generation rarely survive and bloom in the field under natural conditions.

c. Broomsedge seeds mature in the late fall and will not germinate without a period of cold dormancy. A few seedlings appear in old fields the first year of abandonment and do not produce seeds until the fall of the second year. Broomsedge seedlings are relatively drought-resistant and grow best in full sunlight, showing a definite chlorosis and reduced growth in shade. The survival of broomsedge plants of all ages in old fields is high and the species is able to compete successfully with the less drought-resistant asters. Broomsedge shows slightly better growth in soil containing organic matter, especially aster roots, than it does in soil with little organic matter.

5. The above generalizations can be used to explain to a large extent the causes of early old field succession in the Piedmont of North Carolina.

a. Horseweed usually assumes dominance in first-year fields because its seeds are ready to germinate when much farm land is being cultivated for the last time in a season. The low light intensities of fall and winter, and the lack of competition by other plants in first-year fields are favorable for the best growth of the species. Ragweed, a summer annual, may replace horseweed as the dominant of first-year fields if the normal crop of horseweed is eliminated by late fall plowing or by severe erosion.

Horseweed loses its dominance after the first year because it is stunted in the seedling stage by decay products from horseweed roots and also because it cannot compete successfully with the large number of individuals of the same species and with young asters which are abundant in second-year fields.

b. Asters are delayed in assuming dominance until the second year because the seeds do not germinate until the spring following maturity and although the seedlings do not die, they cannot make much growth in competition for light and water with the already established horseweeds. When the first crop of horseweed dies, competition is removed and asters make sufficient growth in spite of the inhibiting effects of decaying horseweed roots to be the dominant species of second-year fields.

Aster sometimes shares dominance with horseweed in first-year fields where the last cultivation of the season takes place in the early summer before all the aster seeds have germinated.

Aster loses its dominance because the seedlings are intolerent of shade and cannot grow to maturity under the old asters, broomsedge and other vegetation, and because the first generation of asters cannot compete with the more drought-resistant broomsedge. Asters may hold dominance past the second year if, for some reason, the lack of soil water does not become critical, but they cannot hold it indefinitely because of the failure of seedlings to grow to maturity.

c. Broomsedge is delayed in assuming dominance until the third year because there is not an adequate supply of seeds present to produce a full stand until the first few invaders are old enough to produce seeds. Environmental conditions are almost as favorable in first- and second-year fields as in third-year fields for the growth of broomsedge seedlings.

6. This study suggests that the particular timing of the events in the life cycles of the first series of invaders and its relation to the season of the year at which secondary succession is initiated may influence the dominant species more than changes in environment. Although the influence of a species on its environment often may produce conditions that keep the species from surviving, those conditions do not always make the environment more favorable for the next invader. The second invader may finally attain dominance in spite of, and not because of, the changes in environment brought about by the first dominant.

There was no evidence from this investigation that toxic substances produced by one living plant influence other plants in such a way as to change the trend of succession. The rate of decay of different kinds of organic matter and the influence of that rate of decay upon the available minerals in the soil does affect the growth of plants and the trend of succession.

LITERATURE CITED

Brenchley, W. E. 1919. Some factors in plant competition. Ann. Appl. Biol. **6**: 142-170.

Benedict, H. M. 1941. The inhibitory effect of dead roots on the growth of bromegrass. Amer. Soc. Agron. Jour. **33**: 1108-1109.

Bonner, J. 1946. Further investigations of toxic substances which arise from guayule plants: relation of toxic substances to the growth of guayule in soil. Bot. Gaz. **107**: 343-351.

Bonner, J. & A. W. Galston. 1944. Toxic substances from the culture media of guayule which may inhibit growth. Bot. Gaz. **106**: 85-198.

Clements, F. E. 1916. Plant succession. Carnegie Inst. Wash. Publ. **242**. 512 pp.

Crafton, W. M. & B. W. Wells. 1934. The old field prisere: an ecological study. Elisha Mitchell Sci. Soc. Jour. **49**: 225-246.

Coile, T. S. 1940. Soil changes associated with loblolly pine succession on abandoned agricultural land of the Piedmont Plateau. Duke Univ. School of Forestry Bul. **5**: 1-85.

Davis, E. F. 1928. The toxic principles of Juglans nigra as identified with synthetic juglone and its toxic effects on tomato and alfalfa plants. Amer. Jour. Bot. **15**: 620.

Gray, R., & J. Bonner. 1948. An inhibitor of plant growth from the leaves of Encelia farinosa. Amer. Jour. Bot. **35**: 52-57.

Hoagland, D. R. & D. I. Arnon. 1938. The water-culture method for growing plants without soil. Calif. Agr. Exp. Sta. Cir. **347**.

Oosting, H. J. 1942. An ecological analysis of the plant communities of Piedmont, North Carolina. Amer. Midl. Nat. **28**: 1-126.

Oosting, H. J. & M. E. Humphreys. 1940. Buried viable seeds in a successional series of old field and forest soils. Torrey Bot. Club Bul. **67**: 253-273.

Oosting, H. J. & P. J. Kramer. 1946. Water and light in relation to pine reproduction. Ecology **27**: 47-53.

Pickering, S. U., & The Duke of Bedford. 1914. The effect of one crop upon another. Jour. Agric. Sci. **6**: 136-151.

1917. The effect of one plant on another. Ann. Bot. **31**: 181-187.

1919. Action of one crop on another. Roy. Hort. Soc. Jour. **43**: 372-380.

Proebsting, E. L. & A. E. Gilmore. 1940. The relation of peach root toxicity to the establishment of peach orchards. Amer. Soc. Hort. Sci. Proc. **38**: 21-26.

Went, F. W. 1942. The dependence of certain annual plants on shrubs in southern Californian deserts. Torr. Bot. Club Bul. **69**: 100-114.

Weaver, J. E. & F. E. Clements. 1938. Plant Ecology. New York. 1-601.

9

Reprinted from *Am. J. Bot.* **59**(10):1033–1040 (1972)

INHIBITION OF NITRIFICATION BY CLIMAX ECOSYSTEMS[1]

ELROY L. RICE AND SUNIL K. PANCHOLY

Department of Botany and Microbiology, University of Oklahoma, Norman

ABSTRACT

Three plots representing two stages of old-field succession and the climax were selected in each of three vegetation types in Oklahoma: oak-pine forest, post oak-blackjack oak forest, and tall grass prairie. Soil samples from the 0–15 and 45–60 cm levels were analyzed every other month for 1 yr for exchangeable ammonium nitrogen and for nitrate. On alternate months numbers of *Nitrosomonas* and *Nitrobacter* were determined in the 0–15 cm level. The amount of ammonium nitrogen was lowest in the first successional stage, intermediate in the second successional stage, and highest in the climax stand. This trend was remarkably consistent throughout all sampling periods, all vegetation types, and both sampling levels in the soil. The amount of nitrate was highest in the first successional stage, intermediate in the second successional stage, and lowest in the climax stand in both sampling levels, all vegetation types, and virtually all sampling periods. The numbers of nitrifiers were high in the first successional stage, generally, and decreased to a very low level in the climax. In fact, there was often no *Nitrobacter* in the climax stands. These results indicate that the nitrifiers are inhibited in the climax so that ammonium nitrogen is not oxidized to nitrate as readily in the climax as in the successional stages. Evidence from other geographic areas and vegetation types strongly supports this conclusion. This would certainly appear to be a logical trend in the evolution of ecosystems because of the increased conservation of nitrogen and energy. The ammonium ion is positively charged and is adsorbed on the negatively charged colloidal micelles, thus preventing leaching below the depth of rooting. On the other hand, nitrate ions are negatively charged, are repelled by the colloidal micelles in the soil, and thus readily leach below the depth of rooting or are washed away in surface drainage. There is growing evidence also that many plant species can use ammonium nitrogen as effectively or more so than nitrate nitrogen. If ammonium nitrogen is used directly, this eliminates four chemical steps because nitrogen which is oxidized to nitrite and then to nitrate must be reduced back to nitrite and then to ammonium nitrogen before it can react with keto-acids in the formation of amino acids. The two reduction reactions require considerable expenditure of energy.

RUSSELL (1914) reported that cropped soil had a much lower total nitrate content than uncropped similar soil, even when the amount taken up by plants was included; and suggested that the lower amount was due to a diminished production in the presence of plants. Richardson (1935, 1938) found in grassland soils at Rothamsted Experimental Station that the level of ammonium nitrogen was several times greater than the level of nitrate nitrogen. He reported, also, that the grasses and other plants absorb the ammonium nitrogen as readily as nitrate and that much of the nitrogen is taken up as ammonia. Theron (1951) reported that the rate of nitrification in a South African grassland soil was lessened by a perennial grass crop. Eden (1951) found that grassland (patana) soils in Ceylon are extremely low in nitrate and that the low nitrification lasts for several years after breaking the land for tea cultivation. Mills (1953), working on crop lands at Kawanda in Uganda (Africa), reported that after three-years resting under elephant grass (*Pennisetum purpureum*), *Chloris*, and *Paspalum* the nitrate content to a soil depth of 6 ft was almost zero. After opening up the areas again, the rate at which nitrate accumulation occurred in the surface horizons varied with the species of grass: *Paspalum*, very slowly with low nitrate for a long time; elephant grass, more rapidly to a higher nitrate figure; *Chloris*, intermediate.

Berlier, Dabin, and Leneuf (1956) found the activity of the nitrifying bacteria to be virtually zero in savanna soils of the Ivory Coast in Africa. They pointed out that the grassy vegetation seems to exert a direct inhibiting action since after clearing the nitrifying activity increases greatly. Greenland (1958) reported that almost no nitrate was found in permanent grassland plots in the savanna in Ghana. He found further that suppression of nitrification under temporary or successional grass plots was less than under the climax grasses. Warren (1965) found that the populations of nitrifiers in the climax purple-veld of South Africa were much lower than in succes-

[1] Received for publication 22 March 1972.

This project was done under grant GB-22859 from the National Science Foundation. We are grateful to Mr. Jack Turner for his able assistance.

TABLE 1. *Results of vegetational analysis of study plots. Data given only for herbs in each plot which made up 3 % or more of total herbaceous composition, but data given for all tree species sampled. P_1 and P_2 were successional stages in old fields, whereas P_3 was a climax stand in each vegetation type*

Species	Tall Grass Prairie			Post Oak–Blackjack Forest			Oak–Pine Forest		
	P_1	P_2	P_3	P_1	P_2	P_3	P_1	P_2	P_3
Herbs (Grasses and forbs)				Per Cent Composition					Sparse, not sampled
Agrostis elliotiana				11.8					
Ambrosia psilostachya				3.1	12.4				
Andropogon gerardi			22.4			8.6			
A. scoparius			42.4			47.6			
A. virginicus					13.9	8.1		71.1	
Antennaria plantaginifolia						3.2			
Aristida oligantha		63.3			9.4				
Coreopsis tinctoria	63.7	6.8							
Cynodon dactylon					7.9		3.0		
Cyperus ovularis				3.5			3.8		
C. strigosus							13.7		
Digitaria sanguinalis				12.9			23.6		
Diodia teres				5.5					
Erigeron canadensis	16.8			4.3			3.0		
E. strigosus		7.9		3.9					
Eupatorium serotinum							20.6		
Gnaphalium purpureum				5.1					
Haplopappus ciliatus	7.1	14.5							
Leptoloma cognatum				8.6	8.3				
Lespedeza stipulacea					19.9			16.7	
Medicago sativa							15.3		
Panicum linearifolium						4.3			
P. scribnerianum						3.2			
P. virgatum			11.2						
Paspalum pubescens				34.1	4.5				
Ptilimnium nuttallii					4.1				
Setaria geniculata and *S. lutescens*								4.1	
Sorghastrum nutans			13.6			3.8			
Sorghum halepense	5.3								
Sporobolus asper						4.3			
Trees				Importance Percentage					
Carya texana						2			5
C. tomentosa									6
Pinus echinata									54
Quercus falcata									2
Q. marilandica						64			5
Q. stellata						26			23
Q. velutina									6
Ulmus alata						7			

sional stages. During the greater part of the year the nitrite oxidizers were almost completely absent in the purple-veld soils. Moreover, there was a gradual decrease in nitrite and nitrate with succession. Jacquemin and Berlier (1956) and Berlier et al. (1956) found that the nitrifying power in forest-covered soils of the lower Ivory Coast was low and that it increased on clearing, even without burning. Nye and Greenland (1960) reported that in the soils of Africa covered by Moist Evergreen Forest the numbers of nitrifiers are exceptionally low. In contrast to the low rate of nitrification in such soils, ammonification proceeds rapidly. In Finland, Viro (1963) reported that most of the available nitrogen in the humus layer from stands in which spruce was dominant, in pure pine stands, and in pure spruce stands, was ammonium nitrogen with the amount of nitrate nitrogen being very low. Data cited by Russell and Russell (1961) and Weetman (1961) indicate that the rate of nitrification is generally very low under spruce and several other conifers and also in forests with a mor type of mulch.

Smith, Bormann, and Likens (1968) found an 18-fold increase in *Nitrosomonas* and a 34-fold increase in *Nitrobacter* after clear-cutting a forest ecosystem in Connecticut.

Boughey et al. (1964) reported that two species

TABLE 2. *Characteristics*[a] *of soils, 0–15 cm*

Vegetation	% sand	% silt	% clay	*p*H	% organic carbon[c]
Tall Grass Prairie					
P_1[b]	59.05	14.26	26.68	6.82	0.561
P_2	65.29	17.94	16.75	6.49	0.589
P_3	78.16	7.17	14.66	6.81	0.885
Post Oak–Blackjack					
P_1	79.40	8.48	13.24	6.23	0.484
P_2	82.88	5.36	11.76	6.58	0.637
P_3	83.20	6.32	10.48	5.91	0.480
Oak–Pine					
P_1	75.06	12.30	12.64	5.52	1.151
P_2	75.22	15.08	9.32	6.06	0.342
P_3	65.52	20.00	14.48	5.42	1.096

[a] Figures are averages of ten replicates.
[b] P_1 and P_2 are successional stages in old fields. P_3—climax vegetation.
[c] Walkley and Black value (Piper, 1942).

of *Hyparrhenia*, grasses abundant in the Rhodesian high-veld savanna, secrete a toxin which suppresses the growth of nitrifying bacteria. In the same year the senior author reported that numerous species of herbaceous plants important in old-field succession in Oklahoma inhibit the growth and nitrifying ability of the nitrifiers (Rice, 1964). The only climax species which he studied, *Andropogon scoparius* and *Erigeron strigosus*, showed little activity against the nitrogen-fixing bacteria, but both were very inhibitory to the nitrifiers. Munro (1966b) found that root extracts of several climax species from the Rhodesian high-veld were more inhibitory to nitrification than several seral species investigated.

We hypothesized, therefore, that many (if not most) soils under climax vegetation are low in nitrate, due to an inhibition of nitrification by climax plant species. This would certainly appear to be a logical successional and evolutionary development of ecosystems because of the increased conservation of nitrogen. The ammonium ion is positively charged and is adsorbed on the negatively charged colloidal micelles, thus preventing leaching below the depth of rooting. On the other hand, the nitrate ions are negatively charged and are repelled by the colloidal micelles in the soil. Therefore, they readily leach below the depth of rooting or are washed away in surface drainage. Experiments were designed to test the hypothesis.

MATERIALS AND METHODS—*Study plots and vegetation analyses*—Three stands, representing two stages of old-field succession and the climax, were selected in each of the following vegetation types in Oklahoma: oak-pine forest, post oak-blackjack oak forest, and tall grass prairie (Duck and Fletcher, 1943; Rice and Penfound, 1959). Stands in the tall grass prairie area are located near Norman in Cleveland and McClain Counties, the post oak-blackjack area stands in Hughes County southeast of Wetumka, and the oak-pine stands in Latimer County north of Wilburton.

The woody vegetation was sampled in June, 1971, using the variable-radius technique at 40 evenly distributed points to determine basal area, and 40 arms-length rectangles of 0.01 acre each were used between points for procuring data on frequency and density (Rice and Penfound, 1955). Any woody plant with a breast-high diameter of 3 in. or more was considered a tree. Relative density, relative frequency, and relative basal area were averaged to give an importance percentage for each tree species in each stand. The herbaceous vegetation was sampled in July, 1971, using the point-frame technique (Whitman and Siggeirsson, 1954) and 1500 basal points were sampled in each stand. Percent composition was calculated for each species in each stand.

Soil analyses—Starting in March, 1971, ten evenly distributed soil samples were taken from the 0–15 cm- and ten from the 45–60 cm-level in each plot every other month for a full year. These were analyzed for ammonium nitrogen by steam distillation with MgO (Bremner, 1965) and for nitrate nitrogen by a specific ion electrode, after extracting the soil with distilled water (1:2 ratio of soil:water) for 1 hr with occasional stirring.

The 0–15-cm samples obtained in March were analyzed additionally for organic carbon by the chromic acid-digestion method (Piper, 1942); for sand, silt, and clay by the hydrometer method of Bouyoucos (1936); and for *p*H by the glass electrode method, using a 1:5 ratio of soil to water.

Starting in April, 1971, eight evenly distributed soil samples were taken from the 0–15 cm level in each plot every other month for a year. These were analyzed for numbers of *Nitrosomonas* (oxidizer of NH_4 to NO_2) and *Nitrobacter* (oxidizer of NO_2 to NO_3) by the most probable number (MPN) method of Alexander and Clark (1965).

RESULTS—*Vegetation of study plots*—The first successional stage investigated (P_1) was in the 1st year after abandonment from cultivation in the oak-pine and tall grass prairie areas and in the 2nd year in the post oak-blackjack area. All P_1 plots were in the pioneer weed stage of succession, Stage 1 of Booth (1941) (Table 1). The second successional stage investigated (P_2) was in the 6th year after abandonment in the tall grass prairie area and was dominated by *Aristida oligantha*, which represented Stage 2, the annual

TABLE 3. *Amounts of ammonium and nitrate nitrogen in 0–15 cm level of research plots. Each number is average of ten analyses*

Date	PPM NH_4^+			PPM NO_3^-		
	P_1	P_2	P_3	P_1	P_2	P_3
			Tall Grass Prairie			
March, 71	3.28	3.98	4.60[bc]	4.42	1.42[a]	1.50[c]
May, 71	2.87	3.71[a]	4.86[bc]	3.11	1.98[a]	1.39[bc]
July, 71	0.94	1.45[a]	2.21[bc]	4.12	2.70[a]	1.77[bc]
September, 71	1.68	5.63[a]	6.68[c]	4.27	2.78[a]	1.78[bc]
November, 71	2.82	6.69[a]	7.83[bc]	2.42	0.91[a]	0.20[bc]
January, 72	2.08	3.10	4.35[c]	2.49	1.31[a]	0.57[bc]
			Post Oak–Blackjack Oak			
March, 71	2.17	3.09[a]	4.49[bc]	2.13	1.62	1.14
May, 71	2.24	3.97[a]	5.45[bc]	2.78	1.53[a]	0.86[bc]
July, 71	1.99	2.90[a]	4.36[bc]	3.43	1.83[a]	1.06[bc]
September, 71	4.93	4.27	2.80[c]	3.21	1.65[a]	1.05[bc]
November, 71	2.87	4.41[a]	4.90[c]	1.67	0.88[a]	0.42[c]
January, 72	1.36	3.04[a]	3.02[c]	1.89	1.04[a]	0.73[bc]
			Oak–Pine			
March, 71	2.40	6.43[a]	7.39[c]	3.92	1.25[a]	1.82[bc]
May, 71	3.36	6.24[a]	7.13[c]	3 09	2.03[a]	1.52[bc]
July, 71	2.62	3.89[a]	4.72[bc]	3.72	2.40[a]	1.62[bc]
September, 71	2.97	3.15	3.01	2.18	1.44	0.07[bc]
November, 71	1.63	3.91[a]	4.37[c]	2.62	2.01[a]	0.79[bc]
January, 72	1.00	3.85[a]	5.43[c]	2.59	0.92[a]	0.35[bc]

[a] Difference between P_1 and P_2 significant at 0.05 level or better.
[b] Difference between P_2 and P_3 significant at 0.05 level or better.
[c] Difference between P_1 and P_3 significant at 0.05 level or better.

grass stage of Booth (Nomenclature follows Fernald, 1950). The P_2 plot in the post oak-blackjack area was in the 8th year after abandonment and was dominated by *Ambrosia psilostachya, Andropogon virginicus, Aristida oligantha,* and *Lespedeza stipulacea.* The P_2 plot in the oak-pine area was abandoned from cultivation for 25 yr and was dominated by *Andropogon virginicus*; a few pine and oak seedlings were present. The chief criteria used in selecting the P_2 plots were availability and a stage of succession somewhere between the pioneer stage and the climax. The climax prairie (P_3) was dominated by *Andropogon gerardi* and *A. scoparius,* with *Panicum virgatum* and *Sorghastrum nutans* as important secondary species. The climax post oak-blackjack oak stand was dominated by *Quercus marilandica* and *Q. stellata,* with ground cover primarily of *Andropogon scoparius* (Table 1). The climax oak-pine stand was dominated by *Pinus echinata* and *Quercus stellata.* This stand had virtually no herbaceous ground cover. There were a few very small and sparse patches consisting of *Andropogon gerardi, A. scoparius, Desmodium laevigatum,* and *Tephrosia virginiana.*

General soil characteristics—All plots had a sandy loam soil except the P_1 plot in the tall grass prairie area which had a sandy clay loam soil (Buckman and Brady, 1960) (Table 2). Moreover, the *p*H was virtually the same in all three plots of each vegetational type (Table 2). The amount of organic carbon often varied considerably in the different plots, with the only consistent trend in relation to succession occurring in the tall grass prairie plots (Table 2).

Ammonia and nitrate in soil—The amount of ammonium nitrogen was lowest in the first successional stage, intermediate in the second successional stage, and highest in the climax stand (Table 3, 4). Moreover, the differences in amounts between P_1 and P_2, P_2 and P_3, and P_1 and P_3 were generally statistically significant. This trend was remarkably consistent throughout all sampling periods, all vegetation types, and both sampling levels in the soil. The only exception to this trend occurred in both the 0–15- and 45–60-cm levels in the post oak-blackjack area in September, and in the 0–15-cm level in the oak-pine area in September.

The amount of nitrate was highest in the first successional stage, intermediate in the second successional stage, and lowest in the climax stand in both sampling levels, all vegetation types, and virtually all sampling periods (Table 3, 4). The chief exceptions occurred in the 45–60-cm level in the oak-pine area in March and September.

TABLE 4. *Amounts of ammonium and nitrate nitrogen in 45–60 cm level of research plots. Each number is average of ten analyses*

Date	PPM NH_4^+ at 45–60 cm			PPM NO_3^- at 45–60 cm		
	P_1	P_2	P_3	P_1	P_2	P_3
			Tall Grass Prairie			
March, 71	1.37	3.40[a]	4.12[c]	4.63	3.68[a]	1.16[bc]
May, 71	1.47	1.89[a]	2.48[bc]	2.49	2.08[a]	1.36[bc]
July, 71	0.76	0.97	1.27	2.67	2.12[a]	1.44[bc]
September, 71	3.32	5.51	5.67[c]	2.66	2.25[a]	1.49[bc]
November, 71	3.18	5.39[a]	5.60[c]	0.85	0.31[a]	0.13[c]
January, 72	1.88	2.95	2.75[c]	1.06	0.92	0.18[bc]
			Post Oak–Blackjack Oak			
March, 71	1.65	4.54[a]	5.19[bc]	2.08	1.63[a]	0.94[c]
May, 71	0.80	1.12	2.65[bc]	1.98	1.40	1.18[c]
July, 71	0.87	1.26	2.14[bc]	2.18	1.66[a]	0.96[bc]
September, 71	3.64	3.01	2.79	2.08	1.63[a]	0.80[bc]
November, 71	1.78	2.95[a]	3.22[c]	0.54	0.45	0.22[c]
January, 72	0.93	1.20	1.55[c]	0.88	0.64[a]	0.60[c]
			Oak–Pine			
March, 71	2.79	5.23[a]	7.06[bc]	2.23	1.00[a]	2.28[b]
May, 71	1.99	5.37[a]	6.05[c]	2.08	1.82	1.58[c]
July, 71	1.89	3.46[a]	4.20[bc]	2.19	2.12[a]	1.06[bc]
September, 71	1.05	1.57	2.10[c]	0.00	0.37	0.00
November, 71	1.33	1.61	1.82[c]	1.39	1.28	0.39[bc]
January, 72	0.98	1.31	2.01[bc]	0.14	0.06	0.04

[a] Difference between P_1 and P_2 significant at 0.05 level or better.
[b] Difference between P_2 and P_3 significant at 0.05 level or better.
[c] Difference between P_1 and P_3 significant at 0.05 level or better.

Again, the differences between P_1 and P_2, P_2 and P_3, and P_1 and P_3 were usually statistically significant.

The consistently high amounts of ammonium nitrogen and very low amounts of nitrate in the climax stands throughout all vegetation types, all sampling periods, and both sampling levels were particularly striking.

Numbers of nitrifiers—The numbers of *Nitrosomonas* per gram of soil were highest in the first successional stage, intermediate in the second successional stage, and lowest in the climax throughout all sampling periods and all vegetation types (Table 5). The one exception was in the tall grass prairie area in the month of April, when numbers were lower than usual in the P_1 and P_2 plots.

The numbers of *Nitrobacter* per gram of soil were generally considerably higher in the first and second successional stages than in the climax (Table 5). In the post oak-blackjack and oak-pine areas the number in the first successional stage was highest with the second successional stage generally having an intermediate number. In the prairie area, the first and second successional stages generally had similar numbers of *Nitrobacter*. The most striking feature was the very low number of *Nitrobacter* which occurred in the climax stands in all vegetation types and all sampling periods, with the exception of the June period in the tall grass prairie. The number was often zero.

The highest counts of both *Nitrosomonas* and *Nitrobacter* occurred in the P_1 plot of the oak-pine area throughout the year. Seasonally, the highest counts occurred in the late spring and summer in all vegetation types, with secondary highs in the fall and early spring.

DISCUSSION—The inverse correlation between the amount of nitrate and the amount of ammonium nitrogen in all plots was striking. The amount of ammonium nitrogen increased from a low in the first successional stage to a high in the climax, whereas the amount of nitrate decreased from a high in the first successional stage to a low value in the climax. When one considers the additional fact that the counts of nitrifiers were high in the first successional stage and low in the climax, the obvious inference is that the nitrifiers are inhibited in some way in the climax plots so that the ammonium nitrogen is not oxidized to nitrate as readily as in the successional stages. It appears that inhibition of nitrification starts during old-field succession and increases in intensity as succession proceeds toward the climax. It is obvious from the general soil data that the

TABLE 5. *Numbers (MPN) of nitrifiers in 0–15 cm level of research plots. Each number is average of four determinations at different locations*

Date	Nitrosomonas/g Soil			Nitrobacter/g Soil		
	P_1	P_2	P_3	P_1	P_2	P_3
			Tall Grass Prairie			
April, 71	111	42[a]	140	25	25	25
June, 71	3012	525[a]	147[bc]	347	417	280
August, 71	334	158	50[c]	62	72	3[bc]
October, 71	817	51	37	24	24	24
December, 71	177	26[a]	32[c]	23	23	18
February, 72	127	51	51	110	270[a]	43[c]
			Post Oak–Blackjack Oak			
April, 71	186	47	22	116	19[a]	5[b]
June, 71	4470	710[a]	32[c]	22	8	0[c]
August, 71	303	207	64[c]	51	1[a]	0[c]
October, 71	216	292	15	50	321	15
December, 71	126	78	15[c]	19	12	9
February, 72	50	15	14	132	41	6[bc]
			Oak–Pine			
April, 71	2770	65[a]	7[bc]	248	9[a]	5[c]
June, 71	3365	188[a]	8[bc]	951	29[a]	0[bc]
August, 71	36286	964[a]	24[bc]	384	45[a]	0[bc]
October, 71	1588	315[a]	5[bc]	661	40[a]	30[c]
December, 71	783	451[a]	30[bc]	192	50[a]	8[bc]
February, 72	3950	57[a]	0[bc]	205	37[a]	0[bc]

[a] Difference between P_1 and P_2 significant at 0.05 level or better.
[b] Difference between P_2 and P_3 significant at 0.05 level or better.
[c] Difference between P_1 and P_3 significant at 0.05 level or better.

low rates of nitrification in the climax plots were not due to *p*H or textural differences. Moreover, the lack of definite trends in amounts of organic carbon in relation to succession in the oak-pine and post oak-blackjack areas indicates that the quantity of organic carbon is not responsible for the low rate of nitrification in the climax plots. The almost complete lack of herbaceous ground cover in the oak-pine climax plot, with the resulting lack of a thoroughly penetrating fibrous root system, indicates that the low level of nitrates in that plot was not due to a greater uptake in comparison to the P_1 and P_2 plots, which had a heavy grass cover with dense fibrous root systems. Data from all vegetation types investigated strongly support our original hypothesis that many soils under climax vegetation are low in nitrate, due to an inhibition of nitrification by climax plant species.

As we indicated in the introduction, evidence from numerous other geographic areas and vegetation types strongly supports our hypothesis. Nye and Greenland (1960), in reviewing many African areas and vegetation types, reported that "Irrespective of the C/N ratio of the soil, its *p*H or moisture regime very little, if any, nitrate nitrogen is found in the soil while the dominant vegetative cover is a grass." Smith et al. (1968) felt that the evidence indicated that the pronounced increase in numbers of nitrifiers resulting from the clear-cutting of a forest ecosystem was due, not to changes in physical conditions, but to elimination of uptake of nitrate by the vegetation, or to a reduction in production of substances inhibitory to the autotrophic nitrifying population. Likens, Bormann, and Johnson (1969) reported a 100-fold increase in nitrate loss in the same ecosystem after cutting. Our own results definitely suggest that the increase in nitrifiers and nitrate were probably not due to elimination of uptake of nitrate by the vegetation.

Lyon, Bizzell, and Wilson (1923) attributed the lower total nitrate in cropped as opposed to uncropped soil (Russell, 1914) to an increased nitrate uptake by microorganisms, stimulated by root excretions with a high C/N ratio. Theron (1951), in discussing the reasons for the low nitrate level in grasslands in Africa, gave two arguments against the idea of Lyon et al. (1923) that the low nitrate was due to the excretion of carbonaceous material by plants: (1) the amount of carbonaceous material required to bring about the result was too great to be excreted by the roots of the plants; and (2) when such carbonaceous material is added to a soil not only the nitrates but also the ammonia are reassimilated by microorganisms, whereas under grass ammonia is found in greater quantities than in cultivated soils. He inferred from (2) above that the plants exerted their influence on the mineralization of nitrogen in the soil by inhibiting the nitrifying process. Our data certainly support Theron's argument because the amount of ammonium nitrogen was consistently very high in the climax plots.

The earliest reported work which attempted experimentally to link plant-produced toxins with low nitrification was that of Stiven (1952). He found that little or no nitrate was produced in pots containing *Trachypogon plumosus* (a bunch grass important in South Africa) roots and soil. A distilled water extract of the roots was found to be very inhibitory to several species of bacteria on agar. Stiven did not test the effect against the nitrifiers, probably because of the difficulty of culturing them. Subsequently, several workers reported inhibition of nitrification by numerous plant species (Boughey et al., 1964; Munro, 1966a, b; Rice, 1964, 1965a, b, 1969).

Basaraba (1964) reported that the addition of vegetable tannins to soil lowered the rate of nitrification. Rice (1965a, b) found that tannins produced by several species of plants were very inhibitory to nitrifying bacteria. Blum and Rice (1969) found 600–800 ppm of tannic acid in the top 5 cm of soil under *Rhus copallina* in a tall grass prairie in Oklahoma throughout the year. Tannic acid was found to a depth of 75 cm with a zone of accumulation in the B-horizon, thus indicating that this compound is stable enough to leach through the soil. Calculations based on data given by Basaraba (1964) indicate that perhaps 150–450 lb of tannin are added per acre of forest each year. These data suggest that tannins alone may constitute a very important inhibitor of nitrification in many vegetation types. Identification of inhibitors is in progress.

There is growing evidence that many plant species can use ammonium nitrogen as effectively or more so than nitrate nitrogen (Allison, 1931; Cramer and Myers, 1948; Ferguson and Bollard, 1969; McFee and Stone, 1968; Moore and Keraitis, 1971; Oertli, 1963; Shen, 1969). Thus, the inhibition of nitrification makes good biological sense.

The inhibition of nitrification has important implications in mineral cycling in ecosystems also. Four biochemical steps are eliminated, and two of these, the reduction of nitrate to nitrite and of nitrite to ammonium nitrogen, require considerable energy. The inhibition of nitrification results, therefore, in a conservation of energy.

When the conservation of nitrogen, the conservation of energy, and the ability of many species to use ammonium nitrogen as effectively or more so than nitrate nitrogen are all considered, it is not surprising that succession proceeds in the direction of inhibition of nitrification. It is also probable that such selective pressures cause ecosystems to evolve in this direction.

LITERATURE CITED

ALEXANDER, M., AND F. E. CLARK. 1965. Nitrifying bacteria, p. 1477–1483. *In* C. A. Black et al. [ed.], Methods of soil analysis, Vol. 2. Amer. Soc. Agron., Madison, Wisconsin.

ALLISON, F. E. 1931. Forms of nitrogen assimilated by plants. Quart. Rev. Biol. 6: 313–321.

BASARABA, J. 1964. Influence of vegetable tannins on nitrification in soil. Plant and Soil 21: 8–16.

BERLIER, Y., B. DABIN, AND N. LENEUF. 1956. Comparaison physique, chimique et microbiologique entre les sols de foret et de savane sur les sables tertiaires de la Basse Côte d'Ivoire. Trans. 6th Int. Congr. Soil. Sci. E: 499–502.

BLUM, U., AND E. L. RICE. 1969. Inhibition of symbiotic nitrogen-fixation by gallic and tannic acid, and possible roles in old-field succession. Bull. Torrey Bot. Club 96: 531–544.

BOOTH, W. E. 1941. Revegetation of abandoned fields in Kansas and Oklahoma. Amer. J. Bot. 28: 415–422.

BOUGHEY, A. S., P. E. MUNRO, J. MEIKLEJOHN, R. M. STRANG, AND M. G. SWIFT. 1964. Antibiotic reactions between African savanna species. Nature 203: 1302–1303.

BOUYOUCOS, G. J. 1936. Directions for making mechanical analyses of soils by the hydrometer method. Soil Sci. 42: 225–229.

BREMNER, J. M. 1965. Inorganic forms of nitrogen, p. 1179–1237. *In* C. A. Black et al. [ed.], Methods of soil analysis, Vol. 2. Amer. Soc. Agron., Madison, Wisconsin.

BUCKMAN, H. O., AND N. C. BRADY. 1960. The nature and properties of soils. Sixth ed. The Macmillan Co., New York.

CRAMER, M., AND J. MYERS. 1948. Nitrate reduction and assimilation in *Chlorella*. J. Gen. Physiol. 32: 92–102.

DUCK, L. G., AND J. B. FLETCHER. 1943. A game type map of Oklahoma. Oklahoma Game and Fish Dept., Oklahoma City, Oklahoma.

EDEN, T. 1951. Some agricultural properties of Ceylon montane tea soils. J. Soil Sci. 2: 43–49.

FERGUSON, A. R., AND E. G. BOLLARD. 1969. Nitrogen metabolism of *Spirodela oligorrhiza*. I. Utilization of ammonium, nitrate and nitrite. Planta 88: 344–352.

FERNALD, M. L. 1950. Gray's manual of botany. 8th ed. American Book Co., New York.

GREENLAND, D. J. 1958. Nitrate fluctuations in tropical soils. J. Agr. Sci. 50: 82–92.

JACQUEMIN, H., AND Y. BERLIER. 1956. Évolution du pouvior nitrifiant d'un sol de basse Côte d'Ivoire sous l'action du climat et de la végétation. Trans. 6th Int. Congr. Soil Sci. C: 343–347.

LIKENS, G. E., F. H. BORMANN, AND N. M. JOHNSON. 1969. Nitrification: Importance to nutrient losses from a cutover forested ecosystem. Science 163: 1205–1206.

LYON, T. L., J. A. BIZZELL, AND B. D. WILSON. 1923. Depressive influence of certain higher plants on the accumulation of nitrates in the soil. J. Amer. Soc. Agron. 15: 457–467.

MCFEE, W. W., AND E. L. STONE, JR. 1968. Ammonium and nitrate as nitrogen sources for *Pinus radiata*, and *Picea glauca*. Soil Sci. Soc. Amer. Proc. 32: 879–884.

MILLS, W. R. 1953. Nitrate accumulation in Uganda soils. East Afr. Agr. J. 19: 53–54.
MOORE, C. W. E., AND K. KERAITIS. 1971. Effect of nitrogen source on growth of eucalypts in sand culture. Aust. J. Bot. 19: 125–141.
MUNRO, P. E. 1966a. Inhibition of nitrite-oxidizers by roots of grass. J. Appl. Ecol. 3: 227–229.
———. 1966b. Inhibition of nitrifiers by grass root extracts. J. Appl. Ecol. 3: 231–238.
NYE, R. H., AND D. J. GREENLAND. 1960. The soil under shifting cultivation. Tech. Commun. 51. Commonwealth Bureau of Soils, Harpenden, England.
OERTLI, J. J. 1963. Effect of the form of nitrogen and *p*H on growth of blueberry plants. Agron. J. 55: 305–307.
PIPER, C. S. 1942. Soil and plant analysis. Interscience Publishers, Inc., New York.
RICE, E. L. 1964. Inhibition of nitrogen-fixing and nitrifying bacteria by seed plants. I. Ecology 45: 824–837.
———. 1965a. Inhibition of nitrogen-fixing and nitrifying bacteria by seed plants. II. Characterization and identification of inhibitors. Physiol. Plant. 18: 255–268.
———. 1965b. Inhibition of nitrogen-fixing and nitrifying bacteria by seed plants. III. Comparison of three species of *Euphorbia*. Proc. Okla. Acad. Sci. 45: 43–44.
———. 1969. Inhibition of nitrogen-fixing and nitrifying bacteria by seed plants. VI. Inhibitors from *Euphorbia supina* Raf. Physiol. Plant 22: 1175–1183.
———, AND W. T. PENFOUND. 1955. An evaluation of the variable-radius and paired-tree methods in the blackjack-post oak forest. Ecology 36: 315–320.
———, AND ———. 1959. The upland forests of Oklahoma. Ecology 40: 593–608.
RICHARDSON, H. L. 1935. The nitrogen cycle in grassland soils. Trans. 3rd Int. Congr. Soil Sci. 1: 219–221.
———. 1938. Nitrification in grassland soils: with special reference to the Rothamsted Park Grass experiment. J. Agr. Sci. 28: 73–121.
RUSSELL, E. J. 1914. The nature and amount of the fluctuations in nitrate contents of arable soils. J. Agr. Sci. 6: 50–53.
———, AND E. W. RUSSELL. 1961. Soil conditions and plant growth. 9th ed. John Wiley and Sons, New York.
SHEN, T. C. 1969. The induction of nitrate reductase and the preferential assimilation of ammonium in germinating rice seedlings. Plant Physiol. 44: 1650–1655.
SMITH, W., F. H. BORMANN, AND G. E. LIKENS. 1968. Response of chemoautotrophic nitrifiers to forest cutting. Soil Sci. 106: 471–473.
STIVEN, G. 1952. Production of antibiotic substances by the roots of a grass (*Trachypogon plumosus* [H. B. K.] Nees) and of *Pentanisia variabilis* (E. Mey.) Harv. (Rubiaceae). Nature 170: 712–713.
THERON, J. J. 1951. The influence of plants on the mineralization of nitrogen and the maintenance of organic matter in the soil. J. Agr. Sci. 41: 289–296.
VIRO, P. J. 1963. Factorial experiments on forest humus decomposition. Soil Sci. 95: 24–30.
WARREN, M. 1965. A study of soil-nutritional and other factors operating in secondary succession in highveld grassland in the neighborhood of Johannesburg. Ph.D. thesis, Univ. of Witwatersrand, Johannesburg, South Africa.
WEETMAN, G. F. 1961. The nitrogen cycle in temperate forest stands (Literature Review). Res. Note #21, Pulp and Paper Res. Inst. Can., Montreal.
WHITMAN, W. C., AND E. I. SIGGEIRSSON. 1954. Comparison of line interception and point contact methods in the analysis of mixed grass range vegetation. Ecology 35: 431–436.

10

Reprinted from *Science* **176:**914–915 (May 26, 1972)

REVEGETATION FOLLOWING FOREST CUTTING: MECHANISMS FOR RETURN TO STEADY-STATE NUTRIENT CYCLING

P. L. Marks, *Section of Ecology and Systematics, Cornell University*

F. H. Bormann, *School of Forestry, Yale University*

Because terrestrial plant communities have always been subjected to various forms of natural disturbances, such as wind storms, fires, and insect outbreaks, it is only reasonable to consider recovery from disturbance as a normal part of community maintenance and repair. Although the structural basis of recovery from disturbance has long been recognized, the functional basis is only now beginning to be understood, largely as a result of the whole ecosystem studies of nutrient cycling at the Hubbard Brook Experimental Forest in New Hampshire (*1–3*).

At Hubbard Brook, experimental clear-cutting and subsequent herbicidal spraying of a 15.6-hectare watershed-ecosystem triggered a chain of events which led to pronounced changes in ecosystem function (*3, 4*). Significant increases in decomposition rates, nitrification, streamwater concentrations of dissolved inorganics (most notably nitrate-nitrogen), total hydrologic runoff, and erosion have been reported (*2, 3, 5, 6*). The combined effect of deforestation and suppression of vegetative regeneration with herbicides caused extreme open or loose cycling of nutrients (nutrient dumping), which were flushed from the forest soils into streams where they caused eutrophication.

The cutting and herbicide experiment raises important and fundamental questions about both the general impact of disturbance on the functioning of ecosystems and the means by which such impact is attenuated by natural recovery processes. Because similar trends are produced by commercial clear-cutting practices in the Northeast (*7*), it is of considerable interest to understand these recovery processes, especially as they are related to nutrient circulation. We present data to show the importance of a fast-growing, short-lived, successional tree species, pin cherry (*Prunus pensylvanica* L.), in reducing the degradative nutrient losses found at Hubbard Brook, and discuss the implications for ecosystem stability and forestry practice. Disturbance is here defined as destruction of vegetation, or parts of vegetation, whether natural or man-induced.

Erosional and nutrient element losses from a forest ecosystem following disturbance are diminished by any form of vegetative regeneration, there being a roughly inverse relationship between the rate of regeneration and the amounts of erosional and nutrient losses

(*8*). To measure the rate of recovery following cutting, we sampled naturally occurring stands of pin cherry in different stages of recovery (1, 4, 6, and 14 years after cutting). To minimize site differences between stands, we sampled stands which were all in north-central New Hampshire, all on the same geologic formation, all subjected to the same disturbance (clear-cutting), all broadly equivalent in drainage and elevation, and, most importantly, all densely stocked with, and dominated by, the same successional species—pin cherry.

We measured the rate of recovery of the ecosystem in terms of amount of biomass, rate of biomass accumulation (by measuring net annual primary production for each stand), rate of canopy closure as indicated by leaf area index (*9*), and rate of accumulation of nutrients (nitrogen, calcium, magnesium, potassium, and sodium) in plant tissues (*10*).

Our approach to biomass and production estimation (*14*) followed harvest techniques based on the allometric relations of individual sample trees (*11, 12*). Three to five replicate chemical analyses were made for each major tissue (leaves, current twigs, older branches, dead branches, roots, stem wood, and stem bark) from each of the three older stands (4, 6, and 14 years), and included macro-Kjeldahl nitrogen and cation analyses by atomic absorption spectrophotometry (*13*). For a particular stand, weights of the different plant parts were multiplied by the appropriate nutrient concentrations to yield estimates of standing crop of each nutrient element per unit area of land surface (in grams per square meter).

Young, very dense stands of pin cherry develop rapidly (Fig. 1). Leaf area indexes equivalent to the upper end of the range for temperate deciduous forests (4.0 to 6.0) (*14*) were attained within 4 years. Reestablishment of a full canopy implies marked increases in transpiration compared to that in the disturbed condition and substantial moderation of soil temperatures during the growing season, the significance of which for nutrient loss is discussed below.

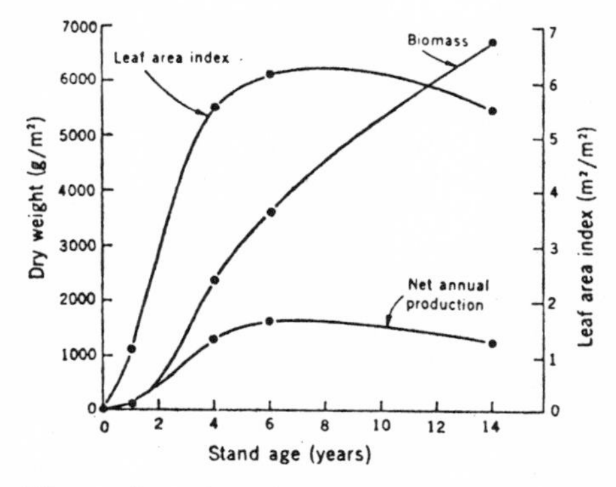

Fig. 1. Relationships of biomass, net annual production, and leaf area index to stand age.

Total net annual production, including estimates of below-ground production (*10*), for the young pin cherry stands is high (*15*) compared to that of other deciduous forests of natural origin, some of considerably older age, including the undisturbed forest at Hubbard Brook. Indeed, the value for the 6-year-old stand, 1650 g/m^2, exceeds the usual range for temperate climax forests (1200 to 1500 g/m^2) (*12, 15*). Accretion of biomass is rapid in these young, dense, successional stands.

Incorporation of nutrient elements into the developing biomass of the three older stands is also rapid, especially for nitrogen and calcium (Fig. 2). We estimate, for example, annual uptake of nitrogen in the 4- and 6-year-old stands to be about 50 percent greater (10.0 compared with 6.5 g/m^2) (*5*) than uptake in the mature, undisturbed ecosystem at Hubbard Brook. It is likely that the high rates of nutrient uptake and growth result from an adaptive capacity of this species to utilize the greater availability of water and nutrients on disturbed sites (*10*).

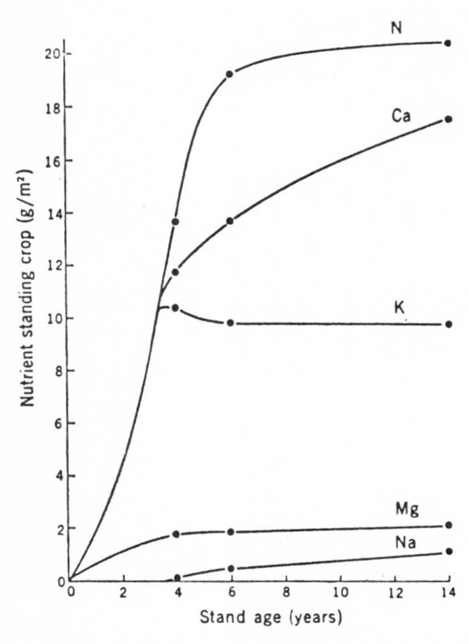

Fig. 2. Relationship of nutrient standing crop to stand age.

Our data indicate that, following severe disturbance such as clear-cutting, the growth and development of dense stands of successional species such as pin cherry may be extremely rapid. Such growth acts to minimize nutrient losses from the ecosystem. This regulation of nutrient cycling is achieved soon after disturbance by a complex interaction involving (i) channeling of water from runoff to evapotranspiration, thereby reducing erosion and nutrient loss; (ii) reduction in rates of decomposition through moderation of the microclimate during the growing season, so that the supply of soluble ions available for loss in drainage water is reduced; and (iii) simultaneous incorporation into the rapidly developing biomass of nutrients that do become available and that otherwise might be lost from the system.

The extent of nutrient loss from a forest ecosystem following clear-cutting, while varying with local conditions, such as type of vegetation, the nature of the cutting, site characteristics (slope, drainage, and so forth), and degree of soil scarification, should nevertheless decrease according to the rate of revegetation. Dense stands of fast-growing, successional species will exhibit pronounced regulation of ecosystem function soon after disturbance, the net effect of which is to move the ecosystem rapidly back toward the steady-state, stable pattern of nutrient circulation that typifies the mature forest (*5, 10, 16*).

The implication of our data is that successional species, specially adapted to exploit disturbed conditions (*10*), ought to be considered integral components of the larger ecosystem, despite the fact that they are typically absent from the terminal, climax community. Successional species have evolved in relation to the pattern of disturbance in the climax community; indeed the design of the life cycle of pin cherry, particularly the storage in soil, longevity, and germination requirements of its seeds, assures that its occurrence is geared closely into the pattern of disturbance in the large system (*10*).

Silvicultural practice that ignores the function of successional, "weed," tree species like pin cherry because they are economically undesirable, may be ecologically unsound. In the important issue of evaluating the effects of clear-cutting practices, the role of successional species in the recovery process is deserving of considerably more basic research.

[*Editor's Note:* The references and notes appear on the following page.]

References and Notes

1. G. E. Likens, F. H. Bormann, N. M. Johnson, R. S. Pierce, *Ecology* **48**, 772 (1967); N. M. Johnson, G. E. Likens, F. H. Bormann, R. S. Pierce, *Geochim. Cosmochim. Acta* **32**, 531 (1968); G. E. Likens and F. H. Bormann, in *Ecosystem Structure and Function*, J. A. Wiens, Ed. (Oregon State Univ. Press, Corvallis, 1971), p. 25; in *Interactions between Land and Water*, F. H. Whitehead, Ed. (International Association for Ecology, London, in press).
2. F. H. Bormann, G. E. Likens, D. Fisher, R. S. Pierce, *Science* **159**, 882 (1968); W. H. Smith, F. H. Bormann, G. E. Likens, *Soil Sci.* **106**, 471 (1968).
3. G. E. Likens, F. H. Bormann, N. M. Johnson, *Science* **163**, 1205 (1969); ———. D. Fisher, R. S. Pierce, *Ecol. Monogr.* **40**, 23 (1970).
4. F. H. Bormann and G. E. Likens, *Sci. Amer.* **223**, 92 (1970).
5. A. Dominski, thesis, Yale University (1971).
6. R. S. Pierce, J. W. Hornbeck, G. E. Likens, F. H. Bormann, in *Proceedings of the IASH-UNESCO Symposium on the Results of Research on Representative and Experimental Basins*, Wellington, New Zealand (1970), p. 311.
7. R. S. Pierce, C. W. Martin, C. C. Reeves, G. E. Likens, F. H. Bormann, unpublished data.
8. F. H. Bormann, G. E. Likens, J. S. Eaton, *BioScience* **19**, 600 (1969).
9. Leaf area index, the dimensionless ratio of the area of leaf surfaces (one side of blade only) above a particular area of ground surface to that area of ground surface, is a convenient measure of canopy closure.
10. P. L. Marks, thesis, Yale University (1971).
11. R. H. Whittaker, *Ecology* **42**, 177 (1961); *ibid.* **43**, 357 (1962); *ibid.* **46**, 365 (1965); G. L. Baskerville, *ibid.*, p. 867; *Forest Sci. Monogr.* 9 (1965); T. Kira and T. Shidei, *Jap. J. Ecol.* **17**, 70 (1967); T. Satoo, in *Analysis of Temperate Forest Ecosystems*, D. E. Reichle, Ed. (Springer-Verlag, New York, 1970), p. 55.
12. D. F. Westlake, *Biol. Rev. Cambridge Phil. Soc.* **38**, 385 (1963); R. H. Whittaker, *Ecology* **47**, 103 (1966); *Communities and Ecosystems* (Macmillan, London, 1970).
13. G. E. Likens and F. H. Bormann, *Yale Univ. Sch. For. Bull.* 79 (1970).
14. R. H. Whittaker and G. M. Woodwell, *Amer. J. Bot.* **54**, 931 (1967).
15. H. W. Art and P. L. Marks, in *Forest Biomass Studies*, H. E. Young, Ed. (Life Sciences and Agricultural Experiment Station, Univ. of Maine, Orono, 1971), p. 1.
16. F. H. Bormann, G. E. Likens, T. G. Siccama, R. S. Pierce, in preparation.

17. Contribution No. 42 of the Hubbard Brook Ecosystem Study. Supported by NSF grants GB-6567, GB-6742, GB-14325, and GB-14289. Fieldwork was done through the cooperation of the Northeastern Forest Experiment Station, Forest Service, U.S. Department of Agriculture, Upper Darby, Pennsylvania. We thank F. T. Ledig, G. K. Voigt, T. G. Siccama, R. S. Pierce, S. Filip, D. M. Smith, G. M. Furnival, T. Delevoryas, and H. J. Lutz for help and advice during the study, and J. S. Eaton, L. N. Miller, R. S. Pierce, T. G. Siccama, and R. H. Whittaker for helpful comments on the manuscript.

13 March 1972

Part IV

INTERPRETATION

Editor's Comments on Papers 11 Through 16

As mentioned earlier, there are two sets of theoretical ideas which are related to the concept of succession. The first, the nature of the community, we have already touched on since it is so central to the concept. The second involves the interpretation of succession as a process in communities. In this part we will consider these ideas further.

The problem of definition and identification of ecological communities has been recurrent throughout the history of ecology. All persons who live in natural landscapes develop some sort of classification system and even for primitive peoples these systems may be exceedingly complex and detailed. A forest is distinguished from a grassland, a pine forest from a hardwood forest and so forth. The problem comes from specifying the boundaries of these units in space and in time. For example, the plant species that comprise a forest usually can tolerate and grow in nonforest environments, although they may not be as robust nor as common there. Thus, as one studies a transect from a forest into a nonforested habitat, an abrupt boundary between forest and nonforest usually is not observed. Rather, forest species gradually become less common, nonforest species more common, while yet other species seem to thrive or be rare in both environments. These communities form a continuum which usually is gradual over the discontinuities of the environment. We find a similar problem in time. Change is gradual, and it usually is difficult to determine when a pine forest becomes a hardwood forest, because pines may persist into the hardwood forest while hardwood species may be present almost from the beginning of the sere.

Clements's concept will be described first because it has been so important in the development of ecology. Frederick Clements [1870–1945] grew up in the grasslands of Nebraska, and his observations of vegetation throughout his boyhood and when he served as Professor of Botany at the University of Nebraska led him to a series of arresting ideas which dominated American ecology for many years. The following short excerpt (Paper 11) from his book *Plant Succession: An Analysis of Development of Vegetation,* describes the vegetation of the region as an organism with a life history comparable to an individual plant. Clements's concept of the vegetation as a superorganism is appealing since we can readily develop the analogy from our personal knowledge of individuals and it is in the spirit of the age as we discussed earlier. The formation "arises, grows, matures, and dies." Thus, the community and formation have unique emergent properties which are greater than the sum of the properties of the parts. Clements's concept of superorganism was carried much further by others, notably by Phillips (1934, 1935a,b), and provides a biological basis to Jans Christian Smuts's philosophy of holism (Smuts, 1926).

An alternative viewpoint is taken by Henry A. Gleason [1882–1975], former Head Curator of the New York Botanical Garden, who stresses that the dynamics of the community can be understood only as the sum of the dynamics of its component species (Paper 12). He concludes his paper with the following remark:

> Plant associations, the most conspicuous illustration of the space relations of plants, depend solely on the coincidence of environmental selection and migration over an area of recognizable extent and usually for a time of considerable duration. A rigid definition of the scope or extent of the association is impossible, and a logical classification of associations into larger groups, or into successional series, has not yet been achieved. [p. 126]

Gleason and Clements represent two dramatically opposite points of view on the nature of the community. As we shall see, the arguments still persist, yet today most ecologists probably accept Gleason's principle of species individuality while also accepting the need for studying communities as a unit. Many ecologists feel that there is little question that communities and ecosystems are real functional systems that have some emergent properties, such as development. Yet it also seems true that, as argued by Gleason, communities often intergrade continuously and their species are not bound together into clearly defined units.

Our lack of certainty about the nature of communities pervades other features of successional study. For example, it is fundamental to another important question—when does succession end? This problem of termination of succession has equally generated considerable discussion and argument. The end of succession is termed the climax, which has a special ecological meaning as applying to the stable endpoint of a succession.

If we could visit a pristine landscape where the vegetation on the uplands was not influenced by man's activities, then we would probably observe a variety of communities. The communities along rivers would differ from those on hills, north slopes might differ from south slopes and so forth. Clements would assert that these different communities would ultimately converge on a single climax pattern or community characteristic of the regional climate; this community is termed the monoclimax. Communities under the influence of environmental factors other than climate (such as fire, soil, man) Clements considers temporary and identifies by special names in his elaborate scheme of classification described in Paper 13.

These Clementsian ideas of the analogy between community succession and development of the individual, and the convergence of all communities in a region to a monoclimax under the influence of regional climate are compelling because of their logical simplicity. They also parallel Davis's concept of cycles of erosion discussed earlier in part one. The reader will note that Clements's climax is analogous to Davis's geological peneplain. But unfortunately, the Clementsian concepts of climax and convergence, while logically interesting, do not often correspond to our observations in nature. Robert Whittaker, coeditor of a volume on the niche in this series (Whittaker and Levin, 1975), and Professor at Cornell University, has probably made the most thorough analysis of the data on climax.

Whittaker's long paper (Paper 14) can be contrasted with Clements's equally long discussion. Whittaker carefully evaluates the various concepts and concludes that the climax concept, while partly subjective and relative, is useful. He presents the idea of the climax as a steady state with both dynamic-functional features, such as productivity, and static structural features, such as species composition. Whittaker offers us a thorough and useful analysis of this problem, and we can leave it at this point. Communities probably never will be described with the same precision as chemical compounds. Communities as real natural units have properties unique to them, but these properties have no mystical characteristics. Rather, they represent a complex dynamic system characterized by the connections, interactions and attributes of the individual components of that community. Ecologists currently are attempting to describe more rigorously this network of relationships.

Thus, the Clementsian concepts of a community as a superorganism and the convergence of successions toward a monoclimax are no longer of much interest to ecologists, except in the historical sense. Rather, ecologists have increasingly taken a systems approach to the natural world. One concept which has been especially useful in this regard has been that of the ecosystem. The word was coined by the British ecologist, A. G. Tansley, who wished to identify a system composed of living beings and their environment. This assemblage was called an ecological system or ecosystem. The concept was popularized by Eugene P. Odum in the first edition of his textbook, *Fundamentals of Ecology,* in 1953. Odum (like Whittaker later), as a student at the University of Illinois, was trained in the midwestern tradition of ecology which developed from Cowles, Clements, Cooper, Shelford and others. Odum's interest in ecosystems came partly through his comparative studies of dynamic old-field successional systems, mentioned in part two, and the very stable marsh communities on the Georgia coastline. From these comparative types of studies, Odum developed a list of attributes of successional, and of stable, mature climax ecosystems, which expands that of Cowles (Paper 2) in part two. Paper 15 describes change in the characteristics of these communities in terms of a strategy. In Odum's conceptual framework, ecosystems may have properties which are unpredicted from a knowledge of the parts. Thus, in Odum's terms, an ecosystem may have a strategy to maintain or recover stability if disturbed.

The conceptual approach to succession of Odum and others is analysed in a recent review by two members of the staff of the Massachusetts Audubon Society, William H. Drury and Ian C. T. Nisbet, in the following paper. Drury and Nisbet (Paper 16) restate many of Gleason's arguments that:

> most of the phenomena of succession should be understood as resulting from the differential growth, differential survival and perhaps dispersal of species adapted to grow at different points on stress gradients. [P. 360]

They conclude:

> we suggest that a complete theory of vegetative succession should be sought at the organismic, physiological or cellular level, and not in emergent properties of populations and communities. [P. 362]

I have chosen to end this part with the papers of Odum and Drury and Nisbet for several reasons. First, they show that the argument over the individualistic concept of succession of a community and the holistic concept continues to the present. Second, the correlation of attributes of ecosystems with succession seems to lead to a series of useful concepts and to hypotheses that should be tested in the field or laboratory. While Odum's long list of correlates are not usually buttressed by series of observations in a variety of ecosystems and, therefore, represent only concepts of what the characteristics of these systems might be, they nevertheless suggest a great number of further questions. Odum and Drury and Nisbet's statements effectively counterbalance one another in their emphasis on orderly development versus complexity, irregularity and diversity in succession. The proper balance in this polarity of order and disorder remains to be struck; I will suggest to the reader only that we keep in view both the real complexities of successions and the common features and implications of successional processes.

REFERENCES

Phillips, J. 1934. Succession, Development, the Climax and the Complex Organism: An Analysis of Concepts. Part I. *Ecology* **22:**554–571.

———. 1935a. Succession, Development, the Climax, and the Complex Organism: An Analysis of Concepts. Part II. *Ecology* **23:**210–246.

———. 1935b. Succession, Development, the Climax, and the Complex Organism: An Analysis of Concepts. Part III. *Ecology* **23:**488–508.

Smuts, J. C. 1926. *Holism and Evolution.* New York: Macmillan.

Whittaker, R. H., and S. A. Levin, eds. 1975. *Niche: Theory and Application.* Benchmark Papers in Ecology Series, vol. 3. Stroudsburg, Pa.: Dowden, Hutchinson & Ross.

11

Reprinted from pp. 3–4 of *Carnegie Inst. Washington Publ. No. 242*:1–512 (1916)

PLANT SUCCESSION: AN ANALYSIS OF THE DEVELOPMENT OF VEGETATION

F. E. Clements

I. CONCEPT AND CAUSES OF SUCCESSION.

The formation an organism.—The developmental study of vegetation necessarily rests upon the assumption that the unit or climax formation is an organic entity (Research Methods, 199). As an organism the formation arises, grows, matures, and dies. Its response to the habitat is shown in processes or functions and in structures which are the record as well as the result of these functions. Furthermore, each climax formation is able to reproduce itself, repeating with essential fidelity the stages of its development. The life-history of a formation is a complex but definite process, comparable in its chief features with the life-history of an individual plant.

Universal occurrence of succession.—Succession is the universal process of formation development. It has occurred again and again in the history of every climax formation, and must recur whenever proper conditions arise. No climax area lacks frequent evidence of succession, and the greater number present it in bewildering abundance. The evidence is most obvious in active physiographic areas, dunes, strands, lakes, flood-plains, bad lands, etc., and in areas disturbed by man. But the most stable association is never in complete equilibrium, nor is it free from disturbed areas in which secondary succession is evident. An outcrop of rock, a projecting boulder, a change in soil or in exposure, an increase or decrease in the water-content or the light intensity, a rabbit-burrow, an ant-heap, the furrow of a plow, or the tracks worn by wheels, all these and many others initiate successions, often short and minute, but always significant. Even where the final community seems most homogeneous and its factors uniform, quantitative study by quadrat and instrument reveals a swing of population and a variation in the controlling factors. Invisible as these are to the ordinary observer, they are often very considerable, and in all cases are essentially materials for the study of succession. In consequence, a floristic or physiognomic study of an association, especially in a restricted area, can furnish no trustworthy conclusions as to the prevalence of succession. The latter can be determined only by investigation which is intensive in method and extensive in scope.

Viewpoints of succession.—A complete understanding of succession is possible only from the consideration of various viewpoints. Its most striking feature lies in the movement of populations, the waves of invasion, which rise and fall through the habitat from initiation to climax. These are marked by a corresponding progression of vegetation forms or phyads, from lichens and mosses to the final trees. On the physical side, the fundamental view is that which deals with the forces which initiate succession and the reactions which maintain it. This leads to the consideration of the responsive processes or functions which characterize the development, and the resulting structures, communities, zones, alternes, and layers. Finally, all of these viewpoints are summed up in that which regards succession as the growth or development and the reproduction of a complex organism. In this larger aspect succession includes both the ontogeny and the phylogeny of climax formations.

Succession and sere.—In the thorough analysis of succession it becomes evident that the use of the term in both a concrete and an abstract sense tends to inexactness and uncertainty. With the recognition of new kinds of succession it seems desirable to restrict the word more and more to the phenomenon itself and to employ a new term for concrete examples of it. In consequence, a word has been sought which would be significant, short, euphonic, and easy of combination. These advantages are combined in the word *sere*, from a root common to both Latin and Greek, and hence permitting ready composition in either. The root *ser-* shows its meaning in Latin *sero*, join, connect; *sertum*, wreath; *series*, joining or binding together, hence sequence, course, succession, lineage. In Greek, it occurs in *εἴρω*, to fasten together in a row, and in *σειρά, σηρά*, rope, band, line, lineage. *Sere* is essentially identical with *series*, but possesses the great advantage of being distinctive and of combining much more readily, as in *cosere*, *geosere*, etc.

Sere and cosere.—A sere is a unit succession. It comprises the development of a formation from the appearance of the first pioneers through the final or climax stage. Its normal course is from nudation to stabilization. All concrete successions are seres, though they may differ greatly in development and thus make it necessary to recognize various kinds, as is shown later. On the other hand, a unit succession or sere may recur two or more times on the same spot. Classical examples of this are found in moors and dunes, and in forest burns. A series of unit successions results, in which the units or seres are identical or related in development. They consist normally of the same stages and terminate in the same climax, and hence typify the reproductive process in the formation. Such a series of unit successions, *i. e.*, of seres, in the same spot constitutes an organic entity. For this, the term *consere* or *cosere* (*cum*, together, *sere; consero*, bind into a whole) is proposed, in recognition of the developmental bond between the individual seres. Thus, while the sere is the developmental unit, and is purely ontogenetic, the cosere is the sum of such units throughout the whole life-history of the climax formation, and is hence phylogenetic in some degree. Coseres are likewise related in a developmental series, and thus may form larger groups, eoseres, etc., as indicated in the later discussion (plate 1, A, B).

Processes in succession.—The development of a climax formation consists of several essential processes or functions. Every sere must be initiated, and its life-forms and species selected. It must progress from one stage to another, and finally must terminate in the highest stage possible under the climatic conditions present. Thus, succession is readily analyzed into initiation, selection, continuation, and termination. A complete analysis, however, resolves these into the basic processes of which all but the first are functions of vegetation, namely, (1) nudation, (2) migration, (3) ecesis, (4) competition, (5) reaction, (6) stabilization. These may be successive or interacting. They are successive in initial stages, and they interact in most complex fashion in all later ones. In addition, there are certain cardinal points to be considered in every case. Such are the direction of movement, the stages involved, the vegetation forms or materials, the climax, and the structural units which result.

[*Editor's Note:* Material has been omitted at this point.]

12

Reprinted from *Torrey Bot. Club Bull.* **53**:7–26 (1962)

The individualistic concept of the plant association *

H. A. Gleason

The continued activity of European ecologists, and to a somewhat smaller extent of American ecologists as well, in discussing the fundamental nature, structure, and classification of plant associations, and their apparently chronic inability to come to any general agreement on these matters, make it evident that the last word has not yet been said on the subject. Indeed, the constant disagreement of ecologists, the readiness with which flaws are found by one in the proposals of another, and the wide range of opinions which have been ably presented by careful observers, lead one to the suspicion that possibly many of them are somewhat mistaken in their concepts, or are attacking the problem from the wrong angle.

It is not proposed to cite any of the extensive recent literature on these general subjects, since it is well known to all working ecologists. Neither is it necessary to single out particular contributions for special criticism, nor to point out what may appear to us as errors in methods or conclusions.

It is a fact, as Dr. W. S. Cooper has brought out so clearly in a manuscript which he has allowed me to read, and which will doubtless be in print before this, that the tendency of the human species is to crystallize and to classify his knowledge; to arrange it in pigeon-holes, if I may borrow Dr. Cooper's metaphor. As accumulation of knowledge continues, we eventually find facts that will not fit properly into any established pigeon-hole. This should at once be the sign that possibly our original arrangement of pigeon-holes was insufficient and should lead us to a careful examination of our accumulated data. Then we may conclude that we would better demolish our whole system of arrangement and classification and start anew with hope of better success.

Is it not possible that the study of synecology suffers at the present time from this sort of trouble? Is it not conceivable that, as the study of plant associations has progressed from its originally simple condition into its present highly organized and complex state, we have attempted to arrange all our facts in ac-

* Contributions from The New York Botanical Garden, No. 279.

cordance with older ideas, and have come as a result into a tangle of conflicting ideas and theories?

No one can doubt for a moment that there is a solid basis of fact on which to build our study of synecology, or that the study is well worth building. It is the duty of the botanist to translate into intelligible words the various phenomena of plant life, and there are few phenomena more apparent than those of their spatial relations. Plant associations exist; we can walk over them, we can measure their extent, we can describe their structure in terms of their component species, we can correlate them with their environment, we can frequently discover their past history and make inferences about their future. For more than a century a general progress in these features of synecology can be traced.

It has been, and still is, the duty of the plant ecologist to furnish clear and accurate descriptions of these plant communities, so that by them the nature of the world's vegetation may be understood. Whether such a description places its emphasis chiefly on the general appearance of the association, on a list of its component species, on its broader successional relations, or on its gross environment, or whether it enters into far greater detail by use of the quadrat method, statistical analysis,[1] or exact environometry, it nevertheless contributes in every case to the advancement of our understanding of each association in detail and of vegetation in all its aspects in general.

It is only natural that we should tend to depart from the various conclusions which we have reached by direct observation or experiment, and to attempt other more general deductions as well. So we invent special terms and methods for indicating the differences between associations and the variation of the plant life within a single community. We draw conclusions for ourselves, and attempt to lay down rules for others as to ways and means of defining single associations, by character species, by statistical studies, by environmental relations, or by successional history. We attempt to classify associations, as individual

[1] Pavillard has cast serious doubt on the efficiency of the statistical method in answering questions of synecology. His argument, based solely on European conditions, needs of course no reply from America, but it may properly be pointed out that the intimate knowledge of vegetational structure obtained in this way may easily lead to a much fuller appreciation of synecological structure, entirely aside from any merits of the actual statistical results.

examples of vegetation, into broader groups, again basing our methods on various observable features and arriving accordingly at various results. We even enter the domain of philosophy, and speculate on the fundamental nature of the association, regard it as the basic unit of vegetation, call it an organism, and compare different areas of the same sort of vegetation to a species.

The numerous conclusions in synecology which depend directly upon observation or experiment are in the vast majority of cases entirely dependable. Ecologists are trained to be accurate in their observations, and it is highly improbable that any have erred purposely in order to substantiate a conclusion not entirely supported by facts. But our various theories on the fundamental nature, definition, and classification of associations extend largely beyond the bounds of experiment and observation and represent merely abstract extrapolations of the ecologist's mind. They are not based on a pure and rigid logic, and suffer regularly from the vagaries and errors of human reason. A geneticist can base a whole system of evolution on his observations of a single species: ecologists are certainly equally gifted with imagination, and their theories are prone to surpass by far the extent warranted by observation.

Let us then throw aside for the moment all our pre-conceived ideas as to the definition, fundamental nature, structure, and classification of plant associations, and examine step by step some of the various facts pertinent to the subject which we actually know. It will not be necessary to illustrate them by reference to definite vegetational conditions, although a few instances will be cited merely to make our meaning clear. Other illustrations will doubtless occur to every reader from his own field experience.

We all readily grant that there are areas of vegetation, having a measurable extent, in each of which there is a high degree of structural uniformity throughout, so that any two small portions of one of them look reasonably alike. Such an area is a plant association, but different ecologists may disagree on a number of matters connected with such an apparently simple condition. More careful examination of one of these areas, especially when conducted by some statistical method, will show that the uniformity is only a matter of degree, and that two sample quadrats with precisely the same structure can scarcely be discovered.

Consequently an area of vegetation which one ecologist regards as a single association may by another be considered as a mosaic or mixture of several, depending on their individual differences in definition. Some of these variations in structure (if one takes the broader view of the association) or smaller associations (if one prefers the narrower view) may be correlated with differences in the environment. For example, the lichens on a tree-trunk enjoy a different environment from the adjacent herbs growing in the forest floor. A prostrate decaying log is covered with herbs which differ from the ground flora in species or in relative numbers of individuals of each species. A shallow depression in the forest, occupied by the same species of trees as the surroundings, may support several species of moisture-loving herbs in the lower stratum of vegetation. In other cases, the variations in vegetational structure may show no relation whatever to the environment, as in the case of a dense patch of some species which spreads by rhizomes and accordingly comes to dominate its own small area. The essential point is that precise structural uniformity[2] of vegetation does not exist, and that we have no general agreement of opinion as to how much variation may be permitted within the scope of a single association.

In our attempts to define the limits of the association, we have but two actually observable features which may be used as a basis, the environment and the vegetation. Logically enough, most ecologists prefer the latter, and have developed a system based on character-species. In northern latitudes, and particularly in glaciated regions, where most of this work has been done, there is a wide diversity in environment and a comparatively limited number of species in the flora. A single association is therefore occupied by few species, with large numbers of individuals of each, and it has not been difficult to select from most associations a set of species which are not only fairly common and abundant, but which are strictly limited to the one association. But in many parts of the tropics, where diversity of environment has been reduced to a minimum by the practical completion of most physiographic processes and by the long-continued cumulation of plant reactions, and where the flora is

[2] It has often occurred to the writer that much of the structural variation in an association would disappear if those taxonomic units which have the same vegetational form and behavior could be considered as a single ecological unit.

extraordinarily rich in species, such a procedure is impracticable or even impossible. Where a single hectare may contain a hundred species of trees, not one of which can be found in an adjacent hectare, where a hundred quadrats may never exhibit the same herbaceous species twice, it is obvious that the method of characteristic species is difficult or impracticable.

It is also apparent that different areas of what are generally called the same association do not always have precisely the same environment. A grove of *Pinus Strobus* on soil formed from decomposed rocks in the eastern states, a second on the loose glacial sands of northern Michigan, and a third on the sandstone cliffs of northern Illinois are certainly subject to different environmental conditions of soil. An association of prairie grass in Illinois and another in Nebraska undoubtedly have considerable differences in rainfall and available water. A cypress swamp in Indiana has a different temperature environment from one in Florida.

Two environments which are identical in regard to physiography and climate may be occupied by entirely different associations. It is perfectly possible to duplicate environments in the Andes of southern Chile and in the Cascade Mountains of Oregon, yet the plant life is entirely different. Duplicate environments may be found in the deserts of Australia and of Arizona, and again have an entirely different assemblage of species. Alpine summits have essentially the same environment at equal altitudes and latitudes throughout the world, apart from local variations in the component rock, and again have different floras. It seems apparent, then, that environment can not be used as a means of defining associations with any better success than the vegetation.

At the margin of an association, it comes in contact with another, and there is a transition line or belt between them. In many instances, particularly where there is an abrupt change in the environment, this transition line is very narrow and sharply defined, so that a single step may sometimes be sufficient to take the observer from one into the other. In other places, especially where there is a very gradual transition in the environment, there is a correspondingly wide transition in the vegetation. Examples of the latter condition are easily found in any arid mountain region. The oak forests of the southern Coast Range in California in many places descend upon the grass-covered foothills by a wide transition zone in which the trees become very gradually fewer and farther apart until they ultimately disappear completely. In Utah, it may be miles from the association of desert shrubs on the lower elevations across a mixture of shrubs and juniper before the pure stands of juniper are reached on the higher altitudes. It is obvious, therefore, that it is not always possible to define with accuracy the geographical boundaries of an association and that actual mixtures of associations occur.

Such transition zones, whether broad or narrow, are usually populated by species of the two associations concerned, but instances are not lacking of situations in which a number of species seem to colonize in the transition zone more freely than in either of the contiguous associations. Such is the case along the contact between prairie and forest, where many species of this type occur, probably because their optimum light requirements are better satisfied in the thin shade of the forest border than in the full sun of the prairie or the dense shade of the forest. Measured by component species such a transition zone rises almost to the dignity of an independent association.

Species of plants usually associated by an ecologist with a particular plant community are frequently found within many other types of vegetation. A single boulder, partly exposed above the ground at the foot of the Rocky Mountains in Colorado, in the short-grass prairie association, may be marked by a single plant of the mountain shrub *Cercocarpus*. In northern Michigan, scattered plants of the moisture-loving *Viburnum cassinoides* occur in the xerophytic upland thickets of birch and aspen. Every ecologist has seen these fragmentary associations, or instances of sporadic distribution, but they are generally passed by as negligible exceptions to what is considered a general rule.

There are always variations in vegetational structure from year to year within every plant association. This is exclusive of mere periodic variations from season to season, or aspects, caused by the periodicity of the component species. Slight differences in temperature or rainfall or other environmental factors may cause certain species to increase or decrease conspicuously in number of individuals, or others to vary in their vigor or luxuriance. Coville describes, in this connection, the remarkable variation in size of an *Amaranthus* in the Death Valley, which was three meters high in a year of abundant rainfall, and its progeny only a decimeter high in the following year of drought.

The duration of an association is in general limited. Sooner or later each plant community gives way to a different type of vegetation, constituting the phenomenon known as succession. The existence of an association may be short or long, just as its superficial extent may be great or small. And just as it is often difficult and sometimes impossible to locate satisfactorily the boundaries of an association in space, so is it frequently impossible to distinguish accurately the beginning or the end of an association in time. It is only at the center of the association, both geographical and historical, that its distinctive character is easily recognizable. Fortunately for ecology, it commonly happens that associations of long duration are also wide in extent. But there are others, mostly following fires or other unusual disturbances of the original vegetation, whose existence is so limited, whose disappearance follows so closely on their origin, that they scarcely seem to reach at any time a condition of stable equilibrium, and their treatment in any ecological study is difficult. The short-lived communities bear somewhat the same relation to time-distribution as the fragmentary associations bear to space-distribution. If our ecological terminology were not already nearly saturated, they might be termed ephemeral associations.

Now, when all these features of the plant community are considered, it seems that we are treading upon rather dangerous ground when we define an association as an area of uniform vegetation, or, in fact, when we attempt any definition of it. A community is frequently so heterogeneous as to lead observers to conflicting ideas as to its associational identity, its boundaries may be so poorly marked that they can not be located with any degree of accuracy, its origin and disappearance may be so gradual that its time-boundaries can not be located; small fragments of associations with only a small proportion of their normal components of species are often observed; the duration of a community may be so short that it fails to show a period of equilibrium in its structure.

A great deal has been said of the repetition of associations on different stations over a considerable area. This phenomenon is striking, indeed, and upon it depend our numerous attempts to classify associations into larger groups. In a region of numerous glacial lakes, as in parts of our northeastern states, we find lake after lake surrounded by apparently the same communities, each of them with essentially the same array of species in about the same numerical proportions. If an ecologist had crossed Illinois from east to west prior to civilization, he would have found each stream bordered by the same types of forest, various species of oaks and hickories on the upland, and ash, maple, and sycamore in the alluvial soil nearer the water. But even this idea, if carried too far afield, is found to be far from universal. If our study of glacial lakes is extended to a long series, stretching from Maine past the Great Lakes and far west into Saskatchewan, a very gradual but nevertheless apparent geographical diversity becomes evident, so that the westernmost and easternmost members of the series, while still containing some species in common, are so different floristically that they would scarcely be regarded as members of the same association. If one examines the forests of the alluvial floodplain of the Mississippi River in southeastern Minnesota, that of one mile seems to be precisely like that of the next. As the observer continues his studies farther down stream, additional species very gradually appear, and many of the original ones likewise very gradually disappear. In any short distance these differences are so minute as to be negligible, but they are cumulative and result in an almost complete change in the flora after several hundred miles.

No ecologist would refer the alluvial forests of the upper and lower Mississippi to the same associations, yet there is no place along their whole range where one can logically mark the boundary between them. One association merges gradually into the next without any apparent transition zone. Nor is it necessary to extend our observations over such a wide area to discover this spatial variation in ecological structure. I believe no one has ever doubted that the beech-maple forest of northern Michigan constitutes a single association-type. Yet every detached area of it exhibits easily discoverable floristic peculiarities, and even adjacent square miles of a single area differ notably among themselves, not in the broader features, to be sure, but in the details of floristic composition which a simple statistical analysis brings out. In other words, the local variation in structure of any association merges gradually into the broader geographical variation of the association-type.

This diversity in space is commonly overlooked by ecologists, most of whom of necessity limit their work to a comparatively small area, not extensive enough to indicate that the small observed floristic differences between associations may be of much significance or that this wide geographical variation is actually in operation. Yet it makes difficult the exact definition of any association-type, except as developed in a restricted locality, renders it almost impossible to select for study a typical or average example of a type, and in general introduces complexities into any attempt to classify plant associations.

What have we now as a basis for consideration in our attempts to define and classify associations? In the northeastern states, we can find many sharply marked communities, capable of fairly exact location on a map. But not all of that region can be thus divided into associations, and there are other regions where associations, if they exist at all in the ordinary sense of the word, are so vaguely defined that one does not know where their limits lie and can locate only arbitrary geographic boundaries. We know that associations vary internally from year to year, so that any definition we may make of a particular community, based on the most careful analysis of the vegetation, may be wrong next year. We know that the origin and disappearance of some are rapid, of others slow, but we do not always know whether a particular type of vegetation is really an association in itself or represents merely the slow transition stage between two others. We know that no two areas, supposed to represent the same association-type, are exactly the same, and we do not know which one to accept as typical and which to assume as showing the effects of geographical variation. We find fragmentary associations, and usually have no solid basis for deciding whether they are mere accidental intruders or embryonic stages in a developing association which may become typical after a lapse of years. We find variation of environment within the association, similar associations occupying different environments, and different associations in the same environment. It is small wonder that there is conflict and confusion in the definition and classification of plant communities. Surely our belief in the integrity of the association and the sanctity of the association-concept must be severely shaken. Are we not justified in coming to the general conclusion, far removed from the prevailing opinion, that an association is not an organism, scarcely even a vegetational unit, but merely a *coincidence?*

This question has been raised on what might well be termed negative evidence. It has been shown that the extraordinary variability of the areas termed associations interferes seriously with their description, their delimitation, and their classification. Can we find some more positive evidence to substantiate the same idea? To do this, we must revert to the individualistic concept of the development of plant communities, as suggested by me in an earlier paper.[3]

As a basis for the presentation of the individualistic concept of the plant association, the reader may assume for illustration any plant of his acquaintance, growing in any sort of environment or location. During its life it produces one or more crops of seeds, either unaided or with the assistance of another plant in pollination. These seeds are endowed with some means of migration by which they ultimately come to rest on the ground at a distance from the parent plant. Some seeds are poorly fitted for migration and normally travel but a short distance; others are better adapted and may cover a long distance before coming to rest. All species of plants occasionally profit by accidental means of dispersal, by means of which they traverse

[3] I may frankly admit that my earlier ideas of the plant association were by no means similar to the concept here discussed. Ideas are subject to modification and change as additional facts accumulate and the observer's geographical experience is broadened. An inkling of the effect of migration on the plant community appeared as early as 1903 and 1904 (Bull. Illinois State Lab. Nat. Hist. **7:** 189.) My field work of 1908 covered a single general type of environment over a wide area, and was responsible for still more of my present opinions (Bull. Illinois State Lab. Nat. Hist. **9:** 35–42). Thus we find such statements as the following: "No two areas of vegetation are exactly similar, either in species, the relative numbers of individuals of each, or their spatial arrangement" (l. c. 37), and again: "The more widely the different areas of an association are separated, the greater are the floral discrepancies. . . . Many of these are the results of selective migration from neighboring associations, so that a variation in the general nature of the vegetation of an area affects the specific structure of each association" (l. c. 41). Still further experience led to my summary of vegetational structure in 1917 (Bull. Torrey Club **44:** 463–481), and the careful quantitative study of certain associations from 1911 to 1923 produced the unexpected information that the distribution of species and individuals within a community followed the mathematical laws of probability and chance (Ecology **6:** 66–74).

distances far in excess of their average journey. Sometimes these longer trips may be of such a nature that the seed is rendered incapable of germination, as in dispersal by currents of salt water, but in many cases they will remain viable. A majority of the seeds reach their final stopping-point not far from the parent, comparatively speaking, and only progressively smaller numbers of them are distributed over a wider circle. The actual number of seeds produced is generally large, or a small number may be compensated by repeated crops in successive years. The actual methods of dispersal are too well known to demand attention at this place.

For the growth of these seeds a certain environment is necessary. They will germinate between folds of paper, if given the proper conditions of light, moisture, oxygen, and heat. They will germinate in the soil if they find a favorable environment, irrespective of its geographical location or the nature of the surrounding vegetation. Herein we find the *crux* of the question. The plant individual shows no physiological response to geographical location or to surrounding vegetation *per se*, but is limited to a particular complex of environmental conditions, which may be correlated with location, or controlled, modified, or supplied by vegetation. If a viable seed migrates to a suitable environment, it germinates. If the environment remains favorable, the young plants will come to maturity, bear seeds in their turn, and serve as further centers of distribution for the species. Seeds which fall in unfavorable environments do not germinate, eventually lose their viability and their history closes.

As a result of this constant seed-migration, every plant association is regularly sowed with seeds of numerous extra-limital species, as well as with seeds of its own normal plant population. The latter will be in the majority, since most seeds fall close to the parent plant. The seeds of extra-limital species will be most numerous near the margin of the association, where they have the advantage of proximity to their parent plants. Smaller numbers of fewer species will be scattered throughout the association, the actual number depending on the distance to be covered, and the species represented depending on their means of migration, including the various accidents of dispersal. This thesis needs no argument in its support. The practical universality of seed dispersal is known to every botanist as a matter of common experience.

An exact physiological analysis of the various species in a single association would certainly show that their optimal environments are not precisely identical, but cover a considerable range. At the same time, the available environment tends to fluctuate from year to year with the annual variations of climate and with the accumulated reactionary effects of the plant population. The average environment may be near the optimum for some species, near the physiological limit of others, and for a third group may occasionally lie completely outside the necessary requirements. In the latter case there will result a group of evanescent species, variable in number and kind, depending on the accidents of dispersal, which may occasionally be found in the association and then be missing for a period of years. This has already been suggested by the writer as a probable explanation of certain phenomena of plant life on mountains, and was also clearly demonstrated by Dodds, Ramaley, and Robbins in their studies of vegetation in Colorado. In the first and second cases, the effect of environmental variation toward or away from the optimum will be reflected in the number of individual plants and their general luxuriance. On the other hand, those species which are limited to a single type of plant association must find in that and in that only the environmental conditions necessary to their life, since they have certainly dispersed their seeds many times into other communities, or else be so far removed from other associations of similar environment that their migration thence is impossible.

Nor are plants in general, apart from these few restricted species, limited to a very narrow range of environmental demands. Probably those species which are parasitic or which require the presence of a certain soil-organism for their successful germination and growth are the most highly restricted, but for the same reason they are generally among the rarest and most localized in their range. Most plants can and do endure a considerable range in their environment.

With the continuance of this dispersal of seeds over a period of years, every plant association tends to contain every species of the vicinity which can grow in the available environment. Once a species is established, even by a single seed-bearing plant, its further spread through the association is hastened, since it no longer needs to depend on a long or accidental migration, and this spread is continued until the species is eventually distributed throughout the area of the association. In general, it may be considered that, other things being equal, those species of wide extent through an association are those of early introduction which have had ample time to complete their spread, while those of localized or sporadic distribution are the recent arrivals which have not yet become completely established.

This individualistic standpoint therefore furnishes us with an explanation of several of the difficulties which confront us in our attempts to diagnose or classify associations. Heterogeneity in the structure of an association may be explained by the accidents of seed dispersal and by the lack of time for complete establishment. Minor differences between neighboring associations of the same general type may be due to irregularities in immigration and minor variations in environment. Geographical variation in the floristics of an association depends not alone on the geographical variation of the environment, but also on differences in the surrounding floras, which furnish the immigrants into the association. Two widely distant but essentially similar environments have different plant associations because of the completely different plant population from which immigrants may be drawn.

But it must be noted that an appreciation of these conditions still leaves us unable to recognize any one example of an association-type as the normal or typical. Every association of the same general type has come into existence and had its structure determined by the same sort of causes; each is independent of the other, except as it has derived immigrants from the other; each is fully entitled to be recognized as an association and there is no more reason for regarding one as typical than another. Neither are we given any method for the classification of associations into any broader groups.

Similar conditions obtain for the development of vegetation in a new habitat. Let us assume a dozen miniature dunes, heaped up behind fragments of driftwood on the shore of Lake Michigan. Seeds are heaped up with the sand by the same propelling power of the wind, but they are never very numerous and usually of various species. Some of them germinate, and the dozen embryonic dunes may thenceforth be held by as many different species of plants. Originally the environment of the dunes was identical and their floristic difference is due solely to the chances of seed dispersal. As soon as the plants have developed, the environment is subject to the modifying action of the plant, and small differences between the different dunes appear. These are so slight that they are evidenced more by the size and shape of the dune than by its flora, but nevertheless they exist. Additional species gradually appear, but that is a slow process, involving not only the chance migration of the seed to the exact spot but also its covering upon arrival. It is not strange that individuals are few and that species vary from one dune to another, and it is not until much later in the history of each dune, when the ground cover has become so dense that it affects conditions of light and soil moisture, and when decaying vegetable matter is adding humus to the sand in appreciable quantities, that a true selective action of the environment becomes possible. After that time permanent differences in the vegetation may appear, but the early stages of dune communities are due to chance alone. Under such circumstances, how can an ecologist select character species or how can he define the boundaries of an association? As a matter of fact, in such a location the association, in the ordinary sense of the term, scarcely exists.

Assume again a series of artificial excavations in an agricultural region, deep enough to catch and retain water for most or all of the summer, but considerably removed from the nearest areas of natural aquatic vegetation. Annually the surrounding fields have been ineffectively planted with seeds of *Typha* and other wind-distributed hydrophytes, and in some of the new pools *Typha* seeds germinate at once. Water-loving birds bring various species to other pools. Various sorts of accidents conspire to the planting of all of them. The result is that certain pools soon have a vegetation of *Typha latifolia*, others of *Typha angustifolia*, others of *Scirpus validus;* plants of *Iris versicolor* appear in one, of *Sagittaria* in another, of *Alisma* in a third, of *Juncus effusus* in a fourth. Only the chances of seed dispersal have determined the allocation of species to different pools, but in the course of three or four years each pool has a different appearance, although the environment, aside from the reaction of the various species, is precisely the same for each. Are we dealing here with several different associations, or with a single association, or with merely embryonic stages of some future association? Under our view, these become merely academic questions, and any answer which may be suggested is equally academic.

But it must again be emphasized that these small areas of vegetation are component parts of the vegetative mantle of the land, and as such are fully worthy of description, of discussion, and of inquiry into the causes which have produced them and into their probable future. It must be emphasized that in citing the foregoing examples, the existence of associations or of successions is not denied, and that the purpose of the two paragraphs is to point out the fact that such communities introduce many difficulties into any attempt to define or classify association-types and successional series.

A plant association therefore, using the term in its ordinarily accepted meaning, represents the result of an environmental sorting of a population, but there are other communities which have existed such a short time that a reasonably large population has not yet been available for sorting.

Let us consider next the relation of migration and environmental selection to succession. We realize that all habitats are marked by continuous environmental fluctuation, accompanied or followed by a resulting vegetational fluctuation, but, in the common usage of the term, this is hardly to be regarded as an example of succession. But if the environmental change proceeds steadily and progressively in one direction, the vegetation ultimately shows a permanent change. Old species find it increasingly difficult or impossible to reproduce, as the environment approaches and finally passes their physiological demands. Some of the migrants find establishment progressively easier, as the environment passes the limit and approaches the optimum of their requirements. These are represented by more and more individuals, until they finally become the most conspicuous element of the association, and we say that a second stage of a successional series has been reached.

It has sometimes been assumed that the various stages in a successional series follow each other in a regular and fixed sequence, but that is frequently not the case. The next vegetation will depend entirely on the nature of the immigration which takes place in the particular period when environmental change reaches the critical stage. Who can predict the future for any one of the little ponds considered above? In one, as the bottom silts up, the chance migration of willow seeds will produce a willow thicket, in a second a thicket of *Cephalanthus* may develop, while a third, which happens to get no shrubby immigrants, may be converted into a miniature meadow of *Calamagrostis canadensis*. A glance at the diagram of observed successions in the Beach Area, Illinois, as published by Gates, will show at once how extraordinarily complicated the matter may become, and how far vegetation may fail to follow simple, pre-supposed successional series.

It is a fact, of course, that adjacent vegetation, because of its mere proximity, has the best chance in migration, and it is equally true that in many cases the tendency is for an environment, during its process of change, to approximate the conditions of adjacent areas. Such an environmental change becomes effective at the margin of an association, and we have as a result the apparent advance of one association upon another, so that their present distribution in space portrays their succession in time. The conspicuousness of this phenomenon has probably been the cause of the undue emphasis laid on the idea of successional series. But even here the individualistic nature of succession is often apparent. Commonly the vegetation of the advancing edge differs from that of the older established portion of the association in the numerical proportion of individuals of the component species due to the sorting of immigrants by an environment which has not yet reached the optimum, and, when the rate of succession is very rapid, the pioneer species are frequently limited to those of the greatest mobility. It also happens that the change in environment may become effective throughout the whole area of the association simultaneously, or may begin somewhere near the center. In such cases the pioneers of the succeeding association are dependent on their high mobility or on accidental dispersal, as well as environmental selection.

It is well known that the duration of the different stages in succession varies greatly. Some are superseded in a very short time, others persist for long or even indefinite periods. This again introduces difficulties into any scheme for defining and classifying associations.

A forest of beech and maple in northern Michigan is lumbered, and as a result of exposure to light and wind most of the usual herbaceous species also die. Brush fires sweep over the clearing and aid in the destruction of the original vegetation. Very soon the area grows up to a tangle of other herbaceous and shrubby species, notably *Epilobium angustifolium*, *Rubus strigosus*, and *Sambucus racemosa*. This persists but a few years before it is overtopped by saplings of the original hardwoods which eventually restore the forest. Is this early stage of fire-weeds and shrubs a distinct association or merely an embryonic phase of the forest? Since it has such a short duration, it is frequently regarded as the latter, but since it is caused by an entirely different type of environmental sorting and lacks most of the characteristic species of the forest, it might as well be called distinct. If it lasted for a long period of years it would certainly be called an association, and if all the forest near enough to provide seeds for immigration were lumbered, that might be the case. Again we are confronted with a purely arbitrary decision as to the associational identity of the vegetation.

Similarly, in the broad transition zone between the oak-covered mountains and the grass-covered foothills in the Coast Range of California, we are forced to deal arbitrarily in any matter of classification. Shall we call such a zone a mere transition, describe the forests above and the grasslands below and neglect the transition as a mere mixture? Or shall we regard it as a successional or time transition, evidencing the advance of the grasslands up the mountain or of the oaks down toward the foothills? If we choose the latter, we must decide whether the future trend of rainfall is to increase, thereby bringing the oaks to lower elevations, or to decrease, thereby encouraging the grasslands to grow at higher altitudes. If we adopt the former alternative, we either neglect or do a scientific injustice to a great strip of vegetation, in which numerous species are "associated" just as surely as in any recognized plant association.

The sole conclusion we can draw from all the foregoing considerations is that the vegetation of an area is merely the resultant of two factors, the fluctuating and fortuitous immigration of plants and an equally fluctuating and variable environment. As a result, there is no inherent reason why any two areas of the earth's surface should bear precisely the same vegetation, nor any reason for adhering to our old ideas of the definiteness and distinctness of plant associations. As a matter of fact, no two areas of the earth's surface do bear precisely the same vegetation, except as a matter of chance, and that chance may be broken in another year by a continuance of the same variable migration and fluctuating environment which produced it. Again, experience has shown that it is impossible for ecologists to agree on the scope of the plant association or on the method of classifying plant communities. Furthermore, it seems that the vegetation of a region is not capable of complete segregation into definite communities, but that there is a considerable development of vegetational mixtures.

Why then should there be any representation at all of these characteristic areas of relatively similar vegetation which are generally recognized by plant ecologists under the name of associations, the existence of which is indisputable as shown by our field studies in many parts of the world, and whose frequent repetition in similar areas of the same general region has led us to attempt their classification into vegetational groups of superior rank?

It has been shown that vegetation is the resultant of migration and environmental selection. In any general region there is a large flora and it has furnished migrating seeds for all parts of the region alike. Every environment has therefore had, in general, similar material of species for the sorting process. Environments are determined principally by climate and soil, and are altered by climatic changes, physiographic processes, and reaction of the plant population. Essentially the same environments are repeated in the same region, their selective action upon the plant immigrants leads to an essentially similar flora in each, and a similar flora produces similar reactions. These conditions produce the well known phenomena of plant associations of recognizable extent and their repetition with great fidelity in many areas of the same region, but they also produce the variable vegetation of our sand dunes and small pools, the fragmentary associations of areas of small size, and the broad transition zones where different types of vegetation are mixed. Climatic changes are always slow, physiographic processes frequently reach stages where further change is greatly retarded, and the accumulated effects of plant reaction often reach a condition beyond which they have relatively little effect on plant life. All of these conspire to give to certain areas a comparatively uniform environment for a considerable period of time, during which continued migration of plants leads to a smoothing out of original vegetational differences and to the establishment of a relatively uniform and static vegetational structure. But other physiographic processes are rapid and soon develop an entirely different environment, and some plant reactions are rapid in their operation and profound in their effects. These lead to the short duration of some plant communities, to the development, through the prevention of complete migration by lack of sufficient time, of associations of few species and of different species in the same environment, and to mixtures of vegetation which seem to baffle all attempts to resolve them into distinct associations.

Under the usual concept, the plant association is an area of vegetation in which spatial extent, describable structure, and distinctness from other areas are the essential features. Under extensions of this concept it has been regarded as a unit of vegetation, signifying or implying that vegetation in general is composed of a multiplicity of such units, as an individual representation of a general group, bearing a general similarity to the relation of an individual to a species, or even as an organism, which is merely a more striking manner of expressing its unit nature and uniformity of structure. In every case spatial extent is an indispensable part of the definition. Under the individualistic concept, the fundamental idea is neither extent, unit character, permanence, nor definiteness of structure. It is rather the visible expression, through the juxtaposition of individuals, of the same or different species and either with or without mutual influence, of the result of causes in continuous operation. These primary causes, migration and environmental selection, operate independently on each area, no matter how small, and have no relation to the process on any other area. Nor are they related to the vegetation of any other area, except as the latter may serve as a source of migrants or control the environment of the former. The effect of these primary causes is therefore not to produce large areas of similar vegetation, but to determine the plant life on every minimum area. The recurrence of a similar juxtaposition over tracts of measurable extent, producing an association in the ordinary use of the term, is due to a similarity in the contributing causes over the whole area involved.

Where one or both of the primary causes changes abruptly, sharply delimited areas of vegetation ensue. Since such a condition is of common occurrence, the distinctness of associations is in many regions obvious, and has led first to the recognition of communities and later to their common acceptance as vegetational units. Where the variation of the causes is gradual, the apparent distinctness of associations is lost. The continuation in time of these primary causes unchanged produces associational stability, and the alteration of either or both leads to succession. If the nature and sequence of these changes are identical for all the associations of one general type (although they need not be synchronous), similar successions ensue, producing successional series. Climax vegetation represents a stage at which effective changes have ceased, although their resumption at any future time may again initiate a new series of successions.

In conclusion, it may be said that every species of plant is a law unto itself, the distribution of which in space depends upon its individual peculiarities of migration and environmental requirements. Its disseminules migrate everywhere, and grow wherever they find favorable conditions. The species disappears from areas where the environment is no longer endurable. It grows in company with any other species of similar environmental requirements, irrespective of their normal associational affiliations. The behavior of the plant offers in itself no reason at all for the segregation of definite communities. Plant associations, the most conspicuous illustration of the space relation of plants, depend solely on the coincidence of environmental selection and migration over an area of recognizable extent and usually for a time of considerable duration. A rigid definition of the scope or extent of the association is impossible, and a logical classification of associations into larger groups, or into successional series, has not yet been achieved.

The writer expresses his thanks to Dr. W. S. Cooper, Dr. Frank C. Gates, Major Barrington Moore, Mr. Norman Taylor, and Dr. A. G. Vestal for kindly criticism and suggestion during the preparation of this paper.

13

Reprinted from *J. Ecol.* **24**:252–284 (1936)

NATURE AND STRUCTURE OF THE CLIMAX

BY FREDERIC E. CLEMENTS

(*Carnegie Institution of Washington, Santa Barbara, California*)

CONTENTS

INTRODUCTION

MORE than a century ago when Lewis and Clark set out upon their memorable journey across the continent of North America (1803–6), they were the first to traverse the great climaxes from deciduous woods in the east through the vast expanse of prairie and plain to the majestic coniferous forest of the north-west. At this time the oak-hickory woodland beyond the Appalachians was almost untouched by the ax except in the neighborhood of a few straggling pioneer settlements, and west of the Mississippi hardly an acre of prairie had known the plow. A few years later (1809), Bradbury states that the boundless prairies are covered with the finest verdure imaginable and will become one of the most beautiful countries in the world, while the plains are of such extent and fertility as to maintain an immense number of animals. It appears probable that at this time no other grassland in the world exhibited such myriads of large mammals belonging to but a few species.

The natural inference has been that the prairies were much modified by the grazing of animals and the fires of primitive man, and this has been reinforced by estimates of the population of each. Seton (1929) concludes that the original number of bison was about 60 million with a probable reduction to 40 million by 1800, and that both the antelope and white-tailed deer were equally abundant, while elk and mule-deer each amounted to not more than 10 million at the maximum. However, these were distributed over a billion or two acres, and the average density was probably never more than a score to the square mile. Estimates of the Indian tribes show the greatest divergence, but it seems improbable that the total population within the grassland ever exceeded a half million. The general habit of migration among the animals further insured that serious effects from overgrazing and trampling were but local or transitory, while the influence of fires set by the Indians was even less significant in modifying the plant cover. As to the forests, those of the north-west were still primeval and in the east they were yet to be changed over wide areas by lumbering and burning on a large scale.

THE CLIMAX CONCEPT

The idea of a climax in the development of vegetation was first suggested by Hult in 1885 and then was advanced more or less independently by several investigators during the next decade or so (cf. Clements, 1916; Phillips, 1935). It was applied to a more or less permanent and final stage of a particular succession and hence one characteristic of a restricted area. The concept of the climax as a complex organism inseparably connected with its climate and often continental in extent was introduced by Clements (1916). According to this view, the climax constitutes the major unit of vegetation and as such forms the basis for the natural classification of plant communities. The relation between climate and climax is considered to be the paramount one, while the intimate bond between the two is emphasized by the derivation of the terms from the same Greek root. In consequence, under this concept climax is invariably employed with reference to the climatic community alone, namely, the formation or its major divisions.

At the outset it was recognized that animals must also be considered members of the climax, and the word *biome* was proposed for the purpose of laying stress upon the mutual roles of plants and animals (Clements, 1916*b*; Clements and Shelford, 1936). With this went the realization that the primary relations to the habitat or ece were necessarily different by virtue of the fact that plants are producents and animals consuments. On land, moreover, plants constitute the fixed matrix of the biome in direct connection with the climate, while the animals bear a dual relation, to plants as well as to climate. The outstanding effect of the one is displayed in reaction upon the ece, of the other in coaction upon plants, which constitutes the primary bond of the biotic community.

Because of its emphasis upon the climatic relation, the term climax has come more and more to replace the word formation, which is regarded as an exact synonym, and this process may have been favored by a tendency to avoid confusion with the geological use. The designation "climatic formation" has now and then been employed, but this is merely to accentuate its nature and to distinguish it from less definite usages. Furthermore, climax and biome are complete synonyms when the biotic community is to be indicated, though climax will necessarily continue to be employed for the matrix when plants alone are considered.

Nature of the climax

This theme has been developed in considerable detail in earlier works (Clements, 1916, 1920, 1928; Weaver and Clements, 1929), as well as in a recent comprehensive treatment by Phillips (1935), and hence a summary account of the major features will suffice in the present place. These may be conveniently grouped under the following four captions, i.e. unity, stabilization and change, origin and relationship, and objective tests.

Unity of the climax

The inherent unity of the climax rests upon the fact that it is not merely the response to a particular climate, but is at the same time the expression and the indicator of it. Because of extent, variation in space and time, and the usually gradual transition into adjacent climates, to say nothing of the human equation, neither physical nor human measures of a climate are adequately satisfactory. By contrast, the visibility, continuity, and sessile nature of the plant community are peculiarly helpful in indicating the fluctuating limits of a climate, while its direct response in terms of food-making, growth and life-form provides the fullest possible integration of physical factors. Naturally, both physical and human values have a part in analyzing and in interpreting the climate as outlined by the climax, but these can only supplement and not replace the biotic indicators.

It may seem logical to infer that the unity of both climax and climate should be matched by a similar uniformity, but reflection will make clear that such is not the case. This is due in the first place to the gradual but marked shift in rainfall or temperature from one boundary to the other, probably best illustrated by the climate of the prairie. In terms of precipitation, the latter may range along the parallel of 40° from nearly 40 in. at the eastern edge of the true prairie to approximately 10 in. at the western border of the mixed grassland, or even to 6 in. in the desert plains and the Great Valley of California. Such a change is roughly 1 in. for 50 miles and is regionally all but imperceptible. The temperature change along the 100th meridian from the mixed prairie in Texas to that of Manitoba and Saskatchewan is even more striking, since only one association is concerned. At the south the average period without killing frost is about 9 months, but at the north it is less than 3, while the mean annual temperatures are 70 and 33° F. respectively. The variation of the two major factors at the extremes of the climatic cycle is likewise great, the maximum rainfall not infrequently amounting to three to four times that of the minimum.

The visible unity of the climax is due primarily to the life-form of the dominants, which is the concrete expression of the climate. In prairie and steppe, this is the grass form, with which must be reckoned the sedges, especially in the tundra. The shrub characterizes the three scrub climaxes of North America, namely, desert, sagebrush, and chaparral, while the tree appears in three subforms, coniferous, deciduous, and broad-leaved evergreen, to typify the corresponding boreal, temperate, and tropical climaxes. The life-form is naturally reflected in the genus, though not without exceptions, since two or more forms or subforms, herb or shrub, deciduous or evergreen, annual or perennial, may occur in the same genus. Hence, the essential unity of a climax is to be sought in its dominant species, since these embody not only the life-form and the genus, but also denote in themselves a definite relation to the climate. Their reactions and coactions are the most controlling both in kind and amount, and thus they determine the conditions under which all the remaining species are associated with them. This is true to a less degree of the animal influents, though their coactions may often be more significant than those of plants.

Stabilization and change

Under the growing tendency to abandon static concepts, it is comprehensible that the pendulum should swing too far and change be overstressed. This consequence is fostered by the fact that most ecological studies are carried out in settled regions where disturbance is the ruling process. As a result, the climax is badly fragmented or even absent over wide areas and subseres are legion. In all such instances it is exceedingly difficult or entirely impossible to strike a balance between stability and change, and it becomes imperative to turn to regions much less disturbed by man, where climatic control is still paramount. It is likewise essential to employ a conceivable measure of time, such as can be expressed in human terms of millennia rather than in eons. No student of past vegetation entertains a doubt that climaxes have evolved, migrated and disappeared under the compulsion of great climatic changes from the Paleozoic onward, but he is also insistent that they persist through millions of years in the absence of such changes and of destructive disturbances by man. There is good and even conclusive evidence within the limitations of fossil materials that the prairie climax has been in existence for several millions of years at least and with most of the dominant species of to-day. This is even more certainly true of forests on the Pacific Coast, owing to the wealth of fossil evidence (Chaney, 1925, 1935), while the generic dominants of the deciduous forests of the Dakota Cretaceous and of to-day are strikingly similar.

It can still be confidently affirmed that stabilization is the universal tendency of all vegetation under the ruling climate, and that climaxes are characterized by a high degree of stability when reckoned in thousands or even millions of years. No one realizes more clearly than the devotee of succession that change is constantly and universally at work, but in the absence of civilized man this is within the fabric of the climax and not destructive of it. Even in a country as intensively developed as the Middle West, the prairie relicts exhibit almost complete stability of dominants and subdominants in spite of being surrounded by cultivation (cf. Weaver and Flory, 1934). It is obvious that climaxes display superficial changes with the season, year or cycle, as in aspection and annuation, but these modify the matrix itself little or not at all. The annuals of the desert may be present in millions one year and absent the next, or one dominant grass may seem prevailing one season and a different one the following year, but these changes are merely recurrent or indeed only apparent. While the modifications represented by bare areas and by seres in every stage are more striking, these are all in the irresistible process of being stabilized as rapidly as the controlling climate and the interference of man permit.

In brief, the changes due to aspection, annuation or natural coaction are superficial, fleeting or periodic and leave no permanent impress, while those of succession are an intrinsic part of the stabilizing process. Man alone can destroy the stability of the climax during the long period of control by its climate, and he accomplishes this by fragments in consequence of a destruction that is selective, partial or complete, and continually renewed.

Origin and relationship

Like other but simpler organisms, each climax not only has its own growth and development in terms of primary and secondary succession, but it has also evolved out of a preceding climax. In other words, it possesses an ontogeny and phylogeny that can be quantitatively and experimentally studied, much as with the individuals and species of plants and animals (*Plant Succession*, 1916, pp. 181, 342). Out of the one has come widespread activity in the investigation of succession, while interest in the other lingers on the threshold, chiefly because it demands a knowledge of the climaxes of more than one continent. With increasing research in these, especially in Europe and Asia, it will be possible to test critically the panclimaxes already suggested (Clements, 1916, 1924, 1929), as well as to determine the origin and relationships of the constituent formations.

This task will also require the services of paleo-ecology for the reconstruction of each eoclimax, which has been differentiated by worldwide climatic changes into the existing units of the panclimax or panformation. As it is, there can be no serious question of the existence of a great hemispheric clisere constituted by the arctic, boreal, deciduous, grassland, subtropical and tropical panclimaxes. Desert formations for the most part constitute an exception and may well be regarded as endemic climaxes evolved in response to regional changes of climate (Clements, 1935).

It is a significant fact that the boreal formations of North America and Eurasia are more closely related than the coniferous ones of the former, but this seeming anomaly is explained by the greater climatic differences that have produced the forests of the Petran and Sierran systems. The five climaxes concerned are relatively well known, and it is possible to indicate their relationships with some assurance, and all the more because of their parallel development on the two great mountain chains. In the case of deciduous forest and grassland, only a single formation of each is present in North America, and the problem of differentiation resolves itself into tracing the origin and relationship of the several associations. It has been suggested that the mixed prairie by virtue of its position, extent and common dominants represents the original formation in Tertiary times, an assumption reinforced by its close resemblance to the steppe climax (Clements, 1935). It is not improbable that the mixed hardwoods of the southern Appalachians bear a similar relation to the associations of the modern deciduous forest (Braun, 1935).

Tests of a climax

As has been previously indicated, the major climaxes of North America, such as tundra, boreal and deciduous forest, and prairie, stand forth clearly as distinct units, in spite of the fact that the prairie was first regarded as comprising two formations, as a consequence of the changes produced by overgrazing. The other coniferous and the scrub climaxes emerge less distinctly because of the greater similarity of life-form within each group, and hence it is necessary to appeal to criteria derived from the major formations just mentioned. This insures uniformity of basis and a high degree of objectivity, both of which are qualities of paramount importance for the natural classification of biomes. In fact, entire consistency in the application of criteria is the best warrant of objective results, though this is obviously a procedure that demands a first-hand acquaintance with most if not all the units concerned and over a large portion of their respective areas.

The primary criterion is that afforded by the vegetation-form, as is illustrated by the four major climaxes. The others of each group, such as coniferous forest or scrub, are characterized also by the same form in the dominants, but this is not decisive as between related climaxes and hence recourse must be taken to the other tests. The value of the life-form is most evident where two climaxes of different physiognomy are in contact, as in the case of the lake forest of pine-hemlock and the deciduous forest of hardwoods. The static view would make the hemlock in particular a dominant of the deciduous formation, but the evidence derived from the vegetation-form is supported by that of phylogeny and by early records of composition and timber-cut to show that two different climaxes are concerned. Secondary forms or subforms rarely if ever mark distinctions between climaxes, but do aid in the recognition of associations. This is well exemplified by the tall, mid and short grasses of the prairie and somewhat less definitely by the generally deciduous character of the Petran chaparral and the typically evergreen nature of the Sierran.

As would be expected, the most significant test of the unity of a formation is afforded by the presence of certain dominant species through all or nearly all of the associations, but often not in the role of dominants by reason of reduced abundance. Here again perhaps the best examples are furnished by prairie and tundra, though the rule applies almost equally well to deciduous forest and only less so to coniferous ones because of a usually smaller number of dominants. For the prairie, the number of such species, or *perdominants*, found in all or all but one or two of the five associations is eight, namely, *Stipa comata*, *Agropyrum smithi*, *Bouteloua gracilis*, *Sporobolus cryptandrus*, *Koeleria cristata*, *Elymus sitanion*, *Poa scabrella* and *Festuca ovina*. Even when a species is lacking over most of an association, as in the case of *Stipa comata* and the true prairie, it may be represented by a close relative, such as *S. spartea*, which is probably no more than a mesic variety of it. As to the three associations of the deciduous forest, a still larger number of dominant species occur to some degree in all; the oaks comprise *Quercus borealis*, *velutina*, *alba*, *macrocarpa*, *coccinea*, *muhlenbergi*, *stellata* and *marilandica*, and the hickories, *Carya ovata*, *glabra*, *alba* and *cordiformis*.

It was the application of this test by specific dominants that led to the recognition of the two climaxes in the coniferous mantle of the Petran and

Sierran cordilleras. The natural assumption was that such a narrow belt could not contain more than one climax, especially in view of its physiognomic uniformity, but this failed to reckon with the great climatic differences of the two portions and the corresponding response of the dominants. The effect of altitude proved to be much more decisive than that of region, dominants common to the montane and subalpine zones being practically absent, though the rule for the same zone in each of the two separate mountain systems. Long after the presence of two climaxes had been established, it was found that Sargent had anticipated this conclusion, though in other terms (1884, p. 8).

As would be inferred, the dominants of related associations belong to a few common genera for the most part. Thus, there are a dozen species of *Stipa* variously distributed as dominants through the grassland, nearly as many of *Sporobolus*, *Bouteloua* and *Aristida*, and several each of *Poa*, *Agropyrum*, *Elymus*, *Andropogon*, *Festuca* and *Muhlenbergia*. In the deciduous forest, *Quercus*, *Carya* and *Acer* are the great genera, and for the various coniferous ones, *Pinus*, *Abies* and *Picea*, with species of *Tsuga*, *Thuja*, *Larix* and *Juniperus* hardly less numerous.

The perennial forbs that play the part of subdominants also possess considerable value in linking associations together, and to a higher degree in the deciduous forest than in the prairie, owing chiefly to the factors of shade and protection. Over a hundred subdominants belonging to two score or more of genera, such as *Erythronium*, *Dicentra*, *Trillium*, *Aquilegia*, *Arisaema*, *Phlox*, *Uvularia*, *Viola*, *Impatiens*, *Desmodium*, *Helianthus*, *Aster* and *Solidago*, range from Nova Scotia or New England beyond the borders of the actual climax to Nebraska and Kansas. Across the wide expanse of the prairie climax, species in common are only exceptional, these few belonging mostly to the composites, notably *Grindelia squarrosa*, *Gutierrezia sarothrae*, *Artemisia dracunculus* and *vulgaris*. On the other hand, the number of genera of subdominants found throughout the grassland is very large.

The greater mobility of the larger mammals in particular renders animal influents less significant than plants as a criterion, but several of these possess definite value and the less mobile rodents even more. The antelope and bison are typical of the grassland climax, the first being practically restricted to it, while jack-rabbits, ground-squirrels and kangaroo-rats are characteristic dwellers in the prairie, as is their chief foe, the coyote.

The remaining criteria are derived from development directly or indirectly, though this is less evident in the case of the ecotone between two associations. Here the mixing of dominants and subdominants indicates their general similarity in terms of the formation, within which range their preferences assign them to different associations. The evidence from primary succession is of value only in the later stages as a rule, since initial associes like the reed-swamp may occur in several climaxes. With subseres, however, all or nearly all the stages are related to the particular climax and such seres denote a corresponding unity in development. This is especially true of all subclimaxes and most evidently in the case of those due to fire. More significant still are postclimaxes in both grassland and forest. For example, the associes of species of *Andropogon*, which is subclimax to the oak-hickory forest, constitutes a postclimax to five out of the six associations of the prairie. On the other hand, the community of *Ulmus*, *Juglans*, *Fraxinus*, etc., found on flood-plains through the region of deciduous forest, forms a common subclimax to the three associations.

In addition to such ontogenetic criteria, phylogeny supplies tests of even greater value. This is notably the case with the two associations of the montane and subalpine coniferous forests of the west, though perhaps the most striking application of this criterion is in connection with the lake forest of pine-hemlock. Though the concrete evidence for such a climax recurs constantly through the region of the Great Lakes to the Atlantic, it is fragmentary and there is no evident related association to the westward. However, the four genera are represented by related species in the two regions, namely, *Pinus strobus* by *P. monticola*, *P. banksiana* by *P. contorta*, *Tsuga canadensis* by *T. heterophylla*, *Larix laricina* by *L. occidentalis*, and *Thuja occidentalis* by *T. plicata*, though the last two genera have changed from a subclimax role in the east to a climax one in the west. As suggested earlier, phylogenetic evidence of still more direct nature is supplied by the mixed prairie with the other enclosing associations and by the remnants of a virgin deciduous forest that exhibits a similar genetic and spatial relation to the associations of this climax (cf. Braun, 1935).

Finally, it is clear that any test will gain in definiteness and accuracy of application whenever dependable records are available with respect to earlier composition and structure. These may belong entirely to the historical period, as in the case of scientific reports or land surveys, they may bridge the gap between the present and the past as with pollen statistics, or they may reach further back into the geological record, as with leaf-impressions or other fossils (Chaney, 1925, 1933; Clements, 1936). Two instances of the scientific record that are of the first importance may be given as examples. The first is the essential recognition by Sargent of the pine-hemlock climax under the name of the northern pine belt (1884), at a time when relatively little of this had been logged, by contrast with 90 per cent. or more at present (cf. also Bromley, 1935). The second is an account, discovered and communicated by Dr Vestal, of the prairies of Illinois as seen by Short in *ca.* 1840. This is of heightened interest since its discovery followed little more than a year after repeated field trips had led to the conclusion that all of Iowa, northern Missouri and most of Illinois were to be assigned to the true prairie,[1] a decision

[1] The true prairie is characterized by the three eudominants, *Stipa spartea*, *Sporobolus asper* and *S. heterolepis*. The presence of tall-grasses in it to-day, particularly *Andropogon furcatus* and *nutans*, is the mark of the disclimax due to the varied disturbances associated with settlement.

confirmed for Illinois by Short's description, and supported by the more general accounts of Bradbury (*ca.* 1815) and Greeley (1860).

CLIMAX AND PROCLIMAX

Essential relations

In accordance with the view that development regularly terminates in the community capable of maintaining itself under a particular climate, except when disturbance enters, there is but one kind of climax, namely, that controlled by climate. This essential relation is regarded as not only inherent in all natural vegetation, but also as implicit in the cognate nature of the two terms. While it is fully recognized that succession may be halted in practically any stage, such communities are invariably subordinate to the true climax as determined by climate alone. From the very meaning of the word, there can not be climaxes scattered along the developmental route with a genuine climax at the end. There is no intention to question the reality of such pauses, but only to emphasize the fact that they are of a different order from the climax.

While it is natural to express new ideas by qualifying an old term, this does not conduce to the clearest thinking or the most accurate usage. Even more undesirable is the fact that the meaning of the original word is gradually shifted until it becomes either quite vague or hopelessly inclusive. At the hands of some, climax has already suffered this fate, and fire, disease, insects, and human disturbances of all sorts are assumed to produce corresponding climaxes (cf. Chapman, 1932). On such an assumption corn would constitute one climax, wheat another, and cotton a third, and it would then become imperative to begin anew the task of properly analyzing and classifying vegetation.

In the light of two decades of continued analysis of the vegetation of North America, as well as the application of the twin concepts of climax and complex organism by workers in other portions of the globe and the strong support brought to them by the rise of emergent evolution and holism (Phillips, 1935), the characterization of the climax as given in *Plant Succession*, in 1916, still appears to be both complete and accurate. "The unit of vegetation, the climax formation, is an organic entity. As an organism, the formation arises, grows, matures and dies. Its response to the habitat is shown in processes or functions and in structures that are the record as well as the result of these functions. Furthermore, each climax formation is able to reproduce itself, repeating with essential fidelity the stages of its development. The life-history of a formation is a complex but definite process, comparable in its chief features with the life-history of an individual plant. The climax formation is the adult organism, of which all initial and medial stages are but stages of development. . . . A formation, in short, is the final stage of vegetational development in a climatic unit. It is the climax community of a succession that terminates in the highest life-form possible in the climate concerned."

To-day this statement would need modification only to the extent of substituting "biome" for climax or formation and "biotic" for vegetational. This characterization has recently been annotated and confirmed by Phillips' masterly discussion of climax and complex organism, as cited above, a treatise that should be read and digested by everyone interested in the field of dynamic ecology and its wide applications.

Proclimaxes

As a general term, proclimax includes all the communities that simulate the climax to some extent in terms of stability or permanence but lack the proper sanction of the existing climate. Certain communities of this type were called potential climaxes in *Plant Succession* (p. 108; 1928, p. 109), and two kinds were distinguished, namely, preclimax and postclimax. To avoid proposing a new term in advance of its need, subclimax was made to do double duty, denoting both the subfinal stage of succession, as well as apparent climaxes of other kinds. This dual usage was criticized by Godwin (1929, p. 144) and partially justified by Tansley in an appended note on the ground just given. However, this discussion made it evident that a new term was desirable and proclimax was accordingly suggested (Clements, 1934). While this takes care of the use of subclimax in the second sense noted above, it is better adapted by reason of its significance to apply to all kinds of subpermanent communities other than the climax proper. However, there is still an important residuum after subclimax, preclimax and postclimax have been recognized, and it is proposed to call these *disclimaxes*, as indicated later.

The proclimax may be defined as any more or less permanent community resembling the climax in one or more respects, but gradually replaceable by the latter when the control of climate is not inhibited by disturbance. Besides its general function, it may be used as a synonym for any one of its divisions, as well as in cases of doubt pending further investigation, such as in water climaxes. The four types to be considered are subclimax, disclimax, preclimax and postclimax.

Subclimax

As the stage preceding the climax in all complete seres, primary and secondary, the subclimax is as universal as it is generally well understood. The great majority of such communities belong to the subsere, especially that following fire, owing to the fact that disturbance is to-day a practically constant feature of most climaxes. Fire and fallow are recurrent processes in cultivated regions generally and they serve to maintain the corresponding subsere until protection or conversion terminates the disturbance. Though the subclimax is just as regular a feature of priseres, these have long ago ended in the climax over most of the climatic area and the related subclimax communities are consequently much restricted in size and widely scattered. Smallness is naturally a characteristic of nearly all subclimaxes, the chief exceptions being due to fire or to fire and logging combined, but by contrast they are often exceedingly numerous.

Because of its position in the succession, the subclimax resembles the preclimax in some respects and in a few instances either term may be properly applied. The distinction between subclimax and disclimax presents some difficulty now and then, as the amount of change necessary to produce the latter may be a matter of judgment. This arises in part also from the structural diversity of formation and association, as a consequence of which the dominants of a particular type of subsere vary in different areas. When there is but a single dominant, as in many burn subclimaxes, no question ensues, but if two or more are present, the decision between subclimax and disclimax may be less simple, as is not infrequent in scrub and grassland.

Examples of the subclimax are legion, the outstanding cases being mostly due to fire, alone or after lumbering or clearing. Most typical are those composed of "jack-pines" or species with closed cones that open most readily after fire. Each great region has at least one of these, e.g. *Pinus rigida*, *virginiana* and *echinata* in the east, *P. banksiana* in the north, *P. murrayana* in the Rocky Mountains, and *P. tuberculata*, *muricata* and *radiata* on the Pacific Slope. *Pinus palustris* and *taeda* play a similar role in the "piney" woods of the Atlantic Gulf region, as does *Pseudotsuga taxifolia* in the north-west. The characteristic subclimaxes of the boreal forest are composed of aspen (*Populus tremuloides*), balsam-poplar (*P. balsamifera*), and paper-birch (*Betula papyrifera*), either singly or in various combinations. Aspen also forms a notable subclimax in the Rocky Mountains, for the most part in the subalpine zone. Prisere subclimaxes are regular features of bogs and muskeags throughout much or all of the boreal and lake forests, the three dominants being *Larix laricina*, *Picea mariana* and *Thuja occidentalis*, often associated as zonal consocies. Where pines are absent in the region of the deciduous forest, two xeric oaks, *Quercus stellata* and *marilandica*, may constitute a subclimax, and this role is sometimes assumed by small trees, *Sassafras*, *Diospyrus* and *Hamamelis* being especially important.

Subclimaxes in the grassland are composed largely of tall-grasses, usually in the form of a consocies. In the true prairie, this part is taken by *Spartina cynosuroides* and in the desert plains by *Sporobolus wrighti*, while *Elymus condensatus* plays a similar role in the mixed prairie and in portions of the bunch-grass prairies. The function of the tall Andropogons is more varied; they are typically postclimax rather than subclimax, though they maintain the latter relation along the fringe of the oak-hickory forest and in oak openings. They occupy a similar position at the margin of the pine subclimax in Texas especially, and hence they are what might be termed "sub-subclimax" in such situations. Beyond the forest and in association with *Elionurus*, *Trachypogon*, etc., they appear to constitute a faciation of the coastal prairie. Chaparral proper is to be regarded as a climax, but with a change of species it extends into the montane and even the subalpine zone and there constitutes a fire subclimax. In the foothills of southern California, the coastal sagebrush behaves in like manner where it lies in contact with the chaparral.

The disposition of seral stages below the subclimax that exhibit a distinct retardation or halt for a longer or shorter period is a debatable matter. It is entirely possible to include them among subclimaxes, but this would again fail in accuracy and definiteness and hence lead to confusion. The decision may well be left to usage by providing a term for such seral or sub-subclimax communities as persist for a long or indefinite period because of continued or recurrent edaphic control or human disturbance. By virtue of its significance, brevity and accord with related terms, the designation "serclimax" is suggested, with the meaning of a seral community usually one or two stages before the subclimax, which persists for such a period as to resemble the climax in this one respect. For reasons of brevity and agreement, the connecting vowel is omitted, but the *e* remains long as in sere.

For the most part, serclimaxes are found in standing water or in saturated soils as a consequence of imperfect drainage. The universal example is the reed-swamp with one or more of several consocies, such as *Scirpus*, *Typha*, *Zizania*, *Phragmites* and *Glyceria*: this is typical of the lower reaches of rivers, of deltas and of certain kinds of lakes, the great tule swamps of California affording outstanding instances. Another type occurs in coastal marshes in which *Spartina* is often the sole or major dominant, while sedge-swamps have a wider climatic range but are especially characteristic of northern latitudes and high altitudes. The Everglades of Florida dominated by *Cladium* constitute perhaps the most extensive example of the general group, though *Carex* swamps often cover great areas and the grass *Arundinaria* forms jungle-like cane-brakes through the south. Among woody species, *Salix longifolia* is an omnipresent consocies of sand-bars and river-sides, but the most unique exemplar is the cypress-swamp of the south, typified by *Taxodium*. In boreal and subalpine districts the distinctive serclimax is the peat-bog, moor or muskeag, more or less regularly associated with other seral communities of *Carex* and usually of *Larix* or *Picea* also in the proper region.

Frequent burning may retard or prevent the development of the normal fire subclimax and cause it to be replaced by a preceding stage. This may be a scrub community or one kept in the shrub form by repeated fires, but along the Atlantic and Gulf Coasts it is usually one of *Andropogon virginicus*, owing to its sufferance of burning. The so-called "balds" of the southern Appalachians are seral communities of heaths or grasses initiated and maintained primarily by fire. Finally, there are serclimaxes of weeds, especially annuals, in cultivated districts, and a somewhat similar community of native annuals is characteristic of wide stretches in the desert region.

Disclimax

As with the related concepts, the significance of this term is indicated by a prefix, *dis-*, denoting separation, unlikeness or derogation, much as in the Greek *dys*, poor, bad. The most frequent examples of this community result from the modification or replacement of the true climax, either as a whole or in part, or from a change in the direction of succession. These ensue chiefly in consequence of a disturbance by man or domesticated animals, but they are also occasionally produced by mass migration. In some cases, disturbance and the introduction of alien species act together through destruction and competition to constitute a quasi-permanent community with the general character of the climax. This type is best illustrated by the *Avena-Bromus* disclimax of California, which has all but completely replaced the bunch-grass prairie.[1] A similar replacement by *Bromus tectorum* has more recently taken place over large areas of the Great Basin, while *Poa pratensis* has during the last half-century steadily invaded the native hay-fields and pastures of the true prairie, an advance first noted by Bradbury in 1809. An even more striking phenomenon is the steadily increasing dominance of *Salsola* over range and crop land in the west, and this is imitated by *Sisymbrium* and *Lepidium* in the north-west. It is obvious that all cultivated crops belong in the same general category, but this point hardly requires consideration.

Probably the example most cited in North America is that of the short-grass plains, which actually represent a reduction of the mixed prairie due to overgrazing, supplemented by periodic drouth. Over most of this association, the mid-grasses, *Stipa*, *Agropyrum*, etc., are still in evidence, though often reduced in abundance and stature, but in some areas they have been practically eliminated. Similar though less extensive partial climaxes of short-grasses characterise pastures in the true and both pasture and range in the coastal prairie, the dominants regularly belonging to *Bouteloua*, *Buchloe* or *Hilaria*. Of essentially the same nature is the substitution of annual species of *Bouteloua*

[1] The grassland climax of North America comprises six well-marked associations (*Plant Indicators*, 1920; *Plant Ecology*, 1929). The mixed prairie, so-called because it is composed of both mid-grasses and short-grasses, is more or less central to the other five and is regarded as ancestral to them. To the east along the Missouri and Mississippi Rivers, it has become differentiated into the true prairie formed by other species of mid-grasses pertaining mostly to the same genera, and this unit is flanked along the western margin of the deciduous forest by a proclimax of tall grasses, chiefly *Andropogon*. Southward the true prairie is replaced by coastal prairie, which in the main occupies the Gulf region of Texas and Mexico and is constituted by similar dominants but of different species. The desert plains are characterized primarily by species of *Bouteloua* and *Aristida*, which range from western Texas to the edge of the deserts of Mexico and Arizona. In the north-west the short-grasses disappear and the Palouse prairie of eastern Washington and adjacent regions is formed by mid-grasses of the bunch-grass life-form, among which *Agropyrum spicatum* is the eudominant. The same life-form signalizes the California prairie, found from the northern part of the state southward into Lower California, but its especial character is derived from endemic species of *Stipa*. As indicated in the discussion, the short-grass plains, composed of *Bouteloua*, *Buchloe*, and *Carex*, are not climatic in nature, and this statement applies likewise to the tall-grass meadows of *Andropogon* mentioned above.

and *Aristida* in the desert plains for perennial ones of the same genera, which is a case of short-grasses being followed by still shorter ones.

In other instances, the effect of disturbance is to produce a community with the appearance of a postclimax, when the life-form concerned is that of an undershrub or tall grass. This is notably the case in the mixed prairie when overgrazing is carried to the point of breaking up the short-grass sod and permitting the dominance of *Artemisia frigida* or *Gutierrezia sarothrae*. In essence, the wide extension of sagebrush (*Artemisia tridentata*) and of creosote-bush (*Larrea tridentata*) is the same phenomenon, though each of these is a climax dominant in its own region. In the case of *Opuntia*, the peculiar life-form suggests an important difference, but the numerous species behave in all significant respects like other shrubs, though with the two advantages of spines and ready propagation.

The communities of tall-grasses formed by species of *Andropogon* originally presented some difficulty, since these naturally have all the appearance of a postclimax to the prairie. Probably the greater number are to be assigned to this type, but the evidence from reconnaissance and record indicates that in the true prairie and especially the eastern portion, *Andropogon furcatus* in particular now constitutes a disclimax due to pasturing, mowing and in some measure to fire also (Clements, 1933). A characteristic disclimax in miniature is to be found in the "gopher gardens" of the alpine tundra, where coaction and reaction have removed the climax dominants of sedges and grasses to make place for flower gardens of perennial forbs. "Towns" of prairie-dogs and kangaroo-rats often produce similar but much more extensive communities.

Selective cutting not infrequently initiates disclimaxes, as may likewise the similar action of other agents such as fire or epidemic disease. The most dramatic example is the elimination of the chestnut (*Castanea dentata*) from the oak-chestnut canopy, but of even greater importance has been the extreme reduction and fragmentation of the lake forest through the overcutting of white pine. Finally, what is essentially a disclimax may result from climatic mass migration, such as in the Black Hills of South Dakota has brought together *Pinus ponderosa* from the montane climax of the Rocky Mountains and *Picea canadensis* from the boreal forest.

Preclimax and postclimax

These related concepts were first advanced in *Plant Succession* (1916, 1928) and have since been discussed in *Plant Ecology* (1929) and in the organisation of the relict method (1934). They are both direct corollaries of the principle of the clisere, the spatial series of climaxes that are set in motion by a major climatic shift, such as that of the glacial epoch with its opposite phases. The clisere is most readily comprehended in the case of high ranges or summits, such as Pikes Peak where the entire series of climaxes is readily visible, and is what Tournefort described on Mount Ararat in his famous journey of 1700.

However, this is but an expression of the continental clisere in latitude, which achieves perhaps its greatest regularity in North America. A similar relation is characteristic of the longitudinal disposition of climaxes in the temperate zone between the two oceans, the portion from deciduous forest through prairie to desert being the most uniform.

With the exception of the two extremes, arctalpine and tropical, each climax has a dual role, being preclimax to the contiguous community of so-called higher life-form and postclimax to that of lower life-form. This may be illustrated by the woodland climax, which is postclimax to grassland and preclimax to montane forest. The arctic and alpine tundras exhibit only the preclimax relation, to boreal and subalpine forest respectively, since a potential lichen climax attains but incomplete expression northward or upward. While the general primary relation is one of water in terms of rainfall and evaporation, temperature constantly enters the situation and at the extremes may be largely controlling, as in the tundra especially. However, in our present imperfect knowledge of causal factors it is simpler and more definite to determine rank by position in the cliseral sequence, each community higher in altitude or latitude being successively preclimax to the preceding one. This relation is likewise entirely consistent in the clisere from deciduous forest to desert, as it is among the associations of the same climax, though in both these cases the zonal grouping may be more or less obscured.

Wherever concrete preclimaxes or postclimaxes occur, either between climaxes or within a single one, they are due to the compensation afforded by edaphic situations. The major examples of the latter are provided by valleys, especially gorges and canyons, long and steep slope-exposures, and by extreme soil-types such as sand and alkali. The seration is a series of communities produced by a graduated compensation across a valley and operating within a formation or through adjacent ones, while the ecocline embraces the differentiation brought about by shifting slope-exposures around a mountain or on the two sides of a high ridge. In the case of such soils as sand or gravel at one extreme and stiff clay at the other, the edaphic adjustment may sometimes appear contradictory. Thus, sand affords a haven for postclimax relicts in the dry prairie and for preclimax ones in the humid forest region, while the effect of heavy soils is just the reverse. However, this is readily intelligible when one recalls the peculiar properties of such soils in terms of absorption, chresard and evaporation (Clements, 1933).

Preclimax

Since they occupy the same general antecedent position with respect to the climax, it is necessary to distinguish with some care between subclimax and preclimax, especially in view of the fact that they often exhibit the same life-form. However, this is not difficult when the priseres and subseres have been investigated in detail, as the actual composition and behavior of the two communities are usually quite different. Moreover, in the first, reaction leads to the entry of the climax dominants with ultimate conversion, while in the second the compensation by local factors is rarely if ever to be overcome within the existing climate, short of man-made disturbance.

Preclimaxes are most clearly marked where two adjacent formations are concerned, either prairie and forest or desert and prairie. Examples of the first kind are found in the grassy "openings" and oak savannahs of the deciduous forest and in the so-called "natural parks" along the margin of the montane and boreal forests. They are also well developed on warm dry slope-exposures or xeroclines in the Rocky Mountains. In the one, compensation is usually afforded by a sandy or rocky soil, in the other by a local climate due to insolation. Desert climaxes regularly bear the proper relation to circumjacent grassland, but this is somewhat obscured by the shrub life-form, which would be expected to characterize the less xeric formation. This may be explained, however, by the wide capacity for adaptation shown by such major dominants as *Larrea tridentata* and *Artemisia tridentata*, a quality that is lacking in most of their associates. Left stranded as relict communities in desert plains and mixed prairie by the recession of the last dry phase, they have profited by the overgrazing of grasses to extend across a territory much larger than that in which they are climax. Here they have all the appearance of a postclimax, especially in the case of *Larrea*, which commonly attains a stature several times that found in the desert. However, since this is the direct outcome of disturbance in terms of grazing, it is better regarded as a disclimax, particularly since the climax grasses still persist in it to some degree.

Within the same formation, the more xeric associations or consociations are preclimax to the less xeric ones. This is the general relation between the oak-hickory and beech-maple associations of the deciduous forest, the former occupying in the latter the warmer drier sites produced by insolation or type of soil. A similar relation may obtain in the case of faciations, the *Quercus stellata-marilandica* community often being a border of marginal preclimax to the more mesic oak-hickory faciations. Such preclimaxes naturally persist beyond the limits of the association proper as relicts in valleys or sandy soils and then assume the role of postclimaxes to the surrounding grassland, a situation strikingly exemplified in the "Cross Timbers" of Texas. In the montane forest of the Rockies, the consociation of *Pinus ponderosa* is preclimax to that of *Pseudotsuga taxifolia*, and a similar condition recurs in all forests where there is more or less segregation of consociations.

In the mixed prairie, fragments of the desert plains occur all along the margin as preclimaxes, the most extensive one confronting the Colorado Valley, where it is at the same time postclimax to the desert. The mixed prairie constitutes relicts of this type where it meets the true prairie. The most frequent examples are provided by *Bouteloua gracilis* and *Sporobolus cryptandrus*, though as with all the short-grasses in this role, grazing has played some part.

Postclimax

As a general rule, postclimax relicts are much more abundant than those that represent preclimaxes, owing in the first place to the secular trend toward desiccation in climate and in the second to the large number of valleys, sandhills and sandy plains, and escarpments in the grassland especially. Postclimaxes of oak-hickory and of their flood-plain associates, elm, ash, walnut, etc., are characteristic features of the true and mixed prairies, holding their own far westward in major valleys but limited as outliers on ridges and sandy stretches to the eastern edge. However, the compensation afforded by the last two is incomplete as a rule and the postclimax is typically reduced to the savannah type. The latter is an almost universal feature where forest, woodland or chaparral touches grassland, owing to the fact that shrinkage under slow desiccation operates gradually upon the density and size of individuals. Savannah is derived from the reduction of deciduous forest along the eastern edge of the prairie, of the aspen subclimax of the boreal forest along the northern, and of the montane pine consociation, woodland or chaparral on the western and southern borders, recurring again on the flanks of the Sierras and Coast Ranges in California. On the south, the unique ability of the mesquite (*Prosopis juliflora*) to produce root-sprouts after fire, its thorniness, palatable pods and resistant seeds have permitted it to produce an extensive savannah that often closely simulates a true woodland climax.

As would be expected, a point is reached in the reduction of rainfall westward in the prairie where sand no longer affords compensation adequate for trees. In general this is along the isohyet of 30 in. in the center and south, and of about 20 in. in the north. Southward from the parallel of 37° the further shrinkage of the oak savannah may be traced in the "shinry", which dwindles from four or five feet to dwarfs only "shin" high. With these are associated tall-grasses, principally *Andropogon* and *Calamovilfa* in the form *gigantea*. To the north of this line, the shin oaks are absent and the tall-grasses make a typical postclimax that extends into Canada, though the compensatory influence of sand is still sufficient to permit an abundance of such low bushes as *Amorpha*, *Ceanothus*, *Artemisia filifolia* and *Yucca*, as well as depauperate hackberry and aspen. In the vast sandhill area of central Nebraska, the tall-grass postclimax attains its best development, which is assumed to reflect the climate when the prairies were occupied by the Andropogons and their associates some millions of years ago. The gradual decrease to the rainfall of the present has led to the tall-grasses finding refuge in all areas of edaphic compensation, not only in sand but likewise on foothills and in valleys, and in addition along the front of the deciduous forest.

Structure of the climax

Community functions

The nature of community functions and their relation to the structure of climax and sere have been discussed in considerable detail elsewhere (*Plant Succession*, 1916, 1928; *Plant Ecology*, 1929; *Bio-ecology*, 1936), and for the present purpose it may well suffice to emphasize the difference in significance between major or primary and minor or secondary functions. The former comprise aggregation, migration, ecesis, reaction, competition, cooperation, disoperation, and coaction, together with the resulting complexes, invasion and succession. Any one of these may have a profound effect upon community structure, but the driving force in the selection and grouping of life-forms and species resides chiefly in reaction, competition, and coaction. Migration deals for the most part with the movement and evolution of units under climatic compulsion, and succession with the development and regeneration of the climax in bare or denuded areas.

In contrast to these stands the group of minor functions that are concerned with numbers and appearance or visibility as it may be termed. The first is annuation, in accordance with which the abundance of any species may fluctuate from dry to wet phases of the various climatic cycles or the growth differ in terms of prominence, the two effects not infrequently being combined. For the grassland, a season of rainfall more or less extreme in either direction often emphasizes one dominant at the expense of others, though the balance is usually redressed by the following year, while in the desert in particular the swing in number of annuals may be from almost complete absence to seasonal dominance, again with one or few species taking the major role. Aspection is mainly the orderly procession of societies through each growing season, more or less modified by changes in number ensuing from annuation. Hibernation and estivation merely affect seasonal appearance and are forms of aspection, with the temporary suspension of coaction effects. While usually applied to the animal members of the biome, it is obvious that plants exhibit certain responses of similar nature. Diurnation is likewise best known in the case of animals, especially nocturnal ones, but it is exhibited also by the vertical movement of plankton and in different form by the opening and closing of flowers and the "sleep" movements of leaves.

Roles of constituent species: dominants

The abundant and controlling species of characteristic life-form were long ago termed dominants (Clements, 1907, 1916), this property being chiefly determined by the degree of reaction and effective competition. In harmony with the concept of the biome, it has become desirable to consider the role of animals likewise; since their influence is seen chiefly in coaction by contrast to the reaction of plants, the term *influent* has been applied to the important species of land biomes (cf. Clements and Shelford, 1936). It is an axiom that the life-form of the dominant trees stamps its character upon forest and woodland, that of the shrub upon chaparral and desert, and the grass form on prairie, steppe and tundra. There are seral dominants as well as climax ones, and these give the respective impresses to the stages of prisere and subsere. Finally, there are considerable differences in rank or territory even among the dominants of each formation. The most important are those of wide range that bind together the associations of a climax; to these the term *perdominant* (*per*, throughout) may well be applied. In contrast to these stand the dominants more or less peculiar to each association, such as beech or chestnut in their respective communities and *Sporobolus asper* in the true and *Stipa comata* in the mixed prairie, for which *eudominant* may be employed.

Subdominants regularly belong to a life-form different from that of the dominants and are subject to the control of the latter in a high degree, as the name indicates. They are best exemplified by the perennial forbs, though biennials and annuals may serve as seral subdominants; all three may be actual dominants in the initial stages of succession and especially in the subsere. The term *codominant* has so far had no very definite status; it is hardly needed to call attention to the presence of two or more dominants, since this is the rule in all cases with the exception of consociation and consocies. In contrast to the types mentioned stands a large number of secondary or accessory species that exhibit no dominance, which may be conveniently referred to as *edominants*, pending more detailed analysis.

Influents

As indicated previously, the designation of *influent* is applied to the animal members of the biome by virtue of the influence or coaction they exert in the community. The significance of this effect depends much upon the life-form and to a large degree upon the size and abundance of the species as well, and is seen chiefly in the coactions involved in food, material, and shelter. Influents may be grouped in accordance with these properties, or they may be arranged with respect to distribution and role in climax or sere, or to time of appearance (Clements and Shelford, 1936). For general purposes it is perhaps most convenient to recognize subdivisions similar to those for dominants and with corresponding terms and significance. Thus, a *perfluent* would occur more or less throughout the formation, while the *eufluent* would be more or less typical or peculiar to an association. *Subfluents* would mark the next lower degree of importance, roughly comparable to that of subdominant, while minute or microscopic influents of still less significance might well be known as *vefluents*.

Climax and seral units

No adequate analysis of vegetation or of the biome is possible without taking full account of development. As the first step, this involves a distinction between climax communities proper and those that constitute the successional movement toward the final stages. The two groups differ in composition, stability, and type of control, but they agree in the possession of dominants, subdominants, and influents. These primary differences made it desirable to recognize two series of communities, viz. climax and seral and to propose corresponding terms, distinguished by the respective suffixes, *-ation* and *-ies* (Clements, 1916). These have gradually come into use as the feeling for dynamic ecology has grown and bid fair to constitute a permanent basis for all such studies. It is not supposed that they embrace all the units finally necessary for a complete system, but their constant application to the great climaxes of North America for nearly two decades indicates that they meet present needs in the matter of analysis.

Not all communities can be certainly placed in the proper category at the outset, but the number of doubtful cases is relatively small and few of these present serious difficulty under combined extensive and intensive research. This statement, however, presupposes an experience sufficiently wide and long to permit distinguishing between climaxes and the various types of proclimax, as well as recognizing the characteristic features of subclimaxes in particular. Comparative studies over a wide region are indispensable and the difficulties will disappear to the degree that this is achieved. While ecotones and mictia necessarily give rise to some questions in this connection, these in turn are resolved by investigations as extensive as they are detailed.

Climax units

In the organization of these, four types of descending rank and importance were distinguished within the formation, namely, association, consociation, society, and clan. Like the formation itself, the first two were based upon the dominant and its life-form, while the last were established upon the subdominant and its different life-form. It was recognized at the time that the association contained within itself other units formed by the dominants (cf. *Plant Indicators*, 1920, pp. 107, 276), and two further divisions, *faciation* and *lociation*, with corresponding seral ones, *facies* and *locies*, were suggested and submitted to Prof. Tansley for his opinion as to their desirability. These have been tested in the course of further field studies and have now and then been used in print, though the complete series was not published until 1932 (cf. Shelford, 1932). The climax group now comprises the following units, viz. association, consociation, faciation, lociation, society, and clan. At the beginning, it was intended to replace society by *sociation* for the sake of greater uniformity in terms, but the former had attained such usage that the idea was relinquished. However, the use of society in quite a different sense by students of social relations, especially among insects, again raises the question of the desirability of such a substitution, in view of the growing emphasis upon bio-ecology (cf. also Du Rietz, 1930; Rübel, 1930).

Association.

Under the climax concept this represents the primary division of the biome or formation, and hence differs entirely from the generalized unit of the plant sociologists, for which the term *community* is to be preferred. Each biome consists regularly of two or more associations, though the lake forest and the desert scrub embody two apparent exceptions, each seeming to consist of one association only. However, these are readily explained by the fact that the western member of the former has been obscured by the expansion of montane and coast forests in the north-west, while one or more additional associations of the desert climax occur to the southward in Mexico, and apparently in South America also.

The number of associations in a particular formation is naturally determined by the number of primary differences and these in turn depend upon the presence of eudominants. Just as the unity of the formation rests upon the wide distribution of several major dominants or perdominants, so the association is also marked by one or more dominants peculiar to it, and often as well by differences in the rank and grouping of dominants held in common. Thus, in the true prairie association, the eudominants are *Stipa spartea*, *Sporobolus asper* and *heterolepis*; for the desert plains, *Bouteloua eriopoda*, *rothrocki* and *radicosa* and *Aristida californica*, while *Stipa comata* and *Buchloe* take a similar part in the mixed prairie. In the deciduous climax, the characteristic dominants of one association are supplied by the beech and hard maple, of a second by chestnut and chestnut-oak, though the oak-hickory association, of wider range and greater complexity, is comparatively poor in eudominants by contrast with the number of species.

The structural and phyletic relations of the associations of a climax are best illustrated by the grassland, which is the most highly differentiated of all North American formations, largely as an outcome of its great extent. The most extensive and varied unit is the mixed prairie, which occupies a generally median position with respect to the other five associations of this climax. Originally, it derived its dominants from three separate regions, *Stipa*, *Agropyrum* and *Koeleria* coming from Holarctica, *Sporobolus* from the south, and the short-grasses from the Mexican plateaux, and it still exhibits the closest kinship with the Eurasian steppe. It contains nearly all the genera that serve as dominants in the related associations, while many of the eudominants of these have all the appearance of direct derivatives from its species, as is shown by *Stipa*, *Sporobolus*, *Poa*, and *Agropyrum*. The evolution of both species and communities is evidently in response to the various subclimates, that of the true prairie being moister, of the coastal warmer as well; the desert plains are hotter and drier, the California prairie marked by winter rainfall and the Palouse by snowfall.

Consociation.

In its typical form the consociation is constituted by a single dominant, but as a matter of convenience the term is also applied to cases in which other dominants are but sparingly present and hence have no real share in the control of the community. It has likewise been convenient to refer in the abstract to each major dominant of the association as a consociation, though with the realization that it occurs more frequently in mixture than by itself. In this sense it may be considered a unit of the association, though the actual area of the latter is to be regarded as divided into definite faciations. Consociation dominants fall into a more or less regular series with respect to factor requirements, especially water content, and often exhibit zonation in consequence. This is a general feature of mixed prairie where *Agropyrum smithi* and *Stipa comata* are the chief mid-grasses, the former occupying swales and lower slopes, the latter upper slopes and ridges.

The consociation achieves definite expression over a considerable area only when the factors concerned fluctuate within the limits set by the requirements of the dominant or when the other dominants are not found in the region. The first case may be illustrated by *Pinus ponderosa* in the lower part of the montane forest and by *Adenostoma fasciculatum* in the Sierran chaparral, while the second is exemplified by *Picea engelmanni* in the Front Range of Colorado, its usual associate, *Abies lasiocarpa*, being absent from the district. In rolling terrain like that of the prairie, each consociation will recur constantly in the proper situation but is necessarily fragmentary in nature. Such behavior is characteristic of dominants with a postclimax tendency, as with *Stipa minor* and *Elymus condensatus* in swales and lower levels of the mixed prairie.

Faciation.

This is the concrete subdivision of the association, the entire area of the latter being made up of the various faciations, except for seral stages or fragments of the several consociations. Each faciation corresponds to a particular regional climate of real but smaller differences in rainfall/evaporation and temperature. It may be characterized by one or two eudominants, such as *Hilaria jamesi* and *Stipa pennata* in the southern mixed prairie, but more often it derives its individuality from a sorting out or a recombination of the dominants of the association. As is evident, the term is formed from the stem *fac-*, show, appear, as seen in *face* and *facies*, and the suffix *-ation,* which denotes a climax unit.

During the past decade, much attention has been given to the recognition and limitation of faciations on the basis of the presence or absence of a eudominant, such as *Hilaria*, *Buchloe*, or *Carex*, or a change in the rank or grouping of common dominants, like *Stipa*, *Agropyrum*, *Sporobolus* or *Bouteloua*. In the prairie this task has been complicated by overgrazing, cultivation and related disturbances, while selective lumbering and fire have added to the difficulties, in the deciduous forest especially. In general, temperature appears to play the leading part in the differentiation of faciations, since they usually fall into a sequence determined by latitude or altitude, though rainfall/evaporation is naturally concerned also. The mixed prairie exhibits the largest number, but it is approached in this respect by the deciduous forest as a consequence of wide extent and numerous dominants. Over the Great Plains from north to south, the successive faciations are *Stipa-Bouteloua*, *Bouteloua-Carex*, *Stipa-Agropyrum-Buchloe*, *Bouteloua-Buchloe*, *Hilaria-Stipa-Bouteloua*, and *Agropyrum-Bouteloua*. However, the short-grass communities are to be regarded as disclimaxes wherever the mid-grasses have been eliminated or nearly so, a condition that fluctuates in relation to dry and wet phases of the climatic cycle.

Lociation.

In its turn, the lociation is the subdivision of the faciation, the term being derived from *locus*, place, as indicating a general locality rather than a large region. Nevertheless, a lociation may occupy a relatively extensive territory up to a hundred miles or more in extent, by comparison with several hundred for the faciation. It is characterized by more or less local differences in the abundance and grouping of two or more dominants of the faciation. These correspond to considerable variations in soil, contour, slope-exposure or altitude, but all within the limits of the faciation concerned. As a consequence, lociations are very often fragmented, recurring here and there as alternes with each other, and frequently with proclimaxes of various types. Like most climax units, they have been modified by disturbance in some degree, and this fact must be constantly kept in mind in the task of distinguishing them from subclimax or disclimax.

A detailed knowledge of the faciation is prerequisite to the recognition of the various lociations in it. The number for a particular faciation naturally depends upon the extent of the latter and the number of dominants concerned. Consequently, lociations are more numerous in the faciations of the mixed prairie and desert plains, of the chaparral and the oak-hickory forest. As would be expected, they are often most distinct in ecotones and in districts where there is local intrusion of another dominant. In correspondence with their local character, it is important to eliminate or diminish superimposed differences through restoration of the original cover by means of protection enclosures and thus render it possible to disclose the true composition.

Society.

This term has had a wide range of application, but by dynamic ecologists it has generally been employed for various groupings of subdominants, of which those constituted by aspects or by layers are the most important. In addition, there is a host of minor communities formed by cryptogams in the ground layer or on host-plants and other matrices. The soil itself represents a major layer, divisible into more or less definite sublayers. Animals regularly assume roles of varying importance in all of these, especially the insects, arachnids and crustacea, and hence most if not all societies comprise both subdominant plants and subinfluent animals. It is doubtful whether animals form true societies independently of their food-plants or those used for materials or shelter, but this is a question that can be answered only after the simplest units, namely, family and colony, have been recognized and coordinated in terms of their coactions.

In view of what has been said previously, it seems desirable to employ society as the general term for all communities of subdominants and subinfluents above the rank of family and colony, much as community is the inclusive term for all groupings of whatsoever rank. This then permits carrying out the suggestion made two decades ago that the major types of societies be set apart by distinctive names. In accordance, it is here proposed to call the aspect society a *sociation* and the layer society a *lamiation*, while the corresponding seral terms would be *socies* and *lamies*. Many of the societies of cryptogams and minute animals would find their place in these, particularly so for those of the surface and soil layers, but many others take part in a miniature sere or *serule*, such as that of a moldering log, and may best receive designations that suggest this relation.

Sociation. Wherever societies are well developed, they regularly manifest a fairly definite seasonal sequence, producing what have long been known as aspects (Pound and Clements, 1898). As phenomena of the growing season, these were first distinguished as early spring or prevernal, vernal proper, estival, and serotinal or autumnal, but there may also be a hiemal aspect, especially for animals, in correspondence with an actual and not merely a calendar winter as in California.

Sociations are determined primarily by the relation between the life cycle of the subdominants and the seasonal march of direct factors, temperature in particular. So far as the matrix of plants is concerned, the constituent species may be in evidence throughout the season, but they give character to it only during the period of flowering, or fruiting in the case of cryptogams. They are present largely or wholly by sufferance of the dominants, and they are to be related to the reactions of these and competition among themselves rather more than to the habitat factors as such. In grassland and desert, they are often more striking than the dominants themselves, sometimes owing to stature but chiefly as an effect of color and abundance, and they may also attain much prominence in woods with the canopy not too dense.

Sociations are usually most conspicuous and best developed in grassland, four or even five distinct aspects occurring in the true prairie from early spring to autumn. In the mixed prairie these are usually reduced to three, and in the desert plains and desert proper, to two major ones, summer and winter, in which however there may be subaspects marked by *sations*, as indicated later. Short seasons due to increasing latitude or altitude afford less opportunity, and the tundra, both alpine and arctic, usually exhibits but two aspects, the sociations however taking a conspicuous role as in the prairie. In woodland the number and character of the sociations depend largely upon the nature of the canopy, and for deciduous forest the flowery sociations regularly belong to the spring and autumn aspects, when the foliage is either developing or disappearing.

It is convenient to distinguish sociations as simple or mixed with respect to the plant matrix in accordance with the presence of a single subdominant or of two or more. However, when animals are included in the grouping, such a distinction appears misleading and may well be dropped. The word "mixed" would be more properly applied to plant-animal societies were it not for the fact that this appears to be the universal condition. Since seasonal insects are legion, many of the societies in which they take part are best denoted as sations.

Lamiation. The term for a layer society is derived from the stem *lam-*, seen in lamina and lamella. As is well known, layers are best developed in forests with a canopy of medium density, so that under the most favorable conditions as many as five or six may be recognized above the soil. In such instances, there are usually two shrub stories, an upper and lower, often much interrupted, followed by tall, medium and low forb layers, and a ground community of mosses, lichens and other fungi, and usually some delicate annuals (cf. Hult, 1881; Grevillius, 1894). The soil population is perhaps best treated as a single unit, though it may exhibit more or less definite sublayers. When the various layers beneath the dominants are distinct, each is regarded as a lamiation, but in many cases only one or two are sufficiently organized to warrant designation, e.g. shrub or tall-herb lamiation.

Layers are often reduced to a single lamiation of low herbs in the climax forest, especially of conifers, and even this may be entirely lacking in dense chaparral. Two or three layers of forbs may be present in true prairie in particular, the upper lamiation being much the most definite and often concealing the grass dominants in the estival aspect, but the structure of grassland generally reflects the greater importance of sociations. Climaxes of sagebrush and desert scrub exhibit no proper lamiations, owing to the interval between the individual dominants, but the herbaceous societies of the interspaces show something of this nature.

As a rule, well-developed lamiations also manifest a seasonal rhythm, corresponding to aspects or to subaspects. These constitute recognizable groupings in the lamiation and for the sake of determination and analysis may be termed *sations*. This word is a doublet of season, both being derived from the root *sa-*, sow, hence grow or appear. Because of the frequent interplay of aspects and layers, the sation may for the present be employed for the subdivision of both sociation and lamiation, and especially where seasonal species of invertebrates play a conspicuous role.

Clan. This is a small community of subordinate importance but commonly of distinctive character. It is marked by a density that excludes all or nearly all competing species, in consequence of types of propagation that agree in the possession of short offshoots. Extension is usually by bulb, corm, tuber, stolon or short rootstock, each of which produces a more or less definite family grouping; in fact, most clans are families developed in the climax matrix and sometimes with a blurred outline in consequence. Clans of a particular species such as *Delphinium azureum* or *Solidago mollis* are dotted throughout the respective sociation, often in large numbers, and contribute a distinctive impress much beyond their abundance. Like all units, but small ones especially, they are subject to much fluctuation with the climatic cycle, as a result of which they may pass into societies or be formed by the shrinkage of the latter.

Seral units.

The concepts of dominance and subdominance apply to the sere as they do to the climax, as does that of influence also, and the corresponding sets of units bear the same general relation to each other. Each of the four major units is the developmental equivalent of a similar community in the climax series and this is likewise true of the various kinds of societies. They constitute the successive stages of each sere, both primary and secondary, including the subclimax, where they often achieve their best expression. It has also been customary to employ seral terms for preclimax and postclimax, and this appears to be the better usage for disclimax and proclimaxes in general. From the fragmentary nature of bare areas and suitable water bodies in particular, seral communities are often but partially developed and one or more units will be lacking in consequence. Thus, the reed-swamp associes is frequently represented by a single one of its several consocies and the minor units are even more commonly absent.

The associes is the major unit of every sere, the number being relatively large in the prisere and small in the subsere. The universal and best understood examples are those of the hydrosere, in which *Lemna, Potamogeton, Nuphar, Nymphaea, Nelumbo* and others form the consocies of the floating stage, and *Scirpus, Typha, Phragmites*, etc., are the dominants of the reed-swamp or amphibious associes. As already indicated, every consocies may occur singly and often does when the habitat offers just the proper conditions for it or the others have failed to reach the particular spot. When the ecial range is wider, various combinations of two or three dominants will appear, to constitute corresponding facies. Locies are less definitely marked as a rule, except in swamps of vast extent, but are to be recognized by the abundance of reed-like dominants of lower stature, belonging to other species of *Scirpus*, to *Heleocharis, Juncus*, etc. Both facies and locies seem to be better developed in sedge-swamp with its larger number of dominants, though the Everglades with the single consocies of *Cladium* form a striking exception.

The tree-swamps of the south-eastern United States contain a considerable number of consocies, such as *Taxodium distichum*, *Nyssa aquatica*, *biflora* and *ogeche*, *Carya aquatica*, *Planera aquatica*, *Persea palustris* and *borbonia*, *Magnolia virginiana*, *Fraxinus pauciflora*, *profunda* and *caroliniana*, and *Quercus nigra*. These are variously combined in several different facies, though a more detailed and exact study of the swamp sere may show the presence of two woody associes, distinguished by the depth or duration of the water. As with other scrub communities, the heath associes of peat-bog and muskeag comprises a large number of dominants and presents a corresponding wealth of facies and locies.

In the hydrosere of the deciduous forest, the typical subclimax is that of the flood-plain associes, composed of species of *Quercus*, *Ulmus*, *Fraxinus*, *Acer*, *Betula*, *Juglans*, *Celtis*, *Platanus*, *Liquidambar*, *Populus* and *Salix* for the most part. There are at least three well-marked facies, namely, northern, central and southern, each with a number of more or less distinct locies. The swamp associes or subclimax of the lake and boreal forests consists of *Larix laricina*, *Picea mariana* and *Thuja occidentalis*, occurring often as consocies but generally in the form of zoned facies. A large number of fire subclimaxes appear in the form of consocies, as with many of the pines, but associes are frequent along the Atlantic Coast, as they are in the boreal climax, where aspens and birches are chiefly concerned. The number of shrubs and small trees that play the part of seral dominants in the deciduous climax is much larger, producing not merely a wide range of associes but of facies and locies as well. More than a dozen genera and a score or so of species are involved, chief among them being *Sassafras*, *Diospyrus*, *Asimina*, *Hamamelis*, *Prunus*, *Ilex*, *Crataegus* and *Robinia*. The subclimax of the xerosere is constituted for the most part by species of *Quercus*, forming an eastern, a south-eastern and a western associes, the last with two well-marked facies, one of *stellata* and *marilandica*, the other of *macrocarpa* and *Carya ovata*.

Among postclimax associes, those of grassland and scrub possess a large number of dominants and exhibit a corresponding variety of facies and locies, together with fairly definite consocies. In the sandhills of Nebraska, the tall-grasses concerned are *Andropogon halli*, *furcatus* and *nutans*, *Calamovilfa longifolia*, *Eragrostis trichodes*, *Elymus canadensis* and *Panicum virgatum*; some of these drop out to the northward and others to the south, thus producing at least three regional facies. The mesquite-acacia associes of the south-west possesses a larger number of dominants and manifests a greater variety of facies through its wide area, and this is likewise true of the coastal sagebrush of California.

With reference to seral societies, it must suffice to point out that these are of necessity poorly developed in the initial stages of both hydrosere and xerosere, as the dominants are relatively few. Even in the reed-swamp, true layers are the exception, being largely restricted to such subdominants as

Alisma, *Pontederia*, *Hydrocotyle* and *Sagittaria*, which are found mostly in borders and intervals. However, in extensive subclimaxes and postclimaxes the situation is quite different. The tall-grass associes of sandhills is often quite as rich in saties and lamies as the true prairie, while the various subclimaxes of the several great forest types may equal the latter in the wealth of subdominants for each season and layer, the actual communities being very much the same.

Serule.

This term, a diminutive of *sere*, has been employed for a great variety of miniature successions that run their short but somewhat complex course within the control of a major community, especially the climax and subclimax. They resemble ordinary seres in arising in bare spots or on matrices of different sorts, such as earth, duff, litter, rocks, logs, cadavers, etc. Parasites and saprophytes play a prominent and often exclusive role in them, and plants and animals may alternate in the dominant parts. The organisms range from microscopic bacteria and worms to mites, larvae and imagoes on the one hand and large fleshy and shelf fungi on the other. The most important of these in terms of coaction and abundance are known as *dominules* (Clements and Shelford, 1936), with *subdominule* and *edominule* as terms for the two degrees of lesser importance. On the same model are formed *associule*, *consociule*, and *sociule* in general correspondence with the units of the sere itself. In addition there are families and colonies of these minute organisms, which are essentially similar to those of the initial stages of the major succession. Up to the present, little attention has been devoted to the development and structure of serules, but they are coming to receive adequate consideration in connection with bio-ecological problems. Many of the coactions, however, have long been the subject of detailed research in the conversion of organic materials.

Rank and correspondence of units

The following table exhibits the actual units of climax and sere, as well as their correspondence with each other. However, for the complete and accurate analysis of a great climax and especially the continental mass of vegetation, it is necessary to invoke other concepts, chiefly that of the proclimax and of communities mixed in space or in time. The several proclimaxes have been characterized (pp. 262–8), and the ecocline and seration briefly defined (p. 267). To these are to be added the *ecotone* and *mictium*, both terms of long standing, the former applied to the mixing of dominants between two units, the latter to the mixed community that intervenes between two seral stages or associes. Finally, there will be the several types of seres in all possible stages of development, the prisere in the form of hydrosere, xerosere, halosere or psammosere in regions less disturbed and a myriad of subseres in those long settled.

TABLE OF CLIMAX AND SERAL UNITS

Climax		Sere
	Eoclimax	
	Panclimax	
	Climax (formation)	
Association		Associes
Consociation		Consocies
Faciation		Facies
Lociation		Locies
Sociation		Socies
Lamiation		Lamies
Sation		Saties
Clan		Colony
		Family
	Serule	
	Associule	
	Consociule	
	Sociule	

As indicated previously, the word *community* is employed as a general term to designate any or all of the preceding units, while *society* may well be used to include those of the second division, i.e. sociation, etc. These are characterized by subdominants in contrast to the dominants that mark the first group. It has also been pointed out that the entire area of the association is divided into faciations and that the consociation is the relatively local expression of complete or nearly complete dominance on the part of a single species. The clan corresponds to the family as a rule, but in some cases resembles the colony in being formed by two species.

Families and colonies may also appear in climax communities, but this is regularly in connection with the serule.

Panclimax and eoclimax

The comprehensive treatment of these concepts is reserved for the succeeding paper in the present series, but it is desirable to characterize them meanwhile. The *panclimax* (παν, all, whole) comprises the two or more related climaxes or formations of the same general climatic features, the same life-form and common genera of dominants. The relationship is regarded as due to their origin from an ancestral climax or *eoclimax* (ἠώς, dawn), of Tertiary or even earlier time, as a consequence of continental emergence and climatic differentiation. In the past, eoclimaxes formed a series of great biotic zones in the northern hemisphere with the pole as a focus, and this zonal disposition or clisere is still largely evident in the arrangement of panclimaxes at the present. It is striking in the case of the arctic tundra and taiga or boreal forest, fairly evident for deciduous forest and prairie-steppe, and somewhat obscure for woodland and chaparral-macchia, while the position of deserts is largely determined by intervening mountain ranges. This is true likewise of grassland in some degree, and taken with the former broad land connection between North and South America explains why both prairie and desert panclimax contain at least one austral formation.

In the light of what has been said earlier, it is readily understood that panclimax and panformation are exact synonyms, as are eoclimax and eoformation. Panbiome and eobiome are the corresponding terms when the biotic community is taken as the basis for research.

Prerequisites to research in climaxes

It would be entirely superfluous to state that the major difficulty in the analysis of vegetation is its complexity, were it not for the fact that it is too often taken as the warrant for the static viewpoint. This was embodied in the original idea of the formation as a unit in which communities were assembled on a physiognomic basis, quite irrespective of generic composition and phyletic relationship. It is not strange that this view and its corollaries should have persisted long past its period of usefulness, since this is exactly what happened with the artificial system of Linnaeus, but the time has come to recognize fully that a natural system of communities must be built just as certainly upon development and consequent relationship as must that of plant families. Complexity is an argument for this rather than against it, and especially in view of the fact that the complexity discloses a definite pattern when the touchstone of development is applied to it.

Though the mosaic of vegetation may appear to be a veritable kaleidoscope in countries long occupied by man, the changes wrought upon it are readily intelligible in terms of the processes concerned. As emphasized previously, the primary control is that of climate, in a descending scale of units that correspond to formation, association, and faciation. Upon this general pattern are wrought the more circumscribed effects of physiography and soil, and both climatic and edaphic figures are overlaid and often more or less completely obscured with a veneer applied by disturbance of all possible kinds. Even above this may be discerned the effect, transient but nonetheless apparent, of such recurrent changes as annuation and aspection. Moreover, the orderly pattern of climate is complicated by great mountain ranges so that such climaxes as tundra and taiga occur far beyond their proper zone, and the effect is further varied by the relative position of the axis.

The migrations of climaxes in the past are a prolific source of fragmentary relicts, the interpretation of which is impossible except in terms of dynamics. This is likewise true of savannah, which represents the shrinkage of forest and scrub under a drying climate and is then usually further modified by fire or grazing. Fragmentation from this and other causes is characteristic of every diversified terrain and reaches its maximum when human utilization enters the scene upon a large scale. Somewhat similar in effect though not in process is the reduction of number of dominants by distance, with the consequence that an association of several may be converted into a consociation of one. Such a shrinkage naturally bears some relation also to climate and physiography, especially as seen in the glacial period, and finds its best illustration in the general poverty of dominants in the coniferous and deciduous climaxes of Europe, by contrast with those of eastern Asia and North America. A similar contrast obtains between the grassland of Asia and of North America, the latter being much richer in dominants, while South America approximates it closely in this respect.

On the part of the investigator, the difficulties in the way of an extensive and thoroughgoing study of climaxes are usually more serious. They arise partly from the handicap too often set by state or national boundaries and partly from the limitations of funds and time. They are also not unrelated to the fact that it is easiest to know a small district well and to assume that it reflects larger ones with much fidelity. As a consequence, it is impossible to lay too much stress upon the need for combining intensive and extensive methods in the research upon climaxes, insofar as their nature, limits and structure are concerned. The detailed development in terms of primary and secondary succession lends itself much more readily to local or regional investigation, but even here a wider perspective is essential to accurate generalization.

REFERENCES

Bradbury, J. "Travels in the Interior of North America, *ca.* 1815." In **Thwaites'** *Early Western Travels*, **5**, 1904.

Bromley, S. W. "The original forest types of southern New England." *Ecol. Mon.* **5**, 61–89, 1935,

Chaney, R. W. "A comparative study of the Bridge Creek flora and the modern redwood forest." *Publ. Carneg. Instn*, No. 349, 1925.

Chaney, R. W. and **E. I. Sanborn.** "The Goshen Flora of West-central Oregon." *Publ. Carneg. Instn*, No. 439, 1933.

Chapman, H. H. "Is the longleaf type a climax?" *Ecology*, **13**, 328–34, 1932.

Clements, F. E. *Plant Physiology and Ecology*, New York, 1907.

Clements, F. E. *Plant Succession*, Washington, 1916.

Clements, F. E. "Development and structure of the biome." *Ecol. Soc. Abs.* 1916.

Clements, F. E. *Plant Indicators*, Washington, 1920.

Clements, F. E. "Phylogeny and classification of climaxes." *Yearb. Carneg. Instn*, **24**, 334–5, 1925.

Clements, F. E. *Plant Succession and Indicators*, New York, 1928.

Clements, F. E. "The relict method in dynamic ecology." This JOURN. **22**, 39–68, 1934.

Clements, F. E. "Origin of the desert climate and climax in North America." This JOURN. **24**, 1936.

Clements, F. E. and **E. S. Clements.** "Climate and climax." *Yearb. Carneg. Instn*, **32**, 203, 1933.

Clements, F. E. and **V. E. Shelford.** *Bio-ecology*, 1936.

Du Rietz, G. E. "Classification and nomenclature of vegetation." *Svensk. Bot. Tid.* **24**, 489, 1930.

Godwin, H. "The subclimax and deflected succession." This JOURN. **17**, 144, 1929.

Greeley, H. *An Overland Journey to California in* 1859, New York, 1860.

Grevillius, A. Y. "Biologisch-physiologische Untersuchungen einiger Schwedischen Hainthälchen." *Bot. Z.* **52**, 147–68, 1894.

Hult, R. "Forsök til analytisk behandling af växformationerna." *Medd. Soc. Faun. Flor. Fenn.* **8**, 1881.

Hult, R. "Blekinges vegetation. Ett bidrag till växformationernas utvecklingshistorie." *Medd. Soc. Faun. Flor. Fenn.* **12**, 161, 1885. *Bot. Zbl.* **27**, 192, 1888.

Lewis, M. and **W. Clark.** Journal, 1803–1806. In **Thwaites'** *The Original Journals of the Lewis and Clark Expedition*, 1904–05.

Phillips, J. "The biotic community." This JOURN. **19**, 1–24, 1931.

Phillips, J. "Succession, development, the climax and the complex organism: an analysis of concepts. Part I." This Journ. **22**, 554–71, 1934.

Phillips, J. "Succession, development, the climax and the complex organism: an analysis of concepts. Part II." This Journ. **23**, 210–46, 1935.

Phillips, J. "Succession, development, the climax and the complex organism: an analysis of concepts. Part III." This Journ. **23**, 488–508, 1935.

Rübel, E. *Pflanzengesellschaften der Erde*, Bern-Berlin, 1930.

Sargent, C. S. *Report on the Forests of North America (exclusive of Mexico)*, Washington, 1884.

Seton, E. T. *Lives of Game Animals*, **3**, New York, 1929.

Shelford, V. E. "Basic principles of the classification of communities and habitats and the use of terms." *Ecology*, **13**, 105–20, 1932.

Short, C. W. "Observations on the botany of Illinois." *West. J. Med. Surg.* **3**, 185, 1845.

Tansley, A. G. "Editorial note." This Journ. **17**, 146–7, 1929.

Tansley, A. G. "The use and abuse of vegetational concepts and terms." *Ecology*, **16**, 284–307, 1935.

Tournefort, J. P. *Relation d'un Voyage du Levant*, Paris, 1717.

Weaver, J. E. and **F. E. Clements.** *Plant Ecology*, New York, 1929.

Weaver, J. E. and **E. L. Flory.** "Stability of climax prairie and some environmental changes resulting from breaking." *Ecology*, **15**, 333–47, 1934.

[*Editor's Note:* Plates VI through XI are not reproduced here.]

14

Reprinted from *Ecol. Mon.* **23**:41–78 (1953) by permission of the publisher, Duke University Press, Durham, N. C.

A CONSIDERATION OF CLIMAX THEORY: THE CLIMAX AS A POPULATION AND PATTERN

R. H. WHITTAKER

Biology Section, Department of Radiological Sciences
General Electric Co., Richland, Wash.*

TABLE OF CONTENTS

INTRODUCTION

The approach to vegetation often described as "dynamic ecology," the central concepts of which are succession and the climax, was developed by Cowles (1899, 1901, 1910, 1911), Clements (1904, 1905, 1916), and Cooper (1913) as a major contribution of American ecology. The antecedents (Cowles 1911, Clements 1916) were almost entirely European (King 1685, Biberg 1749, Dureau de la Malle 1825, Kerner 1863, Hult 1885, Warming 1891, 1895, 1896, Flahault & Combres 1894, Graebner 1895, Meigen 1896, Drude 1896, Schimper 1898, etc.). In spite of these the viewpoint as stated by Clements (1916), especially the "monoclimax" or single climax of a given area, was received with a certain coolness by some authors on the Continent (Gams 1918, Du Rietz 1919, 1921:97, Domin 1923) while being applied with more or less modification by others. Among the Swiss-French or Zürich-Montpellier group, theory and practice have tended to resemble those of the Americans, (Braun-Blanquet & Furrer 1913, Rübel 1913, 1922, Lüdi 1920, 1921, 1923, 1929, 1930, Furrer 1922, Braun-Blanquet & Jenny 1926, Braun-Blanquet & Pavillard 1928, Braun-Blanquet 1928, 1932, 1933, 1951, Lemée 1937-9), but with climax and succession somewhat less emphasized in the conception of vegetation than by American authors. In Braun-Blanquet's treatment (1928, 1932, 1951) the development of vegetation in a given area tends toward the climatically determined end community (*klimatische Schlussgesellschaft*) or climax. A single climax thus exists in a given climatic region, although other permanent communities (*Dauergesellschaften*) determined by other factors in addition to climate may occur there also; these would be termed sub- or proclimaxes in the American monoclimax conception. The grouping of successional communities related to the same climax forms the *Klimax-Komplex*, which corresponds to a geographic territory, the *Klimax-Gebiet*, conceptions approximately equivalent to the climax formation of Clements but free from the organismic analogy of Clements which interprets succession to the climax as growth to maturity of the climax formation as an organism.

Among the Scandinavian and Baltic plant sociologists emphasis of succession in the general interpretation of vegetation has been still less, although such more concrete successions as those of bogs, shores, and moraines have been intensively studied; and dissent from the monoclimax has been more pronounced (Du Rietz 1921, 1924, 1930b, Nordhagen 1928, Lippmaa 1933b, Faegri 1937). In some of the intensive vegetation monographs of this group the climax receives passing mention, if any, although the climax concept may be traced back to Hult (1885) and Warming (1896). While the monoclimax is rejected by Du Reitz, the conception of a major community or dominating coenose (Du Rietz 1930b:344, Faegri 1937, cf. the *Hauptcoenose* of Schmid 1922, 1935) without the assumption of convergence of other communities to this dominating coenose, is related to the prevailing climax discussed below. Nordhagen also rejected the convergence to a single climax, but recognized the major community of a given zone as a *regionale Hauptassoziation*, with which may occur an *edaphische Hauptassoziation* corresponding to the American physiographic climax or subclimax (Nordhagen 1928:526). Lippmaa in Estonia suggests a

* Paper begun while a member of the Department of Zoology, Washington State College, Pullman. The author is indebted to K. Faegri, F. Egler, H. K. Buechner, and H. E. Brewer for critical reading of the manuscript, to W. C. Healy, Jr., and R. H. Titman of the Statistics Unit, Utilities and General Services Department, General Electric Co., Richland for statistical assistance, to the funds for biological and medical research of the State of Washington Initiative Measure No. 171 for support of the Siskiyou Mountains studies, and to library personnel at the University of California, Berkeley, Washington State College, the University of Washington, and General Electric Co., Richland.

similar conception of the climax as the stable community of greatest extent in an area (Lippmaa 1933b:22) but rejects the distinction between climatically determined and edaphically determined permanent communities, since all communities are both edaphically and climatically determined (1933b: 43). Only stable communities are regarded as valid associations, and these are distinguished as undisturbed *primary associations* and disturbed or modified but stable *secondary associations* (Lippmaa 1933a, 1933b, 1935a).

These two major groups of Continental plant sociologists have in some ways followed Warming and Cowles more than Clements; and it might be generalized, with some simplification of the matter, that they have tended to treat succession within the framework of their community classification, while the Americans have tended to treat community units within the framework of a successional conception (*vide* Lüdi 1923:297). A contributor to climax theory from Germany and the Netherlands, Tüxen, has followed Zürich-Montpellier in part (Tüxen 1928), but is also author of the concepts of the paraclimax (Tüxen 1933, 1935) in its more widely used sense corresponding to the edaphic subclimax of Clements (Grabherr 1936) and the climax-group and climax-swarm (Tüxen & Diemont 1937) of polyclimax implication. Another quite distinctive nomenclature of climax and succession has appeared among some of the authors of southern Europe (Del Villar 1929a, 1929b, Gaussen 1933, Cuatrecasas 1934, Ciferri 1936, Trochain 1940) and was carried to its far point by Ciferri with the recognition of eleven climax terms. The climax of an area is conceived by Del Villar as the *conclimax,* a complex or mosaic of climax associations.

Ecological problems of the plains of Russia resemble those of the plains of North America, which influenced the Clementsian monoclimax theory; and a Russian monoclimax conception developed in close relation to Russian soil science and somewhat independently of American ecology (Alechin 1926, Walter 1943). In this conception zonal, extrazonal, and intrazonal or azonal communities correspond to climatic climax, pre- and postclimax, and subclimax communities but are preferred for their freedom from the assumption of succession (Walter 1943:29). The Russian viewpoint was much influenced also by the concept of the steady-state applied to self-maintaining vegetation (Ramensky 1918, Elenken 1921, Iljinski 1921, Alechin 1926, 1927, Stantschinsky 1931). The steady-state conception has been rejected, however, on the basis that "Instead of equilibrium, the dialectical method suggests unity, divergency and the struggle of contradictions (Lenin)" and the further conclusion that "'While studying the succession of cenoses, we have to overcome both the static fossilized conception of European phytocenologists and the somewhat fatalistic theory of successions of American scientists." (Bukovsky 1935:98)—a way of thinking that is suggestive of what has happened in other fields of biology. The distinguished Russian author Alechin has indicated that the later period of Russian phytocenology has been characterized by (1) theoretical study of the fundamental conceptions of the science and (2) reconstruction of Russian phytosociology on the bases of the methodology of Marx and Lenin (Alechin 1946, Roussin 1948), with the apparent assumption that these two pursuits are compatible.

Development of climax conceptions in England was influenced by Warming's concept of the *Schlussverein* (1896) through Moss (1907, 1910), who distinguished climax and seral communities as chief and subordinate associations, and by Crampton's (1911, 1912) concepts of stable and migratory formations. From these sources and a strong influence of Clements and other Americans the British "polyclimax" position developed under the leadership of Tansley (1911, 1920, 1929, 1935, 1939; 1947-8). The British polyclimax majority viewpoint was close to that of some Americans following Nichols (1917, 1923), but the prevailing American view has been that of the monoclimax. Clements' own full formulation of this theory was published in 1936 (also in 1949); and conceptions similar to those of Clements have been stated by Phillips (1934-5), Weaver & Clements (1929), Cain (1939), Dansereau (1946), Braun (1950), etc., and have been generally accepted by ecologists of the "biotic" school (Shelford 1932, Woodbury 1933, Shelford & Olson 1935, Clements & Shelford 1939, Carpenter 1939). The Clementsian monoclimax conception, though widely criticized, has been most characteristic of American ecology in the past and is still an active and influential viewpoint, appearing almost without qualification in current texts (McDougall 1947, Oosting 1948), as well as in more advanced studies. The divergent interpretations of climax and succession in America and elsewhere (Hansen 1921, Scharfetter 1921, Gams 1923, Cooper 1926, Gleason 1927, Godwin 1929, Bourne 1934, Davis 1936, Oberdorfer 1937, Walter 1937, Wood 1939, Tutin 1941, Graham 1941, Beard 1946, Cain 1947, Crocker & Wood 1947, Beadle 1951, and citations above) have led to a degree of confusion in the variety of approaches and units of treatment, indicated in Du Rietz (1930a), Conard (1939), and Cain's climax discussion of 1939.

Something of the nature of this confusion may be illustrated by the profusion of climax terms in the literature. Beginning with Clements' climatic climax, subclimax, preclimax and postclimax, and potential climax (1916), the climax terms have led through the proclimax (1934), disturbance climax or disclimax, serclimax, eoclimax, and panclimax (1936), edaphic and physiographic or topographic climaxes (Nichols 1917, 1923), fire climax (Tansley & Chipp 1926) or pyroclimax (Roberty 1946), aquatic climax (Woodbury 1933), biotic climax and anthropeic climax (Tansley 1935), anthropoclimax (Tüxen & Diemont 1937), or archeological disclimax (Wells 1946), paraclimax (Del Villar 1929, Tüxen 1933), conclimax, anteclimax, and peniclimax (Del Villar 1929), transclimax, metaclimax, and euclimax (Ciferri 1936), deflected climax (Marshall 1934) or plagioclimax (Tansley 1935), pseudoclimax (Ciferri 1936, Carpenter 1938) and quasiclimax (Tansley 1939, Dansereau 1946), co-climax (Wood 1937), pedoclimax (Lemée 1937–9:501), and salt-spray climax (Wells 1939) to, inevitably, the superclimax (Muller 1940). Onward the march of science. A significant difference in climax interpretation has, further, been described in the monoclimax and polyclimax theories (Du Rietz 1930a, Bourne 1934, Tansley 1935, Cain 1939). In the light of the distinguished succession of climax terms it is hoped that the present view will not be known as the anticlimax hypothesis. Such a multiplicity of terms, many of them clearly exceptions to the concept as originally formulated, may imply that the concept is being stretched this way and that to cover evidence for which it is not actually adequate. If an ideal, the climax, must be so modified in application, it may be suspected that the ideal is at fault. The vagueness of usage has led Egler (1947) to abandon the term "climax," regarding it as no longer useful. Usefulness of the term and of the concept it represents should be examined, primarily on the basis of two questions: (1) Can the climax state be meaningfully interpreted, defined, and measured as distinct from succession? (2) Does the concept serve a useful function in synecological research?

In the problems to be discussed, and in the history of the climax concept, two approaches to climax definition may be distinguished. Much of the consideration of climax vegetation, and much of the logic of the monoclimax, have been based on the physiognomy, or structure in terms of growth-forms, of vegetation and its relation to climate. This area of vegetation problems, in the overlap of synecology and geography, has been the subject of an extensive literature from some of the beginnings of plant ecology to the present, indicating the adaptive or epharmonic relation of climax vegetation to environment in terms of both morphological life-form, vegetative form, or growth-form (Humboldt 1806, 1807, Grisebach 1838, 1872, 1877, Kerner 1863, Drude 1890, 1913, Schimper 1898, 1903, Schimper-von Faber 1935, Warming 1895, 1896, 1909, 1933, Rübel 1930, 1936, Stefanoff 1930, Küchler 1949, 1951, Dansereau 1951; review and dissent, Du Rietz 1931) and the Raunkiaer life-forms (Raunkiaer 1910, 1934, 1937, Smith 1913, Paulsen 1915, Hansen 1928, 1930, 1932, Braun-Blanquet 1932, Gelting 1934, Clapham 1935, Jones 1936, 1938, Allan 1937, Adamson 1939, Cain 1950, Dansereau 1951). It will be treated here as a necessary part of climax theory, but not as the primary concern. The other approach is to the climax as a population of plants and animals of different species, with treatment of natural communities in much greater detail than by physiognomy alone. Synecology or biocenology is the science of natural communities or, more broadly conceived, the study of ecosystems (Tansley 1935, 1939: 228) and in its analytical aspects appears to center in problems of populations and productivity in relation to environment. Interpretation of the climax in terms of populations and productivity thus may be sought. It is felt that if the climax has real meaning for analytical synecology, that meaning is primarily in the climax as an object of and basis for research treating populations and the productive or metabolic dynamics of communities. The author has sought, on the basis of theoretical considerations and of his own field work in population analysis of natural communities, to formulate the climax concept in a manner general enough to provide for the evidence available and to give it functional meaning for analytical ecology. He has sought also, in stating the interpretation developed, to draw together citations of previous contributions to climax theory and of some fraction of the world literature of vegetation bearing on the problem as evidence.

EVIDENCE

Some evidence bearing on succession and climax problems may be summarized as a basis for interpretation:

1. Succession. The succession of populations on disturbed or newly exposed sites is one of the best-established phenomena of ecology, circumstantial as most of the evidence is. The most effective demonstrations of the manner in which populations succeed one another are in such short-range, small-scale successions as those of infusions (Woodruff 1912, 1913, Allee 1932), carrion (Fuller 1934, Holdaway 1930), rotting logs and stumps (Shelford 1913, Blackman & Stage 1924, Krogerus 1927, Savely 1939), dung (Mohr 1943), etc. While effective population treatments of succession in the literature are limited, evidence available indicates: (a) that the flow of populations is often quite continuous through and between the stages of succession (data of Cooper 1922a, 1923, 1939, Sampson 1930a, Faegri 1933, Braun 1936, Vaughan & Wiehe 1941, Lüdi 1945, Dansereau 1946, Eggeling 1947, Quarterman 1950, Curtis & McIntosh 1951, Brown & Curtis 1952); (b) that successions of the same general type toward the same climax are variable in rate of change of populations, both between different stages in the same succession and in consequence of local climatic and edaphic differences (Faegri 1933); (c) that successions are variable in composition of the stages, environmental differences causing different successional populations to appear in similar successions in the same area (Watt 1923-5, Kell 1938, Oosting 1942); (d) that the changes may be irregular, shifting back and forth because of environmental fluctuation (data of Weaver & Bruner 1945), (e) that successions are often telescoped (Cooper 1916), with the course of succession variable because of chance differences in what populations enter the succession at what time, whole apparent stages being either skipped or added. Evidence of successional irregularity is familiar wherever stages of succession may be reasonably assumed to appear along a spatial gradient, as around an aging pond or bog, or a meadow being invaded by

forest. A meadow into which the forest is advancing in the Siskiyou Mountains, for example, may be ringed with shrubs; but the shrub stage is highly irregular, appearing in one place as a dense and solid belt, in another as scattered patches of various species, and being absent from other parts of the edge. Where a shrub thicket occurs it may be dominated in one place by Douglas maple *(Acer glabrum* var. *douglasii* (Hook.) Piper) and in another by alder *(Alnus tenuifolia* Nutt.); the patchy shrubs are dominated by various combinations, differing from place to place, of *Corylus rostrata* var. *californica* A. DC., *Holodiscus discolor* (Pursh) Maxim., and *Salix* sp., and with different representations of several minor species. Where the shrubs are absent, young trees *(Abies concolor* Lindl.) are invading the meadow directly; where there are shrubs the tree seedlings may appear ahead of or behind the densest shrubs, or they may be scattered ahead of the whole shrub belt.

Succession may thus be thought to occur, not as series of distinct steps, but as a highly variable and irregular change of populations through time, lacking orderliness or uniformity in detail, though marked by certain fairly uniform over-all tendencies. In its continuity and irregularity, and in the sharing of populations in different combinations by different successions, succession is effectively represented by Cooper's (1926) image, after Vestal, of a braided stream (cf. Alechin 1925).

2. Climax convergence. The conception of convergence of successions to the climax was fundamental to the early formulations of Cowles and Clements and is familiar on the levels of both physiognomy and population. Physiognomic convergence on a world-wide scale in the similar climates of different continents is a basis of the literature recognizing this phenomenon in terms of formations, formation classes (Warming 1909, Brockman-Jerosch & Rübel 1912), formation types (Nichols 1917, Schimper-von Faber 1935, Burtt Davy 1938, Tansley 1939:229, Dansereau 1951), isocies (Gams 1918), formation-groups and panformations (Du Rietz 1930a), homologous formations (Braun-Blanquet 1932), panclimaxes (Clements 1936), biochores (Hesse *et al* 1937, Dansereau 1951), biome types (Allee *et al* 1949, Tischler 1951). Convergence of seres within an area to the same physiognomy is also familiar; most successions in a forest area will lead to forest, most in a grassland area to grass. A degree of convergence appears also on the population level. On similar sites at least, within an area, similar climax populations are likely to be found.

3. Climax patterning. Because different population combinations appear in the different environments of any area, vegetation forms a complex pattern of plant populations. While some studies have attempted to draw all the vegetation types of an area into a successional schema, there are many studies showing the real complexity of patterning of stable vegetation (Stamp 1925, Petrie *et al* 1929, Trapnell 1933, Conard 1935, Meusel 1935, Potts & Tidmarsh 1937, Tansley 1939, Wood 1939, Falk 1940, Daubenmire 1942, Braun 1942, 1950, Crocker & Wood 1947, Watt & Jones 1948, Morison *et al* 1948). Even if the monoclimax theory is accepted, it may be recognized that diversity of species combinations is usually characteristic of the climax (Gilliland 1938).

Various topographic-edaphic effects (Shreve 1915, 1927, Brough *et al* 1924, Rohlena 1927, Cottle 1932, Meusel 1935, Braun 1935b, Bauer 1936, Gillman 1936, Tüxen & Diemont 1937, Potts & Tidmarsh 1937, Raup 1938, Bayer 1938, Steyermark 1940, Cline & Spurr 1942, Oosting 1942, Cobbe 1943, Williams & Oosting 1944, Buechner 1944, Boyko 1945, Egler 1947, Miller 1947, Sears 1947, Lüdi 1948) may result in a mosaic of climax types in an area. Breakdown of the climatic climax in mountain topography was indicated within American ecology as early as 1914 by Shreve (1914a:106), as the same author suggested the breakdown of the association the following year (1915:111). Edaphic factors have been the basis of many objections to the monoclimax theory (Tansley 1916, 1935, Nichols 1917, Bourne 1934, Michelmore 1934, Gillman 1936, Richards 1936, Milne 1937, Walter 1937, Wood 1939, Steyermark 1940); and references to edaphic factors occur throughout the literature, such effects apparently being observed in every vegetational area studied with sufficient intensity. The soil moisture factor is universal in its effect on vegetation, appearing in the form of swamp and marsh, flood-plain and ravine "serclimaxes" and "post-climaxes" (Ainslie 1926, Swanson 1929, Grant 1934, Marshall 1934, Clements 1936, Little 1938, Cain & Penfound 1938, Tansley 1939, Richards 1939, Beaven & Oosting 1939, Trochain 1940, Tolstead 1942, Fairbairn 1943, Eggeling 1947, Zohary & Orshansky 1947, Hotchkiss & Stewart 1947, Tchou Yen-Tcheng 1949, Beard 1949) as well as of alternes, etc., and in physical effects of soil texture and underlying strata on vegetation (McLuckie & Petrie 1926, 1927, Keller 1927, Botke 1928, Patton 1930, Barrington 1930, Shreve & Mallery 1933, McBryde 1933, Richards 1936, Davis 1936, Olmsted 1937, Kielhauser 1939, Shantz & Piemeisel 1940, Oosting 1942:111, McComb & Loomis 1944, Fautin 1946, Dyksterhuis 1948, Wilde *et al* 1948, Livingston 1949, Køie 1950, Platt 1951). Chemical effects of soil (so far as there is any meaning in the distinction between "physical" and "chemical" here) are also widespread (Weiss 1923, Firbas 1924, Keller 1925-6, Lämmermayr 1926, 1927, 1928, Domin 1928, Breien 1932, Schwickerath 1933, Flowers 1934, Soó 1936, Bjorkman 1937, Krist 1940, Johnston 1941, Birrell & Wright 1945, Zohary 1945, Rivas Goday & Box 1947, Wilde *et al* 1948, Pichi-Sermolli 1948a, Billings 1950, Zangheri 1950, Albertson 1950). A few further references may suggest that edaphic effects on climaxes, if not universal, appear on all continents and in all kinds of climates (Stamp & Lord 1923, Shantz & Piemeisel 1924, Shantz 1925, Haman & Wood 1928, Sukachev 1928, Cox 1933, Wilde 1932, 1933, Frey-Wyssling 1933, Killian & Dubuis 1937, Tüxen & Diemont 1937, Falk 1940, Pidgeon 1941, Zohary 1942,

1947, Seifriz 1943, Hou 1944, Marks 1950). The assumption that edaphic effects on vegetation are successional, that with soil maturity differences cease to be effective (Clements 1920, 1928:284-5, Brockman-Jerosch 1925-9, Daubenmire 1947:55) may be rejected as by Beadle (1951) and many others accepting the polyclimax position. Edaphic factors, it would appear, are not special, local, and seral in effect, but are quite general in their contribution to vegetation patterning.

4. Climax continuity. That vegetation types are separated by boundaries called ecotones is a frequent assumption, but equally general is the observation that transition between two types may be very broad and gradual, so that it is impossible to recognize any real discontinuity. Continuous intergradation was indicated for the vegetation of the Faeröes by Ostenfeld (1908), and for the forest and grassland patterns of Bohemian mountains by Domin (1928a). McBryde (1933) has suggested that the vegetational response to a continuous climatic gradient is continuous, and Cooper (1942) has described vegetation of a transitional area in Alaska as a gradating pattern. Data of various authors (Frödin 1921, Linkola 1924, Hansen 1930, 1932, Sampson 1930b, Meusel 1935, Halliday 1935, Horton 1941, Spilsbury & Smith 1947, Marler & Boatman 1952) indicate the continuity of the types described. The author, in an attempt to determine the validity of eastern forest associations and ecotones in the Southern Appalachians, found population transition continuous through and between the traditional associations (Whittaker 1951, 1952). In studies of the Wisconsin forests during the same period by Curtis & McIntosh (1951) continuity was demonstrated and has been further developed for both forest and prairie (Curtis 1951, Brown & Curtis 1951, 1952, Gilbert 1951, Tresner 1951, Randall 1951). In vegetation forming "a complex continuum of populations" (Whittaker 1951) associations have only such subjective meaning as is consistent with the individualistic hypothesis of Gleason (Gleason 1926, 1929, 1939, Raup 1942, Cain 1947, Mason 1947, Egler 1947, Braun 1947, 1950, Whittaker 1951, 1952, Curtis & McIntosh 1951, Brown & Curtis 1952; dissents by Tansley 1920, Allorge following Lenoble 1926, Nichols 1929, Clements *et al* 1929:315, Phillips 1934-5). The "individualistic" conception appeared independently in France (Lenoble 1926, 1928, Fournier 1927), without being made welcome among phytosociologists, and in Australia (Patton 1930) and Russia (Ilínskii & Poselškaia 1929); and the conception of the vegetational mantle as being fundamentally continuous and without boundaries between recognized units was developed in Russia by Ramensky (Ramensky 1924, Roussin 1948). While there is some evidence and theory on the occurrence of discontinuities (Cajander 1909, Du Rietz 1922, 1924, Beauchamp & Ullyott 1932, Gause 1936, 1937, Pfeiffer 1943, Nytzenko 1948, Hairston 1951), it is suggested that usually climax populations change continuously along continuous gradients, observed discontinuities being produced locally by either environmental discontinuities which telescope transitions or incompatibility of growth-forms or species which steepen the rate of change along a part of the gradient.

5. Climax irregularity. In spite of the convergence of climax populations on similar sites, no two stands are quite alike. Patchiness or irregularity is familiar on a smaller scale, within stands, in the herb and shrub layers of many vegetations, as in the prairie quadrat maps of Thornber (1901) and data of Steiger (1930) and the forest quadrat maps of Lippmaa (1935b), in bogs, vegetation of which in composition and micro-relief may form a mosaic of two or more phases (Osvald 1923, 1949, Lewis & Dowding 1926, Ljungqvist 1927-9, Rudolph *et al* 1927, 1928, Nordhagen 1928, Katz 1930, Yoshii & Hayasi 1931, Tansley 1939, Sjörs 1948), and in studies of contagion (Blackman 1935, Ashby 1935, 1948, Clapham 1936, Cole 1946, Archibald 1948, Greig-Smith 1952). Since some positive contagion or clustering appears to be general in natural communities, while negative contagion also occurs, a stand may be conceived (cf. Elton 1949) as a system of superimposed population lattices differing in density, spacing, degree of contagion, and degree of association with one another. Stand-to-stand irregularity is familiar in forests (Cooper 1913, Graham 1941, Maissurow 1941, Stearns 1949) and is indicated by association tabulations (Braun-Blanquet 1915, Allorge 1922, Frey 1922, Szafer *et al* 1924, Braun-Blanquet & Jenny 1926, Nordhagen 1928, 1943, Keller 1930, Mrugowsky 1931, Böcher 1933, Krajina 1933, Klika 1932, 1936, 1939, Schmid 1936, Mikyška 1939, Lüdi 1941, 1943, 1948). Such old and apparently stable vegetation as the Southern Appalachian cove forests (Cain 1943) and tropical and subtropical rain forests (Aubréville 1938, Fraser & Vickery 1938) vary strikingly from stand to stand, probably partly because of site differences, but also because of fluctuation in populations and chance differences in history of the stands. Environmental factors may cause irregularity, as in the California chaparral which is adapted to fire, but in which marked patchiness is produced by irregularity of burning (Bauer 1936), by small-scale soil differences affecting plant populations within a single stand (Kelley 1922, Stewart & Keller 1936), by such special edaphic situations as fallen logs and stumps, rocks and wind-fall pits, the "edaphids" of Sernander (Sernander 1936, Arnborg 1940), by surface water-flow patterns (MacFadyen 1950), by microrelief, as on till (Braun 1936, Chapman 1942) and other soils (Sampson 1930a), solifluction terraces (Troll 1944, Washburn 1947, Watt & Jones 1948, Metcalfe 1950) and other frost-determined landforms—tundra hillocks (Pohle 1908, Cajander 1913, Polunin 1935, Porsild 1938, Sharp 1942, Walter 1943, Troll 1944) and polygons (Zimmermann 1912, Högbom 1914, Summerhayes & Elton 1923, 1928, Huxley & Odell 1924, Elton 1927, Gates 1928, Regel 1932, Polunin 1934, Sørensen 1935, Russell & Well-

ington 1940, Walter 1943, Washburn 1947, Hopkins & Sigafoos 1951), stone stripes (Högbom 1914, Troll 1944), *Strangmoor* (Cajander 1913, Tanttu 1915, Auer 1920, Nordhagen 1928, Troll 1944), frost scars and peat rings (Hopkins & Sigafoos 1951). Irregularity of composition, within and between stands, may be thought a general characteristic of climaxes.

6. Climax instability. Spatial irregularity of populations is closely related to, and may be an expression of, underlying temporal instability. Theoretical considerations, developed primarily for animal populations (Lokta 1925, Volterra 1926, 1931a, 1931b, Bailey 1933a, 1933b, Nicholson & Bailey 1935, Smith 1935, D'Ancona 1939, Thompson 1939, Varley 1947), indicate that cyclic fluctuations of interacting species are to be expected under some conditions and are supported by experiment (Gause 1934a, 1934b, 1935, DeBach & Smith 1941, Utida 1950). Fluctuations of animal populations have been widely observed (Elton 1924, 1942, Uvarov 1931, Duffield 1933, Middleton 1934, Hamilton 1937, MacLulich 1937, Naumov 1939, Clements & Shelford 1939, Carpenter 1940, Dymond 1947, Leopold 1947, Siivonen 1948, Allee *et al* 1949, Shelford 1951), although these may be irregular and environmentally determined, rather than cyclic (Palmgren 1949, Cole 1951). These fluctuations occur within functioning ecosystems and may be thought expressions of, or contributions to, a more general instability of the ecosystems. Plankton communities, with their short life-cycles, offer a picture of instability which may be hidden by the longer life-cycles of major terrestrial organisms. The extreme instability of the plankton, with the populations changing rapidly, often in an apparently unpredictable fashion (Krogh & Berg 1931, Pearsall 1932, Riley 1940, Pennington 1941, Hutchinson 1944, Pennak 1946, 1949, Chandler & Weeks 1945, Verduin 1951) is suggestive, though perhaps exaggerated beyond the instability of terrestrial communities. Instability of marine littoral communities was described by MacGinitie (1939). In vegetation itself, environmental instability may permit only a relatively unstable climax, if it is regarded as such, as in frost-affected arctic terrain (Griggs 1934, 1936, Raup 1941, 1951, Sigafoos 1949, Hopkins & Sigafoos 1951) and active scree slopes (Hess 1910, Allan 1926a, Leach 1930). Climatic fluctuation may result in population change, as has been recorded for the American prairies (Weaver *et al* 1935, 1936, 1939, 1940, 1943, 1944, Albertson & Weaver 1942, 1944, Timmons 1941-2, Pechanek *et al* 1937) and is especially conspicuous in the annual vegetation of arid regions (Paczoski 1917, Alechin 1926, Chipp 1930, 1931, Talbot *et al* 1939, Went 1948, 1949, Went & Westergaard 1949, Zohary & Orshansky 1949, Shreve 1951). Changes in alpine vegetation in a decade were shown by Lüdi (1940). As the climax pattern is altered and shifted by climatic change, such conditions as those described by Blake (1938) of fluctuating climaxes of complementary communities oscillating in time and space and in the observation of Paczoski (1917) of steppe and marsh vegetation alternating in time in the same place, may result. Fires may, in fire-adapted climaxes, produce population changes (Jarrett & Petrie 1929, Bauer 1936, Garren 1943, Sampson 1944, Hopkins *et al* 1948, Lemon 1949, Weaver 1951b). Reproduction may be periodic, depending upon especially favorable conditions, as in the English beech (Watt 1923-5), Finnish pines (Renvall 1912, 1919, Lakari 1915), etc. Stands may tend to stagnate until windfall, fire, or insect attack opens the canopy and permits new reproduction (Cooper 1913, Bergman 1924, Moss 1932, Sernander 1936, Poole 1937, Graham 1939, Maissurow 1941, Meyer & Stevenson 1943, Cline & Spurr 1942, Hough & Forbes 1943, Jones 1945, Stearns 1949). An irregular rhythm, sometimes involving different dominant populations, may thus be implicit in the manner of maintenance of the climax population; and cyclic processes have been observed in various vegetations, depending upon vegetational process (Nilsson 1899, Cajander 1904, Sernander 1910, Osvald 1923, 1949, Tansley 1939:683-695; Watt 1936, 1945, 1947a, 1947b, Tansley 1939:355, Godwin & Conway 1939, Coombe & White 1951), erosion in relation to succession (Bayer 1933), effects of animals (Cockayne 1909, Marler & Boatman 1952), and seasonal environments (Saxton 1922, 1924). From the evidence, particularly that summarized by Watt (1947b) and Jones (1945), it may be judged that a degree of instability, involving both response to environmental fluctuation and internal dynamics of communities, is characteristic of climax populations.

SUCCESSION AND RETROGRESSION

The term "succession" may with equal justice be applied to all vegetational change (Cooper 1926, Gleason 1927) or restricted to the shorter-range development to a climax on a given site (Iaroshenko 1946, Major 1951). The latter convention has been preferred here as permitting a clearer distinction of climax and succession. Fundamental to both succession and climax is the interplay of populations as these affect one another in such ways as either to change directionally in succession or to fluctuate about an average in the climax. In the development of a self-maintaining system of interacting populations, the climax community, from the less orderly seral populations, succession is an evolutionary process in the general sense (Wiener 1948). In the evolution of tides the diverse currents produced by gravitational forces are organized by environment, the ocean basin, into a definite tidal pattern. In organic evolution gene frequencies fluctuate with mutation and population changes, and with environmental selection there is progressive change in frequencies of some genes and change in the adaptive pattern of the species. The varied, experimental adaptive responses of childhood become organized in a manner determined, though not rigidly, in relation to conditions of environment into the personality pattern of the adult. In these, as in ecological succession, there is progressive development of a pattern adapted to environment. The environment directs or fashions the development or evolution; the resulting pattern is an expression of its environment even though quite different patterns may correspond to, and perhaps be equally adapted to, a given environment. Adaptation, the matching of organic pattern with environmental pattern (Whittaker 1952), whether of the organism or community, is thus achieved and maintained through such evolutionary processes.

Certain trends or progressive developments toward the climax may characterize successional evolution. Community productivity tends to increase to a steady-state in the development of a terrestrial community, as in the aging of an aquatic one (Hutchinson & Wollack 1940, Lindeman 1942, Pennington 1943, Hutchinson 1948). Presumably, however, productivity is greater in a late seral stand than in the climax (Clements & Shelford 1939:116); energy activity in the developing community, as in a growing boy, may exceed the sustained level of the mature state. Decrease of productivity in the replacement of forest by heath led Lüdi (1923:295) to distinguish the optimal community of maximal productivity from the terminal community or climax in this succession. Along with increasing productivity, there is usually progressive increase in mass of the community structure (Del Villar 1929) as dominants of "higher" growth-form replace seral dominants, though some seral communities may be replaced by climax communities of smaller or "lower" growth-form. Soil depth increases through many seres, and the soil may develop to maturity in the climax; but this relation is by no means a necessary one, as indicated below. As early seral communities are simple and climax communities usually complex, there may be a progressive increase of community complexity, of diversity of species (Hansen 1930, Bojko 1934) and symmetry of community function (Park 1941), up to the climax. There would seem to be no reason, however, why in some cases, as in an open subclimax stand with rich undergrowth preceding a dense, closed climax stand, a seral stand should not be more complex and diverse than the climax. There may be increasing regularity of population distribution during the succession, as patchy, irregular seral stands are replaced by the more uniform climax stands (Cajander 1909); but there would seem to be no reason why some subclimax stands should not be at least as uniform as the following climax. The fact that species populations in succession may spread as clones or clans around a parent individual might imply that climax populations, when these species have had time to spread until identities of the clones or clans are lost, would show less contagion; but no clear difference in extent of contagion has appeared (Whitford 1949). The exceptions to all such trends through succession suggest considerable wariness in their application, but the exceptions should not obscure the significance of the trends themselves. In general, through the process of succession in terrestrial communities, there is progressive increase in community complexity and diversity, stature or massiveness and productivity, maturity of soil, and relative stability and regularity of populations.

Some vegetational developments are described as "retrogressive succession," but this paradoxical expression, a kind of oxymoron if succession is conceived as by Clements, requires definition if it is to be used with meaning. The possibility of retrogression in communities has been denied by Clements (1916) and Phillips (1934-5), while being defended on various grounds by others (Cowles 1901, 1919, Moss 1910, 1913, Cooper 1916, Gleason 1917, 1927, Nichols 1923, Bourne 1934, Tansley 1935, Buechner 1944). Open energy systems, such as those in which the various evolutionary processes occur, cannot actually run backwards although reversible changes may occur in them. A developing system is, as one author has described the human organism, "an arrow pointed through time in one way" (Wiener 1948:46). Clements' pronouncement that "Retrogression, an actual development backwards, is just as impossible for a sere as it is for a plant." (1928:147) is not in this case a pointless analogy. Changes can occur in these systems which are, by definition, retrogressive, however. Regression, as the term is used in psychiatric psychology, cannot imply an exact return to childhood traits (Conklin 1923:379, Mowrer 1939:61). If it is defined, however, that a loss of "genital" maturity (Freud 1935) or personal creativity or productiveness (Barker *et al* 1941) is to be regarded as regressive, as a turning back in those aspects of the personality, then such a development may be termed regression by definition.

If, in synecology, the mature or climax community is defined in terms of maximum diversity, productivity, soil maturity, stability, etc., then change in the community which involves decrease in one or more of these can be defined as retrogressive. Such changes may differ in character from the reversible changes discussed by Gleason (1927) as retrogression. It should be obvious that such a retrogression is not a return through past time but a turning back at an angle, in a direction different from the original development. Because of the unreliability of some of the criteria of community maturity, the decision as to whether a given change is retrogressive may be necessarily subjective. It may be more reasonable in many cases to speak of "deflected" development (Godwin 1929), indicating a succession in a direction different from that which might otherwise occur but with no implication of retrogressive character. When under severe grazing, however, the productivity of a community is gradually reduced, the soil impoverished and eroded, and the community simplified to a patchy, unstable population of weedy annuals, the change seems meaningfully retrogressive. Retrogression may be in some cases a useful term to describe changes occurring in a community, changes "retrogressive" by definition, and perhaps most significantly involving decrease in community productivity.

CHARACTERISTICS OF THE CLIMAX

Through succession the community develops from one of scattered pioneers utilizing only a fraction of environmental resources available, to a mature community with maximum utilization of resources on a sustained basis (Dansereau 1946). Energy of sunlight, converted into the available energy of organic compounds, is transferred through the several metabolic or energetic levels of the community with progressive degradation of energy which, however, is replaced by continued photosynthetic activity (Lindeman 1942, Allee *et al* 1949). Community activity in a lake, expressed as productivity, is determined, given sufficient light and water, by the inorganic nutrients available for protoplasmic synthesis, especially the one or more present in shortest supply relative to need, and by environmental temperature as it affects rates of activities, as major factors (Rawson 1939, 1942, Deevey 1940, Lindeman 1942). In terrestrial communities amount of water available to support transpiration of plants and other needs is a major determinant along with temperature and nutrients. We find in consequence the decrease in community mass, productivity (Jenny 1941:207), and photosynthetic efficiency (Lindeman 1942) along the moisture gradient from rain forest through other forest types, savannah, woodland, and shrub communities, and grasslands to desert. Along the temperature gradient from tropics to arctic a similar lowering of community productivity and stature appears; and, given a particular level of water supply and temperature, difference in nutrients may determine relative productivity of two terrestrial communities. The relation of environment to structure was expressed in the observation of Dansereau (1951: 219) that "an increasingly favorable combination of temperature and moisture permits increased closing, elevation, and differentiation into strata of the community." In its function, in utilization of energy supply in a manner determined by environment and characteristics of the system, the natural community, like the organism (Haskell 1940, Schrödinger 1946, Weiner 1948) has the characteristics of the counter-entropic, open energy system (D'Ancona 1939, Bertalanffy 1950a, 1951).

One characteristic of open energy systems is the possibility of equifinality (Bertalanffy 1950a, 1950b, 1951). Because material and energy drawn from environment are utilized and organized in a manner determined by the properties of the system itself, as well as in relation to environment, the end point of a development or evolution may be in part independent of its starting point. Essentially identical sea-urchins may develop from a whole embryo or from each of two half-embryos. Similar communities may develop in the conditions of a given site whether the development is a primary succession on exposed rock or a secondary succession following fire or other disturbance. As the final condition of the organism is determined by the organizing effects and interaction of the genes in partial independence of original condition, the final condition of the community is determined by the interaction of populations and their organization into a self-maintaining balance in partial independence of original environment and populations. The fact that similar climax communities may develop from differing initial conditions and may replace widely different seral populations is a striking example of open-system equifinality, and one in which the degree of independence of starting condition and course of development appears much greater than in the organism. The degree of independence of environment appears, in contrast, to be much less; for the developing natural community has no such built-in, determinate pattern as is provided by the genic system of an organism. As the productivity of the natural community is determined directly in relation to environmental conditions, and the populations in which this productivity is expressed are determined by population interactions in relation to environment, the dependence of the final characteristics of the community upon environment is more direct and immediate than in the organism. Thus, while the equifinality of the open system appears in both organism and natural community, the different organization of these two systems implies a contrast in the degree to which the equifinal state is independent of starting point and environment.

As one aspect of climax interpretation it may be said that the equifinal, climax steady-state represents a maximum sustained productivity or level of function for the environmental resources and conditions affecting community function (Lüdi 1921, 1923). Climax structure and function may be thought adapted to maximum productivity under given environmental conditions (Paczoski 1921, Alechin 1926, Sukatschew 1929), though the efficiency of utilization is necessarily determined by the kinds of organic populations making up the community. Given the same environmental conditions in two widely separated sites, as on different continents, there may be differences in productivity determined by the different kinds of organisms available to form the climax communities. The communities which develop are, however, organized wholes or systems showing such community properties as physiognomy, diversity, productivity, organization in terms of metabolic or energetic levels, spatial stratification, daily and seasonal periodicity, etc. As an organized system the community is a functional whole, showing emergent characters on its own level as does the organism (Phillips 1931, 1934-35, Egler 1942, Allee *et al* 1949), though radically different from the organism in its manner of organization. As an organic open system the community may be expected to express in its structure and function the conditions of its environment. Because of the intimate relation of structure and function (in spite of the artificial distinction enforced by language) in adaptation, we may speak of the community's structural-functional pattern, its manner of operating together with the physical basis of its operation to utilize with maximum effectiveness the resources of environment in biological activity.

Some relations of structural-functional patterns of communities to environment are familiar enough in the many types of physiognomic adaptation described by Schimper, Warming, and others. If deciduous trees are better adapted to utilize the productive potential of a continental climate of wet summers and severe winters, evergreen trees better adapted to the maritime temperate climate of wet and moderate winters and dry summers, for this adaptive difference and for historical reasons also (Küchler 1946) we may expect to find the difference in dominance of these growth-forms so strikingly illustrated in the eastern and western forests of North America. It may also be observed that many stands of either area contain, and some may be dominated by, trees of the growth-form characteristic of the other; the difference is one of relative emphasis. Because different species may utilize the resources of the community in a manner complementary as well as competitive, maximum productivity can often be better maintained by a balanced pattern of several growth-forms than by a single growth-form. Individual species in a community tend not to meet in direct competition, as shown for animals (Lotka 1932, Volterra 1926, 1931a, 1931b, Gause 1934a, 1934b, 1935, Gause & Witt 1935, Hutchinson 1941, 1948, Lack 1944, 1945, 1946, Crombie 1945, 1947, Elton 1946, Hairston 1951, Williams 1947, 1951, Bagenal 1951) and indicated for the plant community by Shreve (1915: 112) and Alechin (1926). Stands approaching single-species composition most frequently occur in extreme conditions (Alechin 1926), while stands in favorable tropical environments are diverse (Whitford 1906, 1909, Seifriz 1923, Davis & Richards 1934, Richards 1936, 1939, 1945, Aubréville 1938, Holttum 1941, Beard 1946, Black *et al* 1950). Increasing impoverishment of communities may be expected with increasingly rigorous conditions (Cajander 1909), as in the decrease of plant species toward higher elevations (Tidestrom 1925:14, Davidsson 1946, Fries & Fries 1948), in relation to the moisture gradient (Linkola 1924, Hansen 1930, Whittaker 1952), and in the faunistic trends compiled by Hesse (1937:23-31, 144, 249). The more favorable the environment the greater the diversity of species populations and of growth-forms likely to be present in the community structure.

It is felt that the curves of of Fig. 1 are of some significance in illustrating, for the tree-stratum, the balance among growth-forms and the change in this balance with change of environment. In the most mesic valleys of the Great Smoky Mountains the cove forests (mixed mesophytic association) occur, with diverse stands of broad-leaved decidious trees, among which few are fagaceous. As one moves along the moisture gradient from the cove forests toward drier sites, the cove-forest species become progressively less important while first an abietine, eastern hemlock, increases and declines and then the fagaceous trees, chestnut and oak species, increase to become dominant in intermediate sites. Further along the gradient the pines are increasingly important, finally becoming dominant with decreasing propor-

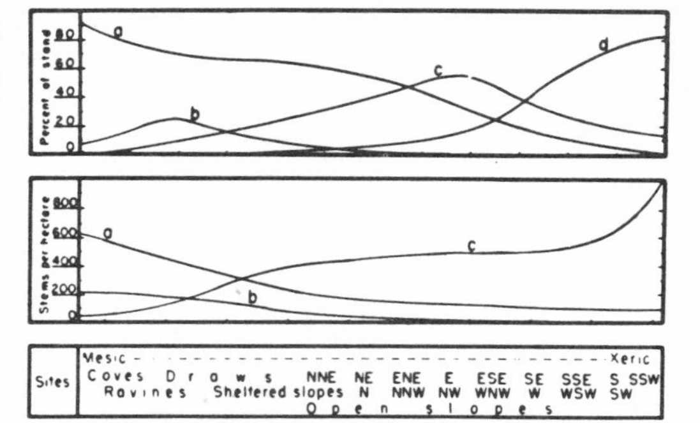

FIG. 1. Distribution of growth-forms of trees in relation to the moisture gradient at low elevations in the Great Smoky Mountains of Tennessee and the Siskiyou Mountains of southwestern Oregon.

Above, in the Great Smoky Mountains National Park, near Gatlinburg, Tenn., 13 stations along the moisture gradient using 37 site-samples for elevations between 1500 and 2500 ft. (data of Whittaker 1951); figures are percentages of stems over 0.5 in. d.b.h. in stand for: **a, deciduous-broadleafs** other than fagaceous and ericaceous species (*Halesia monticola* (Rehd.) Sarg., *Aesculus octandra* Marsh., *Tilia heterophylla* Vent., *Liriodendron tulipifera* L., *Betula alleghenienis* Britt. and *lenta* L., *Acer saccharum* Marsh. and *rubrum* L., *Cladrastis lutea* (Michx. f.) K. Koch, *Cornus florida* L., *Hamamelis virginiana* L., *Nyssa sylvatica* Marsh., *Carya glabra* (Mill.) Sweet and *tomentosa* (Lam.) Nutt., Magnolia spp., etc.); **b,** an **abietine** *Tsuga canadensis* (L.) Carr.); **c, semi-sclerophyllous deciduous,** fagaceous (*Castanea dentata* (Marsh.) Borkh., *Quercus borealis* Michx. f., *montana* Willd., *alba* L., *velutina* Lam., *coccinea* Muench., *marilandica* Muench., *stellata* Wang., and *falcata* Michx.) and ericaceous (*Oxydendrum arboreum* (L.) DC., *Clethra acuminata* Michx.); **d, pines** (*Pinus strobus* L., *rigida* Mill., and *virginiana* Mill.).

Below, in the Siskiyou Mountains, on diorite, Oregon Caves National Monument, Josephine Co., Ore., ten stations along the moisture gradient using 50 site-samples for elevations between 2000 and 3000 ft.; figures are numbers of stems over 0.5 in. d.b.h. per hectare for: **a, needle-leaved evergreens** (*Chamaecyparis lawsoniana* Parl., *Pseudotsuga taxifolia* (Lam.) Britt., *Pinus lambertiana* Dougl., *Taxus brevifolia* Nutt.); **b, deciduous-broadleafs** x.33 (Cornus nuttalli Aud., Corylus rostrata var. *californica* A. DC., *Alnus rhombifolia* Nutt. and *rubra* Bong., *Acer macrophyllum* Pursh and *circinatum* Pursh, *Holodiscus discolor* (Pursh) Maxim., etc.); **c, evergreen-sclerophylls** x.33 *Lithocarpus densiflora* (H. & A.) Rehd., *Quercus chrysolepis* Liebm., *Arbutus menziesii* Pursh, *Castanopsis chrysophylla* (Dougl.) A. DC.), and *Quercus kelloggii* Newb., semi-sclerophyllous deciduous. The populations of deciduous and sclerophyllous species have been reduced to one-third of actual numbers for comparison with the needle-leaved trees as an approximate compensation for the larger size of the latter.

tions of fagaceous trees and small numbers of the cove-forest species toward the xeric extreme. In the Siskiyous, conifers dominate the stand in mesic sites, but with a large deciduous fraction and a smaller evergreen-sclerophyllous one. Toward drier sites there is a progressive decrease of the conifers and deciduous trees relative to the sclerophylls until, on dry south slopes, the community is dominated by sclerophylls with an open coniferous overgrowth and a small fraction of deciduous species. Such a shift in balance among growth-forms is comparable to the progressive shifts in biological spectrum or balance among Raunkiaer life-forms shown along climatic gradients by various authors cited above and along local gradients by Linkola (1924), Cajander (1926), Hansen (1930, 1932), Halliday (1935), along local gradients and through succession by Oosting (1942). Different balances of growth-forms and life-forms may well be equally adapted to the same environment, even within the same area, as mesic sites in the Smokies may support stands dominated either by the conifer, eastern hemlock, or by a mixture of deciduous species. The tendencies in adaptation through growth-forms may be recognized, however; and the structural-functional pattern of the community may involve a balance among growth-forms, a balance changing continuously with change in environment or with local, partial discontinuity.

While the structure of the terrestrial plant community may be thought of in terms of growth-forms, it may also be considered in terms of populations; the structural-functional pattern of a natural community is necessarily one of species populations. "Stability" of these populations has traditionally been an attribute of the climax, but for stability may be substituted the concept of the steady-state, first applied to the plant community by Russian ecologists. If the climax has meaning as a combination of species populations, a degree of balance among these populations should exist in which, as energy and material move through the community, individuals are constantly being lost, but being replaced by reproduction at an equal rate on the average. The balance is a dynamic one, as effectively expressed in the terms *Fliessgleichgewicht* and *équilibre mobile,* an active equilibrium in which the populations tend, by their interplay in relation to environmental limitations, to keep one another within relatively stable limits. As the balance is a balance of populations in relation to one another and environment, difference in environment implies difference in balance (Nicholson 1933). Tabulations from the author's Siskiyou

Table 1. Numbers of stems per hectare of six major tree species on four directions of exposure in the Siskiyou Mountains, southwestern Oregon. Based on ten site-sample counts of 0.1 hectare each for each direction of exposure including all stems over 0.5 inch dbh.; samples taken on slopes with inclinations between 15° and 30° at elevations between 2000 and 3000 feet in the vicinity of the Oregon Caves National Monument on quartz diorite or other non-mafic and non-calcareous rock. Directions of orientation: *NE*=N, NNE, and NE; *E*=E and ENE; *SE*=ESE, SE, and SSE; *S*=S and SSW.

Species	*Pseudotsuga taxifolia*				*Chamaecyparis lawsoniana*				*Lithocarpus densiflora*			
Exposure	NE	E	SE	S	NE	E	SE	S	NE	E	SE	S
Stems per Sample	21	53	13	2					144	138	8	103
	12	16	10	43	51					115	151	19
	35	16	28	6	4				66	125	92	159
	12	17	19	15	10				13	245	77	46
	18	18	19	31	2				29	51	43	60
	57	12	9	16	23					78	41	141
	27	21	3	4	2				35	42	74	114
	19	13	15			2			106	123	16	62
	4	8	7						229	61	100	72
	9	8	8	12					60	196	56	98
Stems per H	214	182	131	129	92	2			682	1174	658	874

Species	*Quercus chrysolepis*				*Arbutus menziesii*				*Castanopsis chrysophylla*			
Exposure	NE	E	SE	S	NE	E	SE	S	NE	E	SE	S
Stems per Sample	28	3	75	17	19	3		10	27	58		8
			35	189		18	7	30	9	32		
	1		13	13		18	5	74	5	14	2	61
		6	2	72		27	12	50	75	15	10	1
	1		106	103		1	5	35				3
				15		4	31	10	27		5	11
			3	17			14	14	35			11
	5	3	23	9			10	106	37			29
	3	2	8	26	11	5	6	128	40		8	8
	3	2	12	37	5	19		22	26	14	2	20
Stems per H	41	16	277	498	35	95	90	479	281	133	27	152

Mountain material may illustrate the difference in populations to be expected in different environments, in this case on four directions of site exposure (Table 1).

A rather high degree of irregularity appears in these populations, occurring in an essentially Mediterranean climate with dry summers and consequent exposure to periodic fires of varying intensity. The differences in populations of the various species on different sites are, however, statistically significant.* The irregularity of these stands is only greater in degree than in some other areas; because of the instability inherent in the climax balance, because of the many chances of dispersal, occupation, etc., the figures can in any area be only average or most probable populations on a particular kind of site. We thus encounter in population problems of climaxes the probability effects in synecology indicated by Egler (1942).

In view of the irregularity and instability of climax populations it may well be asked whether the balance of populations is such that the distinction between climax and succession has meaning for population ecology. Since the climax condition may be defined by the steady-state, we may consider, first that the climax stand is less unstable than the seral stand and, second, that the seral stand differs in that its change is directional, fluctuating about no average. A third useful distinction is that succession usually has a definite starting point or time zero (Jenny 1941, Major 1951) in contrast to the continuing changes in the climax. So far as the steady-state distinction itself is concerned, it is likely that a steady-state of changing and developing character, incompletely stabilized with its balances gradually shifting, exists in many communities. Between the unstable early seral community and the fully stabilized climax may exist a wide area of intermediate conditions (cf. Alechin 1926). For the first distinction based on the steady-state criterion, it may be asked whether a slowly changing seral community, a Douglas-fir forest, say, is not more nearly stable than an herbaceous climax like the prairie. While satisfactory evidence is not at hand, it would seem that even a relative difference between seral and climax communities in extent of fluctuation may easily be broken down. For the second distinction it may be observed that short-term, irregular and rhythmic fluctuations are necessarily part of the seral community's functioning and that long-range, directional trends are necessarily imposed on climax communities by long-range physiographic and climatic changes. The distinction again breaks down, though meaningful as a relative difference. The distinction based on a definite starting point for seral changes must also be qualified; for many of the changes occurring within the climax may have a definite starting point following a sudden climatic change, outbreak of an animal population, fire or wind effect, etc., while some successions, as the filling of a shallow lake and its replacement by forest, convert one type of self-maintaining community into another with no particular starting point for the change.

Evidence that the climax exists as a population phenomenon, as a community significantly less unstable than seral ones, is limited and largely circumstantial in nature. Such evidence includes (1) the observation that in some, climax stands the dominants are reproducing themselves while in other, seral stands they are not but are being replaced by the reproduction of other species; (2) the evidence that some, climax communities have not changed through a fairly long period of record while other, seral communities are more rapidly changing; (3) the observation that there is a fair degree of regularity in the appearance of similar population combinations on similar sites, suggesting that the populations have reached similar climax balances in their similar environments. It must be observed along with the last that some vegetations, as mature as they will become, e.g. the chaparral subject to burning and the unstable forests cited above, may not show any such regularity. It would seem that some communities may reach a climax in terms of growth-form, but never a steady-state of much meaning in terms of populations. While the climax may have surer significance in terms of growth-form, in terms of populations the climax, like the association, is a concept which tends to shrink under critical examination. It does not follow that it shrinks to nothing, and retention of the climax-succession distinction may still be justified.

INTERPRETATION OF EVIDENCE AND CRITICISM OF THE MONOCLIMAX

Before attempting further consideration of climax theory it may be well to compare the interpretations developed with the evidence already given. To be adequate, climax conceptions must provide for, or at least be consistent with, all lines of evidence available.

1. Succession. Evidence of successional characteristics—continuity and irregularity of populations, together with over-all trends—seem to agree well enough with the interpretation suggested. In this irregular continuum of populations our associes and myctia are, like our associations and ecotones, arbitrary and subjective, though justified by usefulness.

2. Climax convergence. If direction of succession and nature of the climax balance are determined in relation to environment, it should be expected that succession in similar environments would lead to climaxes having similar structural-functional patterns. Physiognomic convergence on a world-wide basis would be expected as similar community patterns developed to utilize resources in similar en-

* A test described by Mood (1950:398) indicates that the distribution of Castanopsis, *Quercus chrysolepsis*, and Arbutus, at least, differs significantly from random through the four series. The fact that the sclerophyll populations have more than one peak in the series of stations (e.g. Castanopsis with higher populations on mesic north and xeric south slopes than intermediate east and southeast ones) reflects actual bimodality of these populations, demonstrable with more extensive transect data, rather than sampling error.

vironments. Such convergence should be only partial, however, as observed by Beadle (1951), because of the different kinds of organisms available on different continents. Physiognomic convergence of vegetation of different sites in a single area might well occur in response to climate of the area, but should not be expected in all areas. Population convergence on similar sites in a limited area may also be expected, as a manifestation of community adaptation on the species level, at least in some areas.

3. Climax patterning. Because the sites occupied by communities are varied in any area, different climax populations are to be expected on them, adapted to site conditions. The climax vegetation of an area should consequently form a mosaic or pattern.

4. Climax continuity. Continuous change of site conditions along environmental gradients may be expected to imply continuous change in balance between growth-forms and species populations, little as we understand of the dynamics of such shifting balances. Granting the significance of vegetational discontinuities where they occur, the traditional conception of "a community" extending along a gradient essentially unchanged to some community-wide limit of tolerance where it is replaced by another community is, as a general principle, difficult to conceive dynamically if not meaningless semantically. It is suggested that vegetational change through space is as fundamental, as universal, and as generally continuous as vegetational change through time, even though discontinuities may be imposed on both spatial and temporal change.

5. Climax irregularity. Minor edaphic and climatic differences, chance factors of population entry, reproduction, and interaction, differences in history of past fires, insect attacks, windfalls, etc., may all contribute to the observed differences within a stand and from stand to stand on similar sites.

6. Climax instability. A degree of instability is intrinsic to the climax steady-state, and further instability results from effects of unstable environment on the balance.

As the evidence supports the conception of the climax as a population balance determined by the conditions of its site, some criticism of the monoclimax tradition in American ecology from Clements (1916) to Oosting (1948) and Braun (1950) may be indicated.

Analysis of monoclimax logic reveals that the position is not based upon a single conception, but upon two or more rather disparate ideas or assumptions. The monoclimax position, or any other climax position, must be brought into accord with the diversity of existing vegetation types in any area; and the harmonizing of existing diversity with theoretical unity of the monoclimax is by means of two assumptions. The first of these is the assumption of convergence to identity: "It is a fundamental part of Clements' concept that, given time and freedom, a climax vegetation of the same general type will be produced and stabilized irrespective of earlier site differences." (Cain 1939:150). Since the convergence to identity quite evidently does not occur, a second assumption must be brought into play, the assumption of difference within identity. Different climax populations in an area are assumed to be different lociations or other sub-units of the same climax, and so the diversity of climax vegetation is provided for within the monoclimax (Cain 1939: 153).

It is the first assumption, of convergence to identity, which is the nucleus of the monoclimax theory. Vegetation is thought to develop, and as it develops to modify environment, to climaxes of equal mesophytism on different sites. From the extent of application and the extent of the covergences assumed, the belief seems to have been widespread in American ecology that nothing succeeds like succession. Because a degree of convergence among the seres in an area may be observed, and because in some areas all or most vegetation converges toward identity of dominant growth-form, if not of populations, in the climax, this assumption once seemed a reasonable one. But as a scientific theory judged for present validity, such an idea can scarcely stand against the great body of evidence of edaphic and other special climaxes. Because of the intimate relation of "environment," "vegetation," "soil," and "animal community" as aspects of the ecosystem, the conception of identical vegetation developing in different environments seems in fact an essentially implausible one. It is in opposition to the central ecological conception of organic systems adapted to their whole environments; this first monoclimax assumption seems to state that climax vegetation is both dependent upon environment (regional climate) and independent of environment (local and topographic factors) as if these two aspects of environment were separate in their action on living plants and in the functioning ecosystem. To the extent that the monoclimax theory is based upon the assumption of convergence to identity it may be rejected as both theoretically questionable and contrary to the bulk of existing evidence.

The second assumption of the monoclimax theory is that, since different self-maintaining populations occur on different sites in an area, these must be parts or subdivisions of the climax association. The characteristic terminological complexity of Clementsian ecology arose in part from this necessity of subdividing the climax. Along with the series of climax terms already indicated were the various -ations: the association, consociation, fasciation, and lociation, together with the seration and the more limited sociation, lamiation, and sation (Clements 1936). Further contributions have included the subassociation, the subsociation, (Woodbury 1933), association-segregate (Braun 1935a), presociation (Carpenter 1939), and biociation (Kendeigh 1948). In criticism of Clements himself it may be observed that his response to the problem of vegetation which does not come in simple, natural units was mainly through multiplication of arbitrary units and sub-units. His answer to any new observation or exception, his way of meeting any new scientific challenge, was a new term. As a consequence the climax formation of Clementsian ecology might be described as a terminological jungle. Designation of the different types in a given vegetation pattern as preclimax, postclimax, etc., or as different consociations, lociations, association-segregates, etc., implicitly recognizes the non-convergence of climaxes. Justification of the monoclimax on the basis that north- and south-slope alternes, for example, are different lociations of the same association seems now to be a mere play with undefinable terms. To the extent that the monoclimax theory is based on the assumption of difference within identity it may be rejected as a semantic device which begs the question and obcures the problem.

When both these assumptions break down, as when climaxes which cannot belong to the same association exist in an area, a third provision is at hand—a set of words (pre-, post-, serclimax, etc.) to be applied to some of the climax types which do not fit into the climatic climax. These proclimaxes are often interpreted as arrested seral stages, potentially capable of developing to the climax (Clements 1928:106-7, 1936:262-4, Weaver & Clements 1929: 81, Phillips 1934-5:228). Since, however, they may be fully stabilized for their sites, they are subclimaxes which "only theoretically could be replaced by the climax" (Braun 1950:13). The recognition of a group of -climaxes which are neither actively developing nor true climaxes (Cain 1939:153), both stabilized and somehow developmental, is a position the logic of which does not commend itself. The necessity for these terms implying that some vegetation is climax, but seral also, may further illustrate the ambiguity and internal conflict resulting from imposition of the monoclimax ideal on actual vegetation.

Even apart from the proclimaxes, the monoclimax theory may be seen to be a mixture in indefinite proportions of two major propositions: that the various vegetation types in an area (1) are becoming the same climax or (2) are different parts of the same climax (Phillips 1934-5: 563, 566-7, Cain 1939:150, 153, Oosting 1948:223-4, 225-8, Braun 1950:12). These two assumptions are in a way complementary to one another. The second assumption provides for the failure of the full convergence assumed by the first to occur; the first permits the diversity of climaxes recognized by the second to be interpreted as dynamically related rather than as polyclimaxes. Perhaps it was the hybrid vigor from the union of these two ideas which permitted the monoclimax theory to persist so long in the face of so many criticisms by foreign ecologists and some by Americans. The two assumptions may seem somewhat in conflict since they state that the climaxes of two different sites are the same, but different. This conflict is not itself necessarily a major difficulty, since the assumptions may be reconciled by considering that the convergence is only partial. If the convergence is only partial, however, climaxes of varying degrees of dissimilarity may be expected on different sites; and this is the basis of the polyclimax theory. Some sites, not too greatly different, will support climaxes which differ in populations but not in physiognomy; whether or not these are regarded as belonging to the same association is a matter of terminology and not of theory. Other differences in site will be expressed in differences of physiognomy; and the climaxes will, by convention, be assigned to different formations. In any area a number of climax types will occur and often these will include more than one physiognomic type or formation. Some modification of the polyclimax conception of a mosaic of climax types on different sites may result from the individualistic hypothesis and climax continuity, as indicated below.

The monoclimax theory is felt by the author to be clearly inadequate for the needs of analytical ecology. The view may be expressed that this theory is one of those uncritical, preliminary generalizations which are of the greatest value in stimulating research and furthering the growth of a field of science, but are later found untenable. Thus Clementsian "dynamic ecology" may be evaluated both as a valid and major contribution in its emphasis of process and as a system which was in part artificial and pseudodynamic, interpreting successional processes from situations where they are active to situations where they are not, and thus extrapolating these processes beyond their true effectiveness to the hypothetical monoclimax. It was the great contribution of Clements to have formulated a system, a philosophy of vegetation, which has been a dominating influence on American ecology as a framework for ecological thought and investigation and partly also as a basis for minority dissent. Some negative aspects of Clements' system may be recognized—the superficial verbalism, the tendency to fit evidence by one means or another into the philosophic structure, the thread of non-empiricism which runs through his thought and work—, but his service in providing and relating concepts, integrating ecological knowledge of his time into a conception of vegetation, may also be recognized. The Clementsian system had a certain symmetry about it, it was a fine design if its premises were granted; and for its erection Clements may rank as one of the truly creative minds of the field. Other ecologists of his time probably had clearer conceptions of some ecological problems, and further development of the field may stem less from Clements than from those who dissented from his system. The ecologist wrestling with ecological problems may often wish, however, that the field could be as neatly formulated today as it was by Clements, that the climax problem, for example, was as simple and non-relative as it seemed in Clementsian ecology—before the multiplication of terms was necessary because it was by no means so simple.

FORMULATIONS OF CLIMAX THEORY AND LOGIC

In place of the monoclimax, three major propositions on the nature and structure of climaxes and their relativity may be formulated:

1. The climax is a steady-state of community productivity, structure, and population, with the dynamic balance of its populations determined in relation to its site.

2. The balance among populations shifts with change in environment, so that climax vegetation is a pattern of populations corresponding to the pattern of environmental gradients, and more or less diverse according to diversity of environments and kinds of populations in the pattern.

3. Since whatever affects populations may affect climax composition, this is determined by, or in relation to, all "factors" of the mature ecosystem—properties of each of the species involved, climate, soil and other aspects of site, biotic interrelations, floristic and faunistic availability, chances of dispersal and interaction, etc. There is no absolute climax for any area, and climax composition has meaning only relative to position along environmental gradients and to other factors.

The following secondary or corollary propositions are suggested as of possible significance for synecological research:

A. Propositions of climax determination. Climax composition is determined, as indicated in (3) above, by all factors which are intrinsic to, or act upon, the population on a sustained or repeated basis and do not act with such severity as to destroy the climax population and set new succession in motion. Factors determining climax population will thus include:

1. Characteristics of the populations involved. The balance among populations will necessarily be determined by the kinds of populations entering the community and by the peculiarities of each. The place of a given species in the balance will depend on its ability to maintain a population against environmental resistance, determined by its genetics. Since genetics of species may change along gradients, changing genetics of the species should be part of the background of changing balance among species along gradients. Since a species differs genetically from one place to another, these genetic differences may influence the different places or degrees of importance of a species in climax balances.

2. Climate. All climaxes are adapted to climate (and hence are climatic climaxes); but the climate which acts on and determines a climax population is necessarily the local climate of its site, not the general climate of an area.

3. Site. The climax balance is determined by environment of a specific site, and the climax population has meaning only for a kind of site (Bourne 1934). For the early assumption that climax was independent of site may be substituted the hypothesis that any significant difference in site implies a difference in climax population. As all climax stands occur on sites having some kind of topographic relation to other sites, all climaxes are topographic, as well as climatic, climaxes.

4. Soil. Soil parent-material, as arbitrarily separated from other aspects of site, is a climax determinant; for the traditional assumption that vegetation on any soil parent-material converges to the regional climax may be substituted the hypothesis that any significant edaphic difference, physical or chemical, may imply difference in climax population. All climaxes are edaphic, as well as topographic and climatic, climaxes. The traditional distinctions among climatic climax, pre- and postclimax, physiographic or topographic and edaphic climaxes or paraclimaxes thus break down entirely, as indicated for climatic and edaphic climaxes by Lippmaa (1933b), Bourne (1934), Beard (1946), and Cain (1947). All are part of the climax pattern; all are adapted to climate and all other non-catastrophic factors of environment. The terms topographic and edaphic climaxes may have continued usefulness, however, to indicate that the topographic or edaphic distinctiveness of a site is to be emphasized in relation to the distinctiveness of a particular climax population in the vegetation pattern. The arbitrariness of any distinction between edaphic and topographic climaxes is well illustrated by Platt's (1951) recent work showing that "shale-barren" vegetation, as a recognized type, is determined by some sufficient combination of (1) properties of soil parent material, (2) steepness of inclination, and (3) dryness as affected by direction of exposure.

5. Biotic factors. Natural communities are organic systems of plants and animals in environment; in much of what has been said about the vegetation pattern might be substituted the more awkward phrase natural-community pattern. In the functioning system the balances among plant populations exist in relation to, and are partially determined by, animals acting directly on the plants through consumption and trampling, indirectly through soil, etc. All climaxes are biotic climaxes, balanced in relation to their animal populations. Designation of such cases as the grazing of bison in the Great Plains (Weaver & Clements 1929, Clements & Shelford 1939, Larson 1940), and of rabbits in Britain (Farrow 1916, 1917, Watt 1936, Fenton 1940, Hope-Simpson 1940) and the aspen parkland (Lewis *et al* 1928, Bird 1930), the vegetation of ant and termite colonies (Harshberger 1929b, Martyn 1931, Myers 1936, Burtt 1942, Fries & Fries 1948), prairie-dog towns (Scheffer 1937), and other animal societies, the effects of carp in a lake (Cahn 1929) and of guano deposits (Cockayne 1909, Summerhayes & Elton 1923, 1928, Steffen 1928, Polunin 1935, Russell & Wellington 1940, Rasmussen 1946, Grønlie 1948, Hutchinson 1950), the beaver-meadow complex (Ives 1942b), as biotic or disclimaxes implies no real distinction from other climaxes, but only that effects of animals are rather more evident to us than in other communities.

6. Fire. Periodic burning is an environmental factor to which some climaxes are necessarily adapted (Cooper 1922b, Show & Kotok 1924, Troup 1926, Petrie *et al* 1929, Wells & Shunk 1931, Chapman 1932, Grabherr 1936, Myers 1936, Humbert 1937, 1938, Robyns 1938, Aubréville 1938, Beadle 1940, Trochain 1940, Wells 1942, Rawtischer 1948, Weaver 1951a, Egler 1952). In areas where the fires affecting these climaxes are not set by lightning, they may be lit by native populations; in this as in other influences man, and especially aboriginal man, may be part of the ecosystem in relation to the various factors of which climax vegetation is determined. In the absence of fire the climax populations might well develop to something different; but such an ideal climax is not on the ground subject to measurement. The burning may cause some population fluctuation, and it may then be difficult to draw a distinction between fire (and windfall, etc.) as environmental factors to which some climaxes are adapted and as disturbances introducing still greater instability and initiating successions in others. A continuous series from climaxes fully adapted to fire and scarcely affected by a single burn, through climaxes in which minor changes are produced by each burning and those in which the vegetational structure is altered, but not destroyed, to climaxes which are entirely destroyed by a single fire may be expected. Without attempting to draw a clear line where none exists, it may be thought that, in fire-adapted climaxes, fire either does not destroy the dominant populations or does not cause replacement of the dominant growth-form as in other climaxes.

7. Wind. Although wind is a part of the environment of all climax stands, some stands may show marked effect of wind on composition or physiognomy, especially, toward higher latitudes and altitudes (Crampton 1911, Fries 1913, Braun-Blanquet 1913, 1932, Cockayne 1921, Scharfetter 1921, Szafer *et al* 1924, Brockman-Jerosch 1925-9, Allan 1926b, Cox 1933, Polunin 1934, Russell & Wellington 1940, Issler 1944), and other stands may have windfall permitting reproduction as a normal part of their relation to environment. The combination of sea-wind and salt-spray is particularly effective in producing locally distinctive climaxes along coasts (Crampton 1911, Boodle 1920, Nordhagen 1923, Pavillard 1928, Praeger 1934, Tansley 1939, Davis 1936, Bayer 1938, Wells & Shunk 1938a, 1938b, Wells 1939, 1942, Doutt 1941, Oosting & Billings 1942, Vesey-Fitzgerald 1942, Oosting 1945, Beard 1944b, 1946, Zohary 1947).

8. Other factors. Various other factors may determine locally what self-maintaining or climax populations can exist in a site including, for terrestrial communities, snow-effects (Heer 1835, Vestergren 1902, Smith 1912, Fries 1913, Braun-Blanquet 1913, 1932, Szafer 1924, Watson 1925, Nordhagen 1928, 1936, Lippmaa 1929, 1933b, Harshberger 1929a, Hansen 1930, Trapnell 1933, Cox 1933, Domin 1933, Böcher 1933, Hayashi 1935, Polunin 1935, 1936, Russell & Wellington 1940, Sørensen 1943, Lüdi 1948, Gjaerevoll 1950), fog as it affects the coastal redwoods (Cooper 1917), the fog vegetation of the Peruvian desert (Knuchel 1947), and the mossy or cloud forests of many tropic mountains (Shreve 1914a, 1914b, MacCaughey 1917, 1920, Brown 1919, Seifriz 1923, Holttum 1924, Lane-Poole 1925, Tate 1932, Richards 1936, Beard 1942, 1944b, Pittier & Williams 1945, Beebe & Crane 1947), salt water and tide levels affecting such coastal vegetation as mangrove swamps (Ainslie 1926, Chipp 1927, Marshall 1934, Stehlé 1945, Beard 1946, Dulau & Stehlé 1950, Egler 1952) and salt marshes (Johnson & York 1915, McCrea 1926, Conard 1935, Tansley 1939, 1941, Fontes 1945) although these, not simply seral communities which disappear to be replaced by a climax, may be incompletely stabilized (Vaughan 1910, Yapp *et al* 1917, Walton 1922, Nienburg & Kolumbe 1931, Steiner 1934, Richards 1934, Taylor 1938, Chapman 1938-41, 1940, Nordhagen 1940, Davis 1940, 1942, Purer 1942, Allan 1950, Navalkar 1951) migratory vegetation in the sense of Crampton (1911, 1912, Tansley 1929), communities of shifting equilibrium in the sense of Alechin (1926, 1927).

9. Floristics and faunistics. Climax composition will necessarily be determined by the plant and animal species available in the area. Climax populations in similar environments will vary from place to place for floristic and faunistic reasons, and a recognized climax population type or association will usually have a limited range (Lippmaa 1933b, Bourne 1934, Cain 1947).

10. Chance. Climax composition must, finally, be considerably affected by chances of dispersal and occupation (Palmgren 1929, Lippmaa 1935b) and of population interactions.

B. Propositions on climax relativity. It has been indicated that the climax population has meaning only relative to the environmental conditions of its site, and the inappropriateness of dichotomous logic has been indicated in several connections. The following aspects of climax relativism are suggested:

1. Climax and succession. There are no distinctions between climax and succession or, more concretely, between climax and seral stands, except those of relative instability and relative significance of directional change. This relativity was expressed by Cowles (1901:81), "As a matter of fact we have a variable approaching a variable rather than a constant," and has been indicated by various authors since (Cooper 1913, 1926, Braun-Blanquet & Jenny 1926, Gleason 1927, Braun-Blanquet 1932:322, Tansley 1939:228, Tutin 1941, Graham 1941).

2. Climax and seral species. There is no reason why some species should not be both seral and climax. Climax species may dominate succession as in desert successions (Muller 1940, Shreve 1942, 1951); a species may enter a stand in succession and persist at a different population level into the climax; a species may enter climax stands on one kind of site in an area but only seral stands on another (e.g. the Lake States pines, Eggler 1938, Grant 1934, Nichols 1935), and may enter climax stands in one area but only

seral stands in another (e.g. Douglas-fir, Munger 1940, Sprague & Hansen 1946, Hansen 1947, Merkle 1951). While some species seem clearly seral or climax under specific conditions, for many it is a question of relative position along the time-scale of succession under particular circumstances.

3. Climax and seral types. Types, associations, or stratal communities defined by species may, correspondingly, be seral in one circumstance and climax in another (e.g. pine and Douglas-fir stands, steppe in New Zealand, Zotov 1938, *cf.* Domin 1928a:26), or may be self-maintaining and successional at different times (Tansley 1939:234). Untenable is a familiar kind of logic: Type A (e.g. pines) is being replaced by type B (oak-hickory) on site number 1 (a north slope); therefore, type A growing on site number 2 (a southwest slope) will also ultimately be replaced by type B. Chain-linking of successional observations without regard for site (Type A was seen replacing type B, B replacing C, C replacing D, etc.; therefore, B, C, etc. are all seral to A) is also untenable (Gleason 1927).

C. Propositions on climax recognition. A number of criteria have been used either explicitly or implicitly in the traditional recognition of monoclimaxes (*vide* Cooper 1913:11, 1922b:75, Weaver & Clements 1938:479-80, Oosting 1948:229, Braun 1950:13). It may be profitable to examine some of these which seem no longer tenable in this section and to discuss some which seem applicable in the next section.

1. Unity of growth-form. "The first criterion is that all the climax dominants must belong to the same major life form, since this indicates a similar response to climate and hence, a long association with each other." (Weaver & Clements 1938; *vide* the revealing attempt to exclude *Tsuga canadensis* from the eastern forests for having the wrong growth-form, Clements 1934:64, 1936:258, Weaver & Clements 1938:83). The world-wide occurrence of communities of mixed dominance and of several physiognomic types within a given climatic area may be sufficient commentary.

2. Area of climax. The monoclimax has been thought to be climax of a definite geographic region, so that the climax could be recognized by essential similarity over a large area (Braun 1950) and occurrence of one or more of the dominants throughout the area (Weaver & Clements 1938). Areal extent is irrelevant to achievement of the climax steady-state, however; and there is almost no lower limit on the area of a climax type (Tutin 1941). Such restricted types as the summit balds of mountains (Cain 1930, Camp 1931, 1936, Brown 1941, Daubenmire & Slipp 1943, Merkle 1951), Appalachian shale barrens (Allard 1946, Platt 1951), and southeastern granite flat-rocks (McVaugh 1943), pines on altered andesite in Nevada (Billings 1950), stands as limited as the small marsh in a morainic depression and patch of "alpine rain-forest" at the head of a glacial valley described by Ives (1942a) are, if self-maintaining, climax vegetation. Removal of size restrictions on climax types has, with the variety of populations present, the consequence that distinguishable climax types of the United States are essentially innumerable. While this must certainly be the case, stand types can very well be grouped subjectively into associations-abstract (Nichols 1917, Braun 1950) for some purposes.

3. Convergence on different sites. Convergence of different successions (Cooper 1913, Braun 1950) to similarity of vegetation on different sites (Cooper 1913, 1922b) is a criterion based on the first monoclimax assumption. The convergence is only partial, however, leading to climax vegetation which may be expected to differ on different types of sites. Granting the significance of such partial convergence on differing sites as occurs, it is not a basis of recognizing the self-maintaining condition.

4. Upland position (Nichols 1923). While the vegetation type prevailing on the uplands of an area may be more extensive than other types in the area, it is no more climax than they. Points 2, 3, and 4, may have meaning in relation to the prevailing climax, rather than in recognition of the climax steady-state.

5. Physiography. Convergence was achieved through both biological and physiographic processes in the interpretation of Cowles, but physiographic processes act through too long a period to be directly related to the climax as a biological phenomenon (Domin 1923, Faegri 1937). The consequence of erosion of an area down to a peneplain would be not to produce a uniform environment uniformly occupied by one of the vegetation types already in the area, but to produce a less diverse surface of different climate occupied by other kinds of climax stands. In spite of the impossibility of clearly separating biotic and physiographic (Cowles 1911), autogenic and allogenic (Tansley 1929, 1935) succession, it is the former, biological process, and not the hypothetical result of the latter, physical process, which is the concern of synecology (Michelmore 1934). Since some stands, at least, may reach the climax state in almost any area whether physiographically young, mature, or old, occupation of topographically mature sites (Braun 1950) may bear little relation to climax recognition.

6. Soil maturity (Braun 1950). While relations among climate, vegetation, and soil are recognized, these apply more to vegetation as growth-form than to vegetation as populations. As aspects of the ecosystem, vegetation and soil are, together, related to environment and one another, with neither simply determining the other (Jenny 1941, Major 1951). Soil, vegetation, and environment may consequently be expected to vary together, *pari passu,* through time and space; which is to say that the ecosystemic pattern varies in time and space and hence the coupled aspects of the ecosystem selected for study vary in parallel, though in no simple manner. The alternative definitions of soil maturity, in terms of profile development or of equilibrium with environment, are partially independent (Jenny 1941:48); and profile maturity of soil is not in itself a criterion of the climax state (Tansley 1935). Conditions of

some sites in any area and of most sites in some areas may be such as to prevent soils from developing to profile maturity as, in an extreme case, the "infantile" soils of the arctic (Griggs 1934). Soils in other areas may, in the course of development to the self-maintaining state, become degraded and podsolized so that a less productive vegetation and apparently less mature soil characterize the climax (Bourne 1934).

7. Mesophytism (Cowles 1901, Cooper 1913, Nichols 1917, 1923, Oosting 1948:224). The most mesophytic type of an area is no more "the climax" than other, less mesophytic, self-maintaining stands. The climax-adaptation index of Curtis & MacIntosh (1951) measuring relative mesophytism, etc., of stands is, as indicated by the authors, not simply a measurement of successional status. The course of succession will usually, though not necessarily, lead from apparently less mesophytic to apparently more mesophytic types; but it does not follow that a given xerophytic or hydrophytic stand is becoming more nearly mesophytic. A more mesophytic species is not to be chosen over a less mesophytic one as climax except as it is shown, for a given type of site, that the former replaces the latter. A balance between more and less mesophytic species may exist on sites too dry to support a purely mesophytic stand without implying succession.

8. Tolerance (Graham 1941). Succession will often involve more tolerant species replacing less tolerant ones, but in an area the more tolerant species may be climax on one site and the less tolerant ones on another. There is no reason why such intolerant species as the pines should not form climaxes, either in open stands in which they may reproduce continuously or in denser stands in which they may reproduce only periodically. There is no reason why more and less tolerant species should not form mixed climax stands, provided the stand is open enough to permit the latter to reproduce, or is sufficiently opened at times by fires, windfalls, etc., to permit the less tolerant species to reproduce. In general, in interpreting mixed stands, occurrence together of more and less mesophytic species, or of more and less tolerant species, or of species regarded from other evidence as climax and seral, may or may not mean that succession is occurring in the stand.

9. Higher growth-form (Weaver & Clements 1938: 90, 478). Apart from the difficulty of using criteria of higher and lower growth-forms (Gleason 1927) vegetational stature, at least, will usually increase through the succession into the climax. There is no reason, however, why heath, bog, or grassland should not in some areas replace forest as climax (Graebner 1901, Cajander 1913, Moss 1913, Backman 1920, Lüdi 1920, 1921, 1923, 1929, Katz 1926, Hopkinson 1927, Konovalov 1928, Zinserling 1929, Kleist 1929, Cain 1930, Mikyška 1932, Bayer 1933, Pfaffenberg & Hassenkamp 1934, Braun-Blanquet 1935, Ceballos 1935, Hausrath 1942, Duchaufour 1946-7, 1949, Killian & Moussu 1948, Zach 1950) even though such replacement violates our usual assumptions of trends through succession and of forests as climaxes wherever forests occur. There is no reason why such simple and open communities as those of cliffs (Klika 1932, Davis 1951), hammadas (Cannon 1913, Zohary 1944, 1945), alpine rocks and scree (Hess 1910, Lüdi 1921, Frey 1922, Allan 1926a, Leach 1930, Jenny-Lips 1930, Lippmaa 1933b), and arctic fjaeldmark (Holttum 1922, Polunin 1934, Wager 1938) should not form self-maintaining stands or why the simple algae of a mountain lake should not form a "primitive" climax (Pennak 1951).

10. Relation to successional trends. As indicated for stature and soil, so for productivity, diversity, etc. These features of communities will usually increase through the succession to the climax, but there is no reason why, in a given case, the usual direction should not be reversed. In general, climax status should be determined not by abstract or generalized conceptions of what should be ultimate, but by what populations actually replace other populations and then maintain themselves.

APPLICATION

If the above criteria are to be limited or abandoned, we may seek what criteria of the climax state remain. Although the successional trends have been ruled out for climax recognition apart from site, they, and characteristics of self-maintaining populations, may be the basis of climax decision allowing for site differences.

1. Trends. If, on a number of similar sites, one finds sparse grass among rocks on some, Arctostaphylos heath with thin soil on others, and young red fir forests on a third group, it is reasonable to fit them into series from the first to the third on the basis of increasing community productivity and stature and soil maturity. If in addition the fir stands can be shown to be self-maintaining, it is reasonable to consider them the climax ending the succession. The trends may serve, with limitations in each case already observed, to relate types to one another in relation to succession and the probable climax.

2. Reproduction. In a self-maintaining forest population a J-curve of many young and few old trees (Meyer & Stevenson 1943) will often be encountered, so that there is agreement or accordance (Braun 1950:13) between canopy and reproduction. A climax forest stand may thus often be identified by its J-curve of age or size distribution, a seral stand by marked disparity between the canopy and an undergrowth of species able to grow to canopy stature on the site. Such J-curves, however, may be encountered in some seral stands and are not to be expected in unstable climax stands. While it has been indicated that pre- and postclimaxes, edaphic climaxes, etc., do not represent different kinds of climaxes, climaxes may be distinguished by types of reproduction and constancy or variation in dominance. Stands may reproduce continuously, cyclically, or irregularly, as indicated by Jones (1945) types of structure in virgin forests; and any of these modes of reproduction may occur in stands in which

the dominant species remain the same, while in other kinds of stands the dominant species may alternate cyclically (Watt 1947) or fluctuate and replace one another irregularly (Aubréville 1938).

3. Regularity within the stand. Climax stands often show a marked uniformity on sites of uniform conditions, seral stands are often, in contrast, patchy or irregular. Increasing uniformity in distribution of species in the stand through succession to the climax is the basis of Picchi-Sermolli's "maturity index" (1948a, 1948b), which has, however, as emphasized by the author, only relative, directive significance.

4. Regularity between stands. If the climax population is a balance adapted to environment, it may be expected that similar sites within an area will support similar populations; and such similarity of stands may be evidence of their having reached the climax state. The variation which is to be expected in climax stands may be measurable as a dispersion, indicating in part the degree to which vegetation is stabilized on these sites. When, on similar sites in an area, it is observed that one group of apparently undisturbed stands are very much alike and another group of diverse stands differ from these and one another, it may be reasonable to regard the former as climax and to fit the latter into succession by other criteria.

The phenomenon of stand-to-stand similarity on similar sites may thus be recognized as central to the climax problem. It is an important part of the evidence that the climax as a population steady-state exists, so that similar population balances will occur in similar environments. It is a principal means of recognizing the climax state where it occurs. In it also is the principal utility of the climax concept in research. The existence of stand-to-stand regularity makes possible study of climax populations and climax patterning, study which may exclude succession as an additional variable apart from climax problems as such. Treatment of all stands, seral and climax, together without regard to succession in many areas involves such a bewildering diversity of stands as to defy rational interpretation and relation to environment. Separation of climax and succession permits ordering of the data, in which those stands fully adapted to environment can be related to environment, and those becoming more fully adapted through succession can be related to environment and the time-scale of succession. Treating climax stands alone, their stand-to-stand regularity permits meaningful statements about average or most probable climax populations on a given type of site. Thus in permitting a statement, within probability limits, of what the climax population on a site of known environment but unknown vegetation should be, it meets a prime test of usefulness of a scientific concept or approach: possibility of prediction. It is suggested that a principal justification for retention of the climax concept is in this possibility. If regularity of stands is a principal justification for the concept and for methods using the climax concept to study regularity of stands, a reciprocal interdependence of evidence, concept, and method exists, an interrelation which should be recognized, though it need not invalidate the approach through the climax.

Stand regularity may permit a further approach to climax definition. A climax stand is one which is self-maintaining and relatively permanent on its site (Tansley & Chipp 1926, Tutin 1941, Beard 1944a), but the individual stand is subject to fluctuation and shows chance variation from other stands. Of broader significance than the population of an individual climax stand is the average or most probable population for a type of site. Where data for several similar sites are available, the climax population may be defined as the average population of mature, self-maintaining stands on a type of site as defined and limited (see Table 1). Such an average population is not the absolute climax, which could not be determined if it existed, but is as close to the climax as we can come. Recognizing that there is no ultimate, absolute climax, we may substitute for the climax as an ideal the climax as a measured population; the climax population is the population of climax stands as measured. "The climax" may thus be defined by results of an operation, the counting (and averaging) of populations, though with subjectivity entering the choice of climax stands. In areas much disturbed, where climax stands cannot be measured, the climax may be without meaning as a population, while the alternative definition in terms of physiognomy may be sufficient for most purposes. In other areas largely undisturbed by man the climax populations may be so unstable and variable that averages have little meaning, and study of population changes within the climax communities may be more to the point (*vide* Graham 1941).

Available criteria of the climax state are far from providing for easy, objective identification of climax stands, nor can they be used securely without the judgment and feel for vegetational process of the research worker. In spite of attempts to increase the rigor of climax logic and method there remains an inescapable residuum of subjectivity. It may perhaps be recognized, in the character of the problem, why such must be the case. Not only is the difference between climax and succession relative, but all criteria applicable to the distinction are relative, partly independent of one another, and subject to a sufficient number of exceptions. In dealing with problems of climax and succession one may soon find oneself reasoning in such a network of relativities that some subjective choice of which of these are most significant is unavoidable. It is not surprising that Clements, facing this, sought to banish relativity from ecology, thus spiriting the fundamental difficulty out of sight. It was, however, one of the essential characteristics of the field which was thus apparently done away with.

Yet, in spite of the difficulties, effective methods can be applied in some areas for treatment of the climax as a population.* A basis for some of these may be found in the climax pattern conception. An appropriate method of treating this pattern is that of "gradient analysis" (Whittaker 1951, 1952), analysis of natural communities in terms of population distributions along environmental gradients, an approach for which a theoretical basis is outlined by Major (1951). As a population pattern climax vegetation should be suited to further statistical treatment—averages and means of dispersion of climax stands, correlation with environment and between aspects of the ecosystem, etc.—than have yet been undertaken.

One means of summarizing major features of the pattern is its presentation in terms of the moisture gradient, as in Fig. 1, or in terms of both moisture and elevation (Whittaker 1951, 1952). In many areas the vegetation of a given elevation on soil materials not too diverse can be studied as a pattern of populations and types relative to the moisture gradient (Linkola 1924, Alechin 1926, Arènes 1926-7, McLuckie & Petrie 1927, Phillips 1928, Sampson 1930b, Hansen 1930, 1932, Pessin 1933, Grant 1934, Davis & Richards 1934, Walter 1937, Pidgeon 1938, Horton 1941, Tolstead 1942, Wright & Wright 1948, Morison *et al* 1948), to which vegetational pattern a catena or pattern of soils (Milne 1935, Bushnell 1943, Morison *et al* 1948) may correspond. In many areas several combinations of dominants may be recognized along the gradient; but in northern and mountain forests the canopy may be similar throughout, while the gradient and other site differences find expression in a diverse pattern of undergrowth types, as recognized in the forest site types, etc., of Finnish and other forest ecologists (Cajander 1909, 1926, 1943, Cajander & Ilvessalo 1921, Palmgren 1922, Linkola 1924, 1929, Issler 1926, Klika 1927, Zlatník 1928, Juraszek 1928, Sukachev 1928, Ilvessalo 1929, Kirstein 1929, Sambuk 1930, Borowicki 1932, Heimburger 1934, Meusel 1935, Lippmaa 1935b, Niedziłkowski 1935, Müller 1936, 1938, Mikyška 1937, Arnborg 1940, Mallner 1944, Spilsbury & Smith 1947). In either case, greater or less diversity of climax pattern is to be expected, whether appearing in the "associations" of dominants or only in the undergrowth.

A substitute for the climatic climax to characterize a whole pattern may, finally, be sought in the prevailing climax (Whittaker 1951). It may be determined what dominant populations (or growth-forms) are most numerous in the climax pattern, or what type as defined (or climax physiognomy) occupies the majority of sites in an area. The prevailing climax is thus quantitatively definable in terms of either percentage of climax populations occurring throughout an area or percentage of sites occupied by an arbitrarily defined type. Through the prevailing climax statements can be made about average climax characteristics in an area, correlated with climate, with no assumption of convergence to a monoclimax. In areas of complex environment and complicated vegetation pattern, however, there may be no prevailing climax type or growth-form; and average stand composition for the area may have limited meaning.

* The following applications have been developed in more detail in a manuscript, "A vegetation analysis of the Great Smoky Mountains."

CONCLUSION

In the light of what has been said and what is, in any case, familiar to practicing ecologists, the difficulty, complexity, relativity, and subjective involvements of climax theory and practice may be evident. Vegetation presents a field of phenomena notably lacking in fixed points of reference, lines of division, invariable rules, and easy definitions. The spatial pattern of natural communities is so complex that it is beyond reasonable possibility to achieve understanding of it in full detail. Neither do we need to seek such knowledge; two approaches on different levels can be applied—extensive, geographic study seeking generalized statements about major features of the pattern, and intensive, analytical treatment of local areas—to obtain different aspects of understanding desired. It is felt that the climax concept, in terms of growth-forms or populations, may contribute to both of these.

No completely rigorous definition of the climax and its distinction from succession has been found, and apparently none need be expected. If the retention of the climax-succession distinction is to be justified, presumably it must be not because the distinction is sharp and invariable, but because the distinction, relative as it is, has some real significance and usefulness. While "climax" and "succession" are only words referring to different degrees of instability or of approach to a steady-state no longer changing directionally, the distinction appears significant; and in the absence of these terms it would be necessary to refer to more or less stable, or more or less mature stands, substituting other expressions. In spite of all the difficulties the distinction has been found useful in many research problems; in many areas, at least, it has been found to work. Granting the usefulness of the concept, the more clearly it can be formulated the more serviceable a research tool it may be. It has been an objective of this study to show approaches toward less subjectivity and greater rigor, to indicate research possibilities of quantitative approaches to the climax free from some assumptions made in the past.

The interpretation developed, particularly that of the climax as a population pattern, represents a third climax hypothesis in addition to the traditional monoclimax and polyclimax theories, a conception which may be designated the climax pattern hypothesis. This conception is in part intermediate to or synthetic of the monoclimax and polyclimax, since the diversity of climax stands is recognized, but these are regarded, not as two or more discrete climax associations, but as parts of a single, often continuously gradating climax pattern. Vegetation is conceived as, even more fundamentally than the polyclimax "mosaic of plant communities whose distribution is determined by a corresponding mosaic of habitats" recognized by Tansley (1939:216, cf. Domin 1929), a pattern of populations, variously related to one another, corresponding to the pattern of environmental gradients. Both the monoclimax and the polyclimax assume the existence of associations as valid, discrete units of vegetation, whether one or more than one of these may exist in a given area. The climax pattern hypothesis does not require the subjective grouping of both environments and stands into pieces of the mosaic; it is thus independent of the community-unit assumption and distinct from both the traditional theories. Evaluation of this concept against the monoclimax and polyclimax, as of these against one another, may be based on the criteria of simplicity of assumption, closeness of fit or correspondence to field observation, and methodological implication and productiveness.

The difference between the monoclimax and polyclimax has been recognized as semantic in part (Cain 1939, 1947) since ecologists may, in following one or the other, describe the same stands or stand types by different terms, or even by the same terms with different successional implications. The semantic aspect of the difference between these or among them and the climax pattern hypothesis by no means implies indifference of choice. Judged by simplicity of assumption, the polyclimax has the advantage over the monoclimax in freedom from the assumption of convergence in spite of difference in environment; and the climax pattern hypothesis may have the advantage over the polyclimax theory in freedom from the assumption that climaxes represent valid and distinct units or associations. Judged by correspondence to field data, the monoclimax is clearly least adequate, since the relation between the single ideal climax and the actual diversity of stands is abstract and arbitrary, and in fact somewhat remote. The polyclimax conception corresponds better to the diversity of stands, but implies that these belong to a definite number of distinct climaxes, while the climax pattern hypothesis allows in addition for the continuity of climax types and the diversity of stands only arbitrarily classifiable into a particular number of climax associations. The series—monoclimax, polyclimax, climax pattern—is thus regarded by the author as one of decreasing degree of abstraction and increasing closeness of fit to the actual pattern of vegetation, while the prevailing climax is a definable abstraction of higher order from the climax pattern to replace the undefinable climatic climax.

Judged in terms of methodological implication, the monoclimax is again least adequate, since it implies as a research approach the subjective choice of a climatic climax and the relation of other types in an area to this on a largely speculative basis. The polyclimax theory serves better for the study of actual successions observed to climax associations recognized; it permits both greater realism in the separation of concrete from supposititious successions and greater effectiveness in the relation of climax to environment, though retaining the community-unit assumption and the resulting mosaic conception. The climax pattern hypothesis, free from this assumption, permits the direct relation of community gradients to environmental gradients, of the whole diversity of climax stands in an area to the whole range of environments. The pattern conception and the emphasis of continuity lead to research methods relating populations of species and growth-forms to environmental gradients, to the approach of gradient analysis which is fundamentally different from the traditional approach through units. In methodological implication the climax pattern hypothesis differs more from the traditional two theories than these from each other; it is thus regarded by the author as both a synthesis of the other theories with open-system and population interpretations and an advance beyond them, a logical and natural development from the polyclimax. The difference among climax conceptions is, in any case, far from being a semantic quibble or straw-man tactic (Egler 1951); it is one of the most important decisions to be made by synecologists and one for which quite adequate bases of choice exist.

It is the author's hope that some convergence of the divergent schools of ecology might be possible on the basis of this and other approaches which in part synthesize, in part develop beyond differences among schools. It is surely to be regretted that plant ecology has so largely lacked, since the early great figures in the field, the world-wide perspective which is to be sought in all fields of science and is an achievement of many. Contact is not what it should be even between American and British ecologists (Godwin 1949), and the two major groups of Continental ecologists have been equally divided from one another (Faegri 1937). The alienation of American and Continental ecology in matters other than method is all too familiar. One basis of this may be in some of the interpretive inadequacies rather characteristic of traditional American plant ecology, recently pointed out by Egler (1951). The difficulty may also, however, lie in the difference in relative emphasis of succession and climax and relative size of the "fundamental units" used: the series—American, British, Swiss-French, and Scandinavian—is, on the whole, one of decreasing emphasis of succession and climax and decreasing magnitude of principal vegetational units. It may be observed that these two concepts which separate American and Continental ecology, the monoclimax and the association, are both subjective; and it would scarcely seem that these subjective concepts constitute grounds for divorce between ecological schools. An intermediate position can be conceived, one free from the overemphasis of climax and succession among many American ecologists and the underemphasis among many Europeans, free from the overemphasis of terminology and classification of units in both, and more concerned with dynamic process within the community and with vegetational problems on a world-wide scale. This best common denominator of American and Continental ecology is essentially the position occupied for some time by British ecology. The viewpoint developed here differs from the British tradition in the full abandonment of the climatic climax (except in the definable form of the prevailing climax) and the conception of natural communities and environment as forming together an ecosystemic pattern subject to analysis. Perhaps in the development of analytical approaches to natural communities and of concepts more closely related to analysis and its results, the differences among schools in classificatory and descriptive concepts and abstract interpretation may lose their present importance.

SUMMARY

1. A brief review of the development of climax theory in different schools of ecology indicates the confusion existing and the need for reconsideration of the concept.

2. Some of the evidence bearing on the problem may be summarized as: 1) continuity and irregularity of populations in succession, and 2) convergence, 3) patterning, 4) continuity, 5) irregularity, and 6) instability of climaxes. Adequate conceptions of climax and succession must be consistent with these lines of evidence and sufficiently general in formulation to provide for climax phenomena on a world-wide and not a regional basis.

3. Common to succession and the climax is the interplay of populations, which in succession is expressed in directional change and in the climax as fluctuation about an average. In the progressive development from originally unorganized populations of an organized community pattern adapted to environment, succession is an evolutionary process in a general sense. Through the course of succession there tends to be progressive increase in community complexity and diversity, stature or massiveness and productivity, maturity of soil, and relative stability and regularity of populations; but exceptions to all these trends may be noted. Certain successions involving marked decrease in productivity and others of the above may be termed, by definition, retrogressive.

4. The climax may be interpreted as a partially stabilized community steady-state adapted to maximum sustained utilization of environmental resources in biological productivity. As a functioning system developing in relation to environment, the natural community has a structural-functional pattern adapted to the whole pattern of environmental factors in which it exists. Similar or convergent community patterns tend to develop in adaptation to similar environments. Change in environment implies change in the community pattern; and along a continuous environmental gradient, community composition will usually change continuously.

5. These interpretations appear consistent with the evidence. The climatic climax or monoclimax theory is regarded as untenable.

6. Three principal propositions of climax theory are offered:

1) The climax is a steady-state of community productivity, structure, and population, with the dynamic balance of its populations determined in relation to its site.

2) The balance among populations shifts with change in environment, so that climax vegetation is a pattern of populations corresponding to the pattern of environmental gradients, and more or less diverse according to diversity of environments and kinds of populations in the pattern.

3) Since whatever affects populations may affect climax composition, this is determined by, or in relation to, all "factors" of the mature ecosystem—properties of each of the species involved, climate, soil, and other aspects of site, biotic interrelations, floristic and faunistic availability, chances of dispersal and interaction, etc. There is no absolute climax for any area, and climax composition has meaning only relative to position along environmental gradients and to other factors.

Secondary propositions are developed from these. All climaxes are topographic or physiographic, edaphic, and biotic, as well as climatic; many are determined in relation to fire, wind-effects, snow, salt-spray, etc.; and all may be affected by "chance" factors. Distinctions between climax and succession, climax and seral species, and climax and seral types or stands are necessarily relative. Various of the traditional criteria of climax recognition (unity of dominant growth-form, area of climax, physiography, soil maturity, upland position, mesophytism, tolerance, higher growth-form) are wholly or partially irrevelant for the recognition of the climax state.

7. Useful aids to recognition of the climax state are: successional trends, with limitations and for particular sites, observed reproduction, regularity within stands, and stand-to-stand regularity. The last is central to climax theory as an evidence that a climax steady-state exists, as a means of recognizing the climax state, and as a basis of the principal utility of the climax in research. The climax may be defined by the average population of mature, self-maintaining stands on a type of site as defined and limited. Subjectivity is not eliminated from such a definition nor, from the nature of the problem, can it be. The climax may often be analyzed and represented as a pattern of populations along the moisture gradient, or in relation to this and other gradients. In some areas a prevailing climax may be defined, as a replacement of the climatic climax, by the percentages of dominant species in the pattern or by the percentage of sites occupied by a type as defined.

8. For all its subjectivity and relativity, the climax concept has real meaning and usefulness. The climax pattern hypothesis developed here represents a third climax conception, preferred over the traditional monoclimax and polyclimax theories, and a basis for the research approach of "gradient analysis." It is to be hoped that, with the recognition of the subjectivity of the association and climatic climax and increasing use of analytical approaches, some convergence among the divergent schools of ecology may be possible.

LITERATURE CITED

Adamson, R. S. 1939. The classification of life-forms of plants. Bot. Rev. **5:** 546-561.

Ainslie, J. R. 1926. The physiography of southern Nigeria and its effect on the forest flora of the country. Oxford Forestry Mem. **5:** 1-36.

Albertson, F. W. & J. E. Weaver. 1942. History of the native vegetation of western Kansas during seven years of continuous drought. Ecol. Monog. **12:** 23-51.

———. 1944. Nature and degree of recovery of grassland from the great drought of 1933 to 1940. Ecol. Monog. **14:** 393-479.

Albertson, N. 1950. Das grosse südliche Alvar der Insel Öland. Eine pflanzensoziologische Übersicht. Svensk Bot. Tidskr. **44:** 269-331.

Alechin, W. W. 1925. Assoziationskomplexe und Bildung ökologischer Assoziationsreihen. Bot. Jahrb. **59:** 30-40.

———. 1926. Was ist eine Pflanzengesellschaft? Ihr Wesen und ihr Wert als Ausdruck des sozialen Lebens der Pflanzen (Autorisierte Übersetzung aus dem Russischen von Selma Ruoff). Repert. Spec. Novarum Regni. Veg. Beih. **37:** 1-50.

———. 1927. Die Alluvion der Flusstäler in Russland. Repert. Spec. Novarum Regni. Veg. Beih. **47:** 1-79.

———. 1946. L'histoire de la phytosociologie russe et ses particularités (Russian, *fide* Roussin 1948). Outscheny Zapisky Moskowsk. Gosoud. Ouniversit. Wip. **103:** 85-95.

Allan, H. H. 1926a. Notes on the study of the open communities of high-mountain areas in New Zealand. In Tansley and Chipp, "Aims and methods in the study of vegetation" pp. 366-372.

———. 1926b. A remarkable New Zealand scrub association. Ecology **7:** 72-76.

———. 1937. A consideration of the "biological spectra" of New Zealand. Jour. Ecol. **25:** 116-152.

Allan, P. F. 1950. Ecological bases for land use planning in Gulf Coast marshlands. Jour. Soil & Water Conserv. **5:** 57-62, 85.

Allard, H. A. 1946. Shale barren associations on Massanutten Mountain, Virginia. Castanea **11:** 71-124.

Allee, W. C. 1932. Animal life and social growth. Baltimore: Williams & Wilkins. 159 pp.

Allee, W. C., A. E. Emerson, O. Park, T. Park & K. P. Schmidt. 1949. Principles of animal ecology. Philadelphia: W. B. Saunders. 837 pp.

Allorge, P. 1922. Les associations végétales du Vexin français. Rev. Gén. de Bot. **34:** 71-79, etc. (12 sections).

Archibald, E. E. A. 1948. Plant populations. I. A new application of Neyman's contagious distribution. Ann. Bot. N. S. **12:** 221-235.

Arènes, J. 1926-27. Étude phytosociologique sur la chaîne de la Sainte-Baume et la Provence. Soc. Bot. de France Bull. **73:** 1016-1022, **74:** 65-85.

Arnborg, T. 1940. Der Vallsjö-Wald, ein nordschwedischer Urwald. Acta Phytogeog. Suecica **13:** 128-154.

Ashby, E. 1935. The quantitative analysis of vegetation. With an appendix by W. L. Stevens. Ann. Bot. **49:** 779-802.

———. 1948. Statistical ecology. II—A reassessment. Bot. Rev. **14:** 222-234.

Aubréville, A. 1938. La forêt coloniale (Les forêts de l'Afrique Occidentale Française). Acad. des. Sci. Colon. Paris, Ann. **9:** 1-244.

Auer, V. 1920 Über die Entstehung der Stränge auf den Torfmooren. Acta Forest. Fenn. **12**(2): 1-145.

Backman, A. L. 1920. Torymarksundersökningar i mellersta Österbotten. (German summary.) Acta Forest. Fenn. **12**(1): 1-190, 1-22.

Bagenal, T. B. 1951. A note on the papers of Elton and Williams on the generic relations of species in small ecological communities. Jour. Anim. Ecol. **20:** 242-245.

Bailey, V. A. 1933a. The quantitative theory of interaction between different species of animals. Roy. Soc. N. S. Wales Jour. & Proc. 1932, **66:** 387-393.

———. 1933b. On the interaction between several species of hosts and parasites. Roy. Soc. London Proc. Ser. A **143:** 75-88.

Barker, R., T. Dembo & K. Lewin. 1941. Frustration and regression: An experiment with young children. Iowa Univ. Studies in Child Welfare **18**(1): 1-314.

Barrington, A. H. M. 1930. Burma forest soils. Jour. Ecol. **18:** 145-150.

Bauer, H. L. 1936. Moisture relations in the chaparral of the Santa Monica Mountains, California. Ecol. Monog. **6:** 409-454.

Bayer, A. W. 1933. The relationship of vegetation to soil erosion in the Natal thornveld. So. African Jour. Sci. **30:** 280-287.

———. 1938. An account of the plant ecology of the coastbelt and midlands of Zululand. Natal Mus. Ann. **8:** 371-454.

Beadle, N. C. W. 1940. Soil temperatures during forest fires and their effect on the survival of vegetation. Jour. Ecol. **28:** 180-192.

———. 1951. The misuse of climate as an indicator of vegetation and soils. Ecology **32:** 343-345.

Beard, J. S. 1942. Montane vegetation in the Antilles. Caribbean Forester **3**(2): 61-74.

———. 1944a. Climax vegetation in tropical America. Ecology **25:** 127-158.

———. 1944b. The natural vegetation of the island of Tobago, British West Indies. Ecol. Monog. **14:** 135-163.

———. 1946. The natural vegetation of Trinidad. Oxford Forestry Mem. **20:** 1-152.

———. 1949. The natural vegetation of the Windward and Leeward Islands. Oxford Forestry Mem. **21:** 1-192.

Beauchamp, R. S. A. & P. Ullyott. 1932. Competitive relationships between certain species of fresh-water triclads. Jour. Ecol. **20:** 200-208.

Beaven, G. F. & H. J. Oosting. 1939. Pocomoke Swamp: A study of a cypress swamp on the eastern shore of Maryland. Torrey Bot. Club Bull. **66:** 367-389.

Beebe, W. & J. Crane. 1947. Ecology of Rancho Grande, a subtropical cloud forest in northern Venezuela. Zoologica [New York] **32:** 43-60.

Bergman, H. F. 1924. The composition of climax plant formations in Minnesota. Mich. Acad. Sci. Papers **3:** 51-60.

Bertalanffy, L. von. 1950a. The theory of open systems in physics and biology. Science **111:** 23-29.

———. 1950b. An outline of general system theory. British Jour. Philos. Sci. **1:** 134-165.

———. 1951. General system theory: A new approach

to unity of science. I. Problems of general system theory. Human Biol. **23**: 302-312.

Biberg, I. J. 1749. Oeconomia Naturae. Linné Amoen. Acad. **2**: 1-52 (*fide* Clements 1928:9).

Billings, W. D. 1950. Vegetation and plant growth as affected by chemically altered rocks in the western Great Basin. Ecology **31**: 62-74.

Bird, R. D. 1930. Biotic communities of the aspen parkland of central Canada. Ecology **11**: 356-442.

Birrell, K. S. & A. C. S. Wright. 1945. A serpentine soil in New Caledonia. New Zeal. Jour. Sci. & Technol. **27A**: 72-76.

Bjorkman, G. 1937. Floran i trakten av Äpartjåkkos magnesit-fält. Svenska Vetensk. Akad. Skr. **33**: 1-36.

Black, G. A., T. Dobzhansky, & C. Pavan. 1950. Some attempts to estimate species diversity and population density of trees in Amazonian forests. Bot. Gaz. **111**: 413-425.

Blackman, G. E. 1935. A study by statistical methods of the distribution of species in grassland associations. With an appendix by R. S. Bartlett. Ann. Bot. **49**: 749-777.

Blackman, M. W. & H. H. Stage. 1924. Succession of insects in dying, dead and decaying hickory. N. Y. State Col. Forestry, Syracuse Univ. Bull. **24**: 1-269.

Blake, S. T. 1938. The plant communities of western Queensland and their relationships, with special reference to the grazing industry. Roy. Soc. Queensland Proc. **49**: 156-204.

Böcher, T. W. 1933. Studies on the vegetation of the east coast of Greenland between Scoresby Sound and Angmagssalik (Christian IX.s Land). Medd. om. Grønland. **104**(4): 1-132.

Boodle, L. A. 1920. The scorching of foliage by seawinds. Min. Agr. & Fisheries Jour. [G. Brit.] **27**: 479-486.

Borowicki, S. 1932. Zespoły florystyczne lasu bukowego w Kątach (Wielkopolska). (German summary.) Acta Soc. Bot. Poloniae **9** (Suppl.): 57-94.

Botke, J. 1928. Undurchlässige Horizonte im Boden und die Pflanzenvegetation. Rec. des Trav. Bot. Néerland **25a**: 50-57.

Bourne, R. 1934. Some ecological conceptions. Empire Forestry Jour. **13**: 15-30.

Boyko, H. (Bojko) 1934. Die Vegetationsverhältnisse im Seewinkel. Versuch einer pflanzensoziologischen Monographie des Sand- und Salzsteppengebietes östlich vom Neusiedler See. II. A) Allgemeines. B) Die Gesellschaften der Sandsukzession. Bot. Centbl. Beihefte, Abt. 2, **51**: 600-747.

———. 1945. On forest types of the semi-arid areas at lower altitudes. Palestine Jour. Bot., Ser. R, **5**: 1-21.

Braun, E. Lucy. 1935a. The undifferentiated deciduous forest climax and the association-segregate. Ecology **16**: 514-519.

———. 1935b. The vegetation of Pine Mountain, Kentucky. An analysis of the influence of soils and slope exposure as determined by geological structure. Amer. Midland Nat. **16**: 517-565.

———. 1936. Forests of the Illinoian till plain of southwestern Ohio. Ecol. Monog. **6**: 89-149.

———. 1942. Forests of the Cumberland Mountains. Ecol. Monog. **12**: 413-447.

———. 1947. Development of the deciduous forests of eastern North America. Ecol. Monog. **17**: 211-219.

———. 1950. Deciduous forests of eastern North America. Philadelphia: Blakiston. 596 pp.

Braun-Blanquet, J. 1913. Die Vegetationsverhältnisse der Schneestufe in den Rätisch-Lepontischen Alpen. Ein Bild des Pflanzenlebens an seinen äussersten Grenzen. Schweiz. Naturf. Gesell. Denkschr. **48**: 1-347.

———. 1915. Les Cévennes méridionales (Massif de l'Aigoual). Étude phytogéographique. Biblioth. Univ., Arch. des Sci. Phys. et Nat., Ser. 4, **39**: 72-80 etc., **40**: 39-63 etc. (10 sections).

———. 1928. Pflanzensoziologie, Grundzüge der Vegetationskunde. Berlin: Springer. 330 pp. 2nd edn. 1951. Wien: Springer. 631 pp.

———. 1932. Plant sociology, the study of plant communities. Authorized English translation of "Pflanzensoziologie" by G. D. Fuller & H. S. Conard. New York: McGraw-Hill. 439 pp.

———. 1933. L'association végétale climatique, unité phytosociologique, et le climax du sol dans le Midi méditerranéen. Soc. Bot. de France Bull. **80**: 715-722.

———. 1935. Wald und Bodenentwicklung im Schweizerischen Nationalpark. Zesde Internat. Bot. Congr. Amsterdam Proc. **2**: 64-66.

Braun-Blanquet, J. & E. Furrer. 1913. Remarques sur l'étude des groupements de plantes. Soc. Languedoc Geog. Bull. **36**: 20-41.

Braun-Blanquet, J. & H. Jenny. 1926. Vegetations-Entwicklung und Bodenbildung in der alpinen Stufe der Zentralalpen. Schweiz. Naturf. Gesell. Denkschr. **63**: 181-349.

Braun-Blanquet, J. & J. Pavillard. 1928. Vocabulaire de sociologie végétale. 3rd ed. Montpellier: Roumégous et Déhan. 23 pp.

Breien, K. 1932. Vegetasjonen på skjellsandbanker i indre Østfold. Nyt Mag. Naturvidensk. **72**: 131-282.

Brockmann-Jerosch, H. 1925-29. Die Vegetation der Schweiz. Geobot. Landesaufn. Schweiz Beitr. **12**: 1-499

Brockmann-Jerosch, H. & E. Rübel. 1912. Die Einteilung der Pflanzengesellschaften nach ökologisch-physiognomischen Gesichtspunkten. Leipzig: Engelmann. 72 pp.

Brough, P., J. McLuckie, & A. H. K. Petrie. 1924. An ecological study of the flora of Mount Wilson. I. The vegetation of the basalt. Linn. Soc. N. S. Wales Proc. **49**: 475-498.

Brown, D. M. 1941. Vegetation of Roan Mountain: A phytosociological and successional study. Ecol. Monog. **11**: 61-97.

Brown, R. T. & J. T. Curtis. 1951. A northern conifer-hardwood forest continuum. Ecol. Soc. Amer. Bull. **32**: 56 (Abstract).

———. 1952. The upland conifer-hardwood forests of northern Wisconsin. Ecol. Monog. **22**: 217-234.

Brown, W. H. 1919. Vegetation of Philippine Mountains. The relation between the environment and physical types at different altitudes. Bur. Sci. Publ. 13. Manila: Bur. of Printing. 434 pp.

Buechner, H. K. 1944. The range vegetation of Kerr County, Texas, in relation to livestock and white-tailed deer. Amer. Midland Nat. **31**: 697-743.

Bukovsky, V. 1935. To the criticism of the basic problems and concepts of biocenology. (Russ. with Engl. summ.) Voprosy Ekologii i Biotsenologii, Leningrad Univ. **2**: 74-99.

Burtt, B. D. 1942. Some East African vegetation communities. Ed. by C. H. N. Jackson, with a foreword by W. H. Potts. Jour. Ecol. **30**: 65-146.

Burtt Davy, J. 1938. The classification of tropical woody vegetation-types. Oxford Univ. Imp. Forest. Inst. Paper **13**: 1-85.

Bushnell, T. M. 1943. Some aspects of the soil catena concept. Soil Sci. Soc. Amer. Proc. 1942, **7**: 466-476.

Cahn, A. R 1929. The effect of carp on a small lake: The carp as a dominant. Ecology **10**: 271-274.

Cain, S. A. 1930. An ecological study of the heath balds of the Great Smoky Mountains. Butler Univ. Bot. Studies **1**: 177-208.

———. 1939. The climax and its complexities. Amer. Midland Nat. **21**: 146-181.

———. 1943. The Tertiary character of the cove hardwood forests of the Great Smoky Mountains National Park. Torrey Bot. Club Bull. **70**: 213-235.

———. 1947. Characteristics of natural areas and factors in their development. Ecol. Monog. **17**: 185-200.

———. 1950. Life-forms and phytoclimate. Bot. Rev. **16**: 1-32.

Cain, S. A. & W. T. Penfound. 1938. Aceretum rubri: the red maple swamp forest of central Long Island. Amer. Midland Nat. **19**: 390-416.

Cajander, A. K. 1904. **Beiträge** zur Kenntniss der Vegetation der **Hochgebirge** zwischen Kittilä und Muonio. Fennia **20**: **9** (*fide* Clements 1928: 151).

———. 1909. Ueber **Waldtypen**. Acta Forest. Fenn. **1**(1): 1-175.

———. 1913. Studien über die Moore Finnlands. Acta Forest. Fenn. **2**(3): 1-208.

———. 1926. The theory of forest types. Acta Forest. Fenn. **29**(3): 1-108.

———. 1943. Wesen und Bedeutung der Waldtypen. Intersylva **3**: 169-209.

Cajander, A. K. & Y. Ilvessalo. 1921. Über Waldtypen. II. Acta Forest. Fenn. **20**(1): 1-77.

Camp, W. H. 1931. The grass balds of the Great Smoky Mountains of Tennessee and North Carolina. Ohio Jour. Sci. **31**: 157-164.

———. 1936. On Appalachian trails. N. Y. Bot. Gard. Jour. **37**: 249-265.

Cannon, W. A. 1913. Botanical features of the Algerian Sahara. Carnegie Inst. Wash. Publ. **178**: 1-81.

Carpenter, J. R. 1938. An ecological glossary. Norman, Okla.: Univ. Oklahoma Press. 306 pp.

———. 1939. The biome. Amer. Midland Nat. **21**: 75-91.

———. 1940. Insect outbreaks in Europe. Jour. Anim. Ecol. **9**: 108-147.

Ceballos, L. 1935. Regresión del encinar (*Quercetum ilicis*) en los terrenos graniticos próximos a Avila. Soc. Españ. de Hist. Nat. Bol. **35**: 407-421.

Chandler, D. C. & O. B. Weeks. 1945. Limnological studies of western Lake Erie. V. Relation of limnological and meteorological conditions to the production of phytoplankton in 1942. Ecol. Monog. **15**: 435-457.

Chapman, A. G. 1942. Forests of the Illinoian till plain of southeastern Indiana. Ecology **23**: 189-198.

Chapman, H. H. 1932. Is the longleaf type a climax? Ecology **13**: 328-334.

Chapman, V. J. 1938-41. Studies in salt-marsh ecology. Jour. Ecol. **26**: 144-179 (Sections I to III), **27**: 160-201 (IV & V), **28**: 118-152 (VI & VII), **29**: 69-82 (VIII).

———. 1940. Succession on the New England salt marshes. Ecology **21**: 279-282.

Chipp, T. F. 1927. The Gold Coast forest: a study in synecology. Oxford Forestry Mem. **7**: 1-94.

———. 1930. The vegetation of the central Sahara. Geog. Jour. **76**: 126-137.

———. 1931. The vegetation of northern tropical Africa. Scottish Geog. Mag. **47**: 193-214.

Ciferri, R. 1936. Studio geobotanico dell'Isola Hispaniola (Antille). Atti d'Ist. Bot. Univ. Pavia, Ser. 4, **8**: 1-336.

Clapham, A. R. 1935. Review of Raunkiaer's "The life forms of plants and statistical plant geography." Jour. Ecol. **23**: 247-249.

———. 1936. Over-dispersion in grassland communities and the use of statistical methods in plant ecology. Jour. Ecol. **24**: 232-251.

Clements, F. E. 1904. The development and structure of vegetation. Nebr. Univ., Bot. Surv. of Nebr. 7, Studies in the vegetation of the state, 3, Lincoln, Nebr.: Bot. Seminar. 175 pp.

———. 1905. Research methods in ecology. Lincoln, Nebr.: Univ. Publ. Co. 334 pp.

———. 1916. Plant succession: An analysis of the development of vegetation. Carnegie Inst. Wash. Publ. **242**: 1-512.

———. 1920. Plant indicators: The relation of plant communities to process and practice. Carnegie Inst. Wash. Publ. **290**: 1-388.

———. 1928. Plant succession and indicators. A definitive edition of plant succession and plant indicators. New York: Wilson. 453 pp.

———. 1934. The relict method in dynamic ecology. Jour. Ecol. **22**: 39-68.

———. 1936. Nature and structure of the climax. Jour. Ecol. **24**: 252-284.

———. 1949. Dynamics of vegetation. Selections from the writings of Frederick E. Clements, compiled and edited by B. W. Allred and Edith S. Clements. New York: Wilson. 296 pp.

Clements, F. E. & V. E. Shelford. 1939. Bio-ecology. New York: Wiley & Sons. 425 pp.

Clements, F. E., J. E. Weaver & H. C. Hanson. 1929. Plant competition: An analysis of community functions. Carnegie Inst. Wash. Publ. **398**: 1-340.

Cline, A. C. & S. H. Spurr. 1942. The virgin upland forest of central New England: A study of old growth stands in the Pisgah Mountain section of southwestern New Hampshire. Harvard Forest Bull. **21**: 1-51.

Cobbe, T. J. 1943. Variations in the Cabin Run forest, a climax area in southwestern Ohio. Amer. Midland Nat. **29**: 89-105.

Cockayne, L. 1909. The ecological botany of the Subantarctic Islands of New Zealand. **In** C. Chilton,

"The Subantarctic Islands of New Zealand" **1**: 182-235. New Zealand: Philos. Inst. of Canterbury.

———. 1921. The vegetation of New Zealand. In "Die Vegetation der Erde." Leipzig: Engelmann, 364 pp.

Cole, L. C. 1946. A theory for analyzing contagiously distributed populations. Ecology **27**: 329-341.

———. 1951. Population cycles and random oscillations. Jour. Wildlife Mangt. **15**: 233-252.

Conard, H. S. 1935. The plant associations of central Long Island: A study in descriptive plant sociology. Amer. Midland Nat. **16**: 433-516.

———. 1939. Plant associations on land. Amer. Midland Nat. **21**: 1-27.

———. 1938. Climax group and climax swarm. Ecology. **19**: 315-316.

———. 1951. The background of plant ecology. A translation from the German "The plant life of the Danube Basin" by A. Kerner (1863). Ames, Iowa: State College Press. 238 pp.

Conklin, E. S. 1923. The definition of introversion, extroversion and allied concepts. Jour. Abnorm. Psychol. & Soc. Psychol. **17**: 367-382.

Coombe, D. E. & F. White. 1951. Notes on calcicolous communities and peat formation in Norwegian Lappland. Jour. Ecol. **39**: 33-62.

Cooper, W. S. 1913. The climax forest of Isle Royale, Lake Superior, and its development. Bot. Gaz. **55**: 1-44, 115-140, 189-235.

———. 1916. Plant successions in the Mount Robson Region, British Columbia. Plant World **19**: 211-238.

———. 1917. Redwoods, rainfall and fog. Plant World **20**: 179-189.

———. 1922a. The ecological life history of certain species of *Ribes* and its application to the control of the white pine blister rust. Ecology **3**: 7-16.

———. 1922b. The broad-sclerophyll vegetation of California: An ecological study of the chaparral and its related communities. Carnegie Inst. Wash. Publ. **319**: 1-124.

———. 1923. The recent ecological history of Glacier Bay, Alaska: II. The present vegetation cycle. Ecology **4**: 223-246.

———. 1926. The fundamentals of vegetational change. Ecology **7**: 391-413.

———. 1939. A fourth expedition to Glacier Bay, Alaska. Ecology **20**: 130-155.

———. 1942. Vegetation of the Prince William Sound region, Alaska; with a brief excursion into post-Pleistocene climatic history. Ecol. Monog. **12**: 1-22.

Cottle, H. J. 1932. Vegetation on north and south slopes of mountains in southwestern Texas. Ecology **13**: 121-134.

Cowles, H. C. 1899. The ecological relations of the vegetation on the sand dunes of Lake Michigan. I. Geographical relations of the dune floras. Bot. Gaz. **27**: 95-117, 167-202, 281-308, 361-391.

———. 1901. The physiographic ecology of Chicago and vicinity; a study of the origin, development, and classification of plant societies. Bot. Gaz. **31**: 73-108, 145-182.

———. 1910. The fundamental causes of succession among plant associations. Brit. Assoc. Adv. Sci. Rept. 1909: 668-670.

———. 1911. The causes of vegetative cycles. Bot. Gaz. **51**: 161-183.

———. 1919. Review of Clements' "Plant Succession." Bot. Gaz. **68**: 477-478.

Cox, C. F. 1933. Alpine plant succession on James Peak, Colorado. Ecol. Monog. **3**: 299-372.

Crampton, C. B. 1911. The vegetation of Caithness considered in relation to the geology. Published under the auspices of the Committee for the Survey and Study of British Vegetation. 132 pp.

———. 1912. The geological relations of stable and migratory plant formations. Scottish Bot. Rev. **1**: 1-17, 57-80, 127-146.

Crocker, R. L. & J. G. Wood. 1947. Some historical influences on the development of the South Australian vegetation communities and their bearing on concepts and classification in ecology. Roy. Soc. So. Austral. Trans. **71**: 91-136. (Reviewed by Tansley, 1948).

Crombie, A. C 1945. On competition between different species of graminivorous insects. Roy. Soc. London Proc. Ser. B. **132**: 362-395.

———. 1947. Interspecific competition. Jour. Anim. Ecol. **16**: 44-73.

Cuatrecasas, J. 1934. Observaciones geobotánicas en Colombia. Mus. Nac. de Cien. Nat. Trab. [Madrid] Ser. Bot. **27**: 1-144.

Curtis, J. T. 1951. A Wisconsin prairie continuum based upon presence data. Ecol. Soc. Amer. Bull. **32**: 56 (Abstract).

Curtis, J. T. & R. P. McIntosh. 1951. An upland forest continuum in the prairie-forest border region of Wisconsin. Ecology **32**: 476-496.

D'Ancona, U. 1939. Der Kampf ums Dasein. Eine biologisch-mathematische Darstellung der Lebensgemeinschaften und biologischen Gleichgewichte. Übersetz von L. Holzer. Abh. Exakten Biol., Berlin **1**: 1-196.

Dansereau, P. 1946. L'érablière laurentienne. II. Les successions et leurs indicatuers. Canadian Jour. Res. Sect. C, Bot. Sci. **24**: 235-291.

———. 1951. Description and recording of vegetation upon a structural basis. Ecology **32**: 172-229.

Daubenmire, R. F. 1942. An ecological study of the vegetation of southeastern Washington and adjacent Idaho. Ecol. Monog. **12**: 53-79.

———. 1947. Plants and environment: A textbook of plant autecology. New York: Wiley & Sons. 424 pp.

Daubenmire, R. F. & A. W. Slipp. 1943. Plant succession on talus slopes in northern Idaho as influenced by slope exposure. Torrey Bot. Club Bull. **70**: 473-480.

Davidsson, I. 1946. Notes on the vegetation of Árskógsströnd, North Iceland. Acta Nat. Island. **1**(4): 1-20.

Davis, C. 1936. Plant ecology of the Bulli District. I. Stratigraphy, physiography and climate; general distribution of plant communities and interpretation. Linn. Soc. N. S. Wales, Proc. **61**: 285-297.

Davis, J. H., Jr. 1940. The ecology and geologic rôle of mangroves in Florida. Carnegie Inst. Wash. Publ. **517**: 303-412.

———. 1942. The ecology of the vegetation and topography of the sand keys of Florida. Carnegie Inst. Wash. Publ. **524**: 113-195.

Davis, P. H. 1951. Cliff vegetation in the eastern Mediterranean. Jour. Ecol. **39**: 63-93.

Davis, T. A. W. & P. W. Richards, 1934. The vegetation of Moraballi Creek, British Guiana: An ecological study of a limited area of tropical rain forest. II. Jour. Ecol. **22:** 106-155.

De Bach, P. & H. S. Smith. 1941. Are population oscillations inherent in the host-parasite relation? Ecology **22:** 363-369.

Deevey, E. S. 1940. Limnological studies in Connecticut. V. A contribution to regional limnology. Amer. Jour. Sci. **238:** 717-741.

Domin, K. 1923. Is the evolution of the earth's vegetation tending towards a small number of climatic formations? Acta Bot. Bohemica **2:** 54-60.

———. 1928a. The plant associations of the valley of Radotín. Preslia, Věstník Českoslov. Bot. Společ. **7:** 3-68.

———. 1928b. The relations of the Tatra Mountain vegetation to the edaphic factors of the habitat: A synecological study. Acta Bot. Bohemica **6/7:** 133-164.

———. 1929. Some problems of plant ecology. Internat. Congr. Plant Sci., Ithaca, Proc. 1926, **1:** 497-524.

———. 1933. Die Vegetationsverhältnisse der Bucegi in den rumänischen Südkarpathen. Veröff. Geobot. Inst. Rübel, Zürich **10:** 96-144.

Doutt, J. K. 1941. Wind pruning and salt spray as factors in ecology. Ecology **22:** 195-196.

Drude, O. 1890. Handbuch der Pflanzengeographie. Stuttgart: Engelhorn. 582 pp.

———. 1896. Deutschlands Pflanzengeographie. Stuttgart (*fide* Moss 1910).

———. 1913. Die Ökologie der Pflanzen. Braunschweig: Vieweg & Sohn. 308 pp.

Duchaufour, P. 1946-47. Le sol et la flore forestière en quelques points des secteurs parisien et ligérien. Rev. des Eaux et Forêts **84:** 701-722, **85:** 16-38.

———. 1949. Ecological researches on the French Atlantic oak forest. Vegetatio **1:** 340-342.

Duffield, J. E. 1933. Fluctuations in numbers among freshwater crayfish, *Potamobius pallipes* Lereboullet. Jour. Anim. Ecol. **2:** 184-196.

Dulau, L. & H. Stehlé. 1950. L'évolution éco-phytosociologique du littoral de l'Anse Dumont à la Guadeloupe (Antilles Françaises). Mus. Natl. d'Hist. Nat. Paris Bull. **22:** 488-501.

Dureau de la Malle, A. J. C. A. 1825. Mémoire sur l'alternance ou sur ce problème: la succession alternative dans la réproduction des espèces végétales vivant en société, est-elle une loi générale de la nature? Ann. Sci. Nat. I. **5:** 353-381.

Du Rietz, G. E. 1919. Review of Clements' "Plant Succession." Svensk Bot. Tidskr. **13:** 117-121.

———. 1921. Zur methodologischen Grundlage der modernen Pflanzensoziologie. Wien: Holzhausen. 272 pp.

———. 1922. Die Grenzen der Assoziationen. Eine Replik an John Frödin. Bot. Notiser **1922:** 90-96.

———. 1924. Studien über die Vegetation der Alpen, mit derjenigen Skandinaviens verglichen. Veröff. Geobot. Inst. Rübel Zürich **1:** 31-138.

———. 1930a. Classification and nomenclature of vegetation. Svensk. Bot. Tidskr. **24:** 489-503.

———. 1930b. Vegetationsforschung auf soziationsanalytischer Grundlage. Handb. Biol. Arbeitsmeth. ed. E. Abderhalden XI, **5:** 293-480.

———. 1931. Life forms of terrestrial flowering plants. Acta Phytogeog. Suecia **3:** 1-95.

Dyksterhuis, E. J. 1948. The vegetation of the western Cross Timbers. Ecol. Monog. **18:** 325-376.

Dymond, J. R. 1947. Fluctuations in animal populations with special reference to those in Canada. Roy. Soc. Canada Trans., 3rd Ser., Sect. 5, **41:** 1-34.

Eggeling, W. J. 1947. Observations on the ecology of the Budongo rain forest, Uganda. Jour. Ecol. **34:** 20-87.

Eggler, W. A. 1938. The maple-basswood forest type in Washburn County, Wisconsin. Ecology **19:** 243-263.

Egler, F. E. 1942. Vegetation as an object of study. Phil. of Sci. **9:** 245-260.

———. 1947. Arid southeast Oahu vegetation, Hawaii. Ecol. Monog. **17:** 383-435.

———. 1951. A commentary on American plant ecology, based on the textbooks of 1947-1949. Ecology **32:** 673-695.

———. 1952. Southeast saline everglades vegetation, Florida, and its management. Vegetatio **3:** 213-265.

Elenken, A. A. 1921. La loi de l'équilibre mobile dans les symbioses et les associations des plantes. Isv. Glav. Bot. Sad., Leningrad **20:** 75-121.

Elton, C. S. 1924. Periodic fluctuations in the numbers of animals: Their causes and effects. Jour. Expt. Biol. **2:** 119-163.

———. 1927. The nature and origin of soil-polygons in Spitsbergen. Geol. Soc. London Quart. Jour. **83:** 163-194.

———. 1942. Voles, mice and lemmings; problems in population dynamics. New York: Oxford Univ. Press. 496 pp.

———. 1946. Competition and the structure of ecological communities. Jour. Anim. Ecol. **15:** 54-68.

———. 1949. Population interspersion: an essay on animal community patterns. Jour. Ecol. **37:** 1-23.

Faegri, K. 1933. Über die Längenvariationen einiger Gletscher des Jostedalsbre und die dadurch bedingten Pflanzensukzessionen. Bergens Mus. Aarbok, Naturv. rekke 1933(7): 1-255.

———. 1937. Some recent publications on phytogeography in Scandinavia. Bot. Rev. **3:** 425-456.

Fairbairn, W. A. 1943. Classification and description of the vegetation types of the Niger Colonie, French West Africa. Oxford Univ. Imp. Forest Inst. Paper **23:** 1-38.

Falk, P. 1940. Further observations on the ecology of central Iceland. Jour. Ecol. **28:** 1-41.

Farrow, E. P. 1916. On the ecology of the vegetation of Breckland. II. Factors relating to the relative distributions of *Calluna*-heath and grass-heath in Breckland. Jour. Ecol. **4:** 57-64.

———. 1917. On the ecology of the vegetation of Breckland. III. General effects of rabbits on the vegetation. Jour. Ecol. **5:** 1-18.

Fautin, R. W. 1946. Biotic communities of the northern desert shrub biome in western Utah. Ecol. Monog. **16:** 251-310.

Fenton, E. W. 1940. The influence of rabbits on the vegetation of certain hill-grazing districts of Scotland. Jour. Ecol. **28:** 438-449.

Firbas, F. 1924. Studien über den Standortscharakter auf Sandstein und Basalt (Ansiedlung und Lebens-verhältnisse der Gefasspflanzen in der Felsflur des Rollbergs in Nordböhmen). Bot. Centbl. Beih., Abt. 2, **40:** 253-409.

Flahault, C. & P. Combres. 1894. Sur la flore de la Camargue et des alluvions du Rhône. Soc. Bot. de France Bull. **41:** 37-58.

Flowers, S. 1934. Vegetation of the Great Salt Lake Region. Bot. Gaz. **95:** 353-418.

Fontes, F. C. 1945. Algumas características fitosociológicas dos "salgados" de Sacavém. Soc. Broteriana Bol. **19**(2a Ser.): 789-813.

Fournier, P. 1927. Qu'est-ce que l'association du Hêtre? Soc. Bot. de France Bull. **74:** 416-429.

Fraser, L. & J. W. Vickery. 1938. The ecology of the upper Williams River and Barrington Tops Districts. II. The rain-forest formations. Linn. Soc. N. S. Wales Proc. **63:** 139-184.

Freud, S. 1935. A general introduction to psychoanalysis. New York: Boni and Liveright. 406 pp.

Frey, E. 1922. Die Vegetationsverhältnisse der Grimselgegend im Gebiet der zukünftigen Stauseen. Ein Beitrag zur Kenntnis der Besiedlungsweise von kalkarmen Silikatfels- und Silikatschuttböden. Naturf. Gesell. in Berne Mitt. **1921:** 85-281.

Frey-Wyssling, A. 1933. Over de zandsteppen von Kota Pinang ter Oostkust van Sumatra. Trop. Nat. **22**(4): 69-72.

Fries, R. E. & T. C. E. Fries. 1948. Phytogeographical researches on Mt. Kenya and Mt. Aberdare, British East Africa. Svenska Vetensk. Handl. Ser. 5, **25**(5): 1-83.

Fries, T. C. E. 1913. Botanische Untersuchungen im nördlichsten Schweden. Akad. Abh., Uppsala. (*fide* authors.)

Frödin, J. 1921. Quelques associations de lande dans le Bohuslän nord-ouest. Bot. Notiser 1921: 81-97.

Fuller, M. E. 1934. Insect inhabitants of carrion: a study in animal ecology. Austral. Council. Sci. & Indus. Res. Bull. **82:** 1-62.

Furrer, E. 1922. Begriff und System der Pflanzensukzession. Naturf. Gesell. Zürich Vierteljahrsschr. **67:** 132-156.

Gams, H. 1918. Prinzipienfragen des Vegetationsforschung. Ein Beitrag zur Begriffsklärung und Methodik der Biocoenologie. Naturf. Gesell. Zürich Vierteljahrsschr. **63:** 293-493.

———. 1923. Die Waldklimate der Schweizeralpen, ihre Darstellung und ihre Geschichte. Naturf. Gesell. Basel Verh. **35**(1): 262-276.

Garren, K. H. 1943. Effects of fire on vegetation of the southeastern United States. Bot. Rev. **9:** 617-654.

Gates, R. R. 1928. Notes on the tundra of Russian Lapland. Jour. Ecol. **16:** 150-160.

Gause, G. F. 1934a. The struggle for existence. Baltimore: Williams & Wilkins. 163 pp.

———. 1934b. Experimental analysis of Vito Volterra's mathematical theory of the struggle for existence. Science **79:** 16-17.

———. 1935. Vérifications expérimentales de la théorie mathématique de la lutte pour la vie. Paris: Hermann. 61 pp.

———. 1936. The principles of biocoenology. Quart. Rev. Biol. **11:** 320-336.

———. 1937. Experimental populations of microscopic organisms. Ecology **18:** 173-179.

Gause, G. F. & A. A. Witt. 1935. Behavior of mixed populations and the problem of natural selection. Amer. Nat. **69:** 596-609.

Gaussen, H. 1933. Géographie des plantes. Paris: Colin. 222 pp.

Gelting, P. 1934. Studies on the vascular plants of East Greenland between Franz Joseph Fjord and Dove Bay (Lat. 73° 15′-76° 20′ N). Medd. om Grønland **101**(2): 1-337.

Gilbert, M. L. 1951. Herb composition of the hardwood forests of southern Wisconsin. Ecol. Soc. Amer. Bull. **32:** 57 (Abstract).

Gilliland, H. B. 1938. The vegetation of Rhodesian Manicaland. Jour. So. African Bot. **4:** 73-99.

Gillman, C. 1936. East African vegetation types. Jour. Ecol. **24:** 502-505.

Gjaerevoll, O. 1950. The snow-bed vegetation in the surroundings of Lake Torneträsk, Swedish Lappland. Svensk Bot. Tidskr. **44:** 387-440.

Gleason, H. A. 1917. The structure and development of the plant association. Torrey Bot. Club Bull. **44:** 463-481.

———. 1926. The individualistic concept of the plant association. Torrey Bot. Club Bull. **53:** 7-26.

———. 1927. Further views on the succession-concept. Ecology **8:** 299-326.

———. 1929. Plant associations and their classification: A reply to Dr. Nichols. Internat. Congr. Plant Sci., Ithaca Proc. 1926, **1:** 643-646.

———. 1939. The individualistic concept of the plant association. Amer. Midland Nat. **21:** 92-110.

Godwin, H. 1929. The sub-climax and deflected succession. Jour. Ecol. **17:** 144-147.

———. 1949. Review of H. J. Oosting, "The study of plant communities: an introduction to plant ecology." Jour. Ecol. **37:** 174-175.

Godwin, H. & V. M. Conway. 1939. The ecology of a raised bog near Tregaron, Cardiganshire. Jour. Ecol. **27:** 313-359.

Grabherr, W. 1936. Die Dynamik der Brandflächenvegetation auf Kalk- und Dolomitböden des Karwendels. Bot. Centbl. Beih., Abt. B, **55:** 1-94.

Graebner, P. 1895. Studien über die norddeutsche Heide, Versuch einer Formationsgliederung. Bot. Jahrb. **20:** 500-654.

———. 1901. Die Heide Norddeutschlands. In "Die Vegetation der Erde" 5 (*fide* Clements 1928: 159-164).

Graham, S. A. 1939. Forest insect populations. Ecol. Monog. **9:** 301-310.

———. 1941. Climax forests of the Upper Peninsula of Michigan. Ecology **22:** 355-362.

Grant, M. L. 1934. The climax forest community in Itasca County, Minnesota, and its bearing upon the successional status of the pine community. Ecology **15:** 243-257.

Greig-Smith, P. 1952. The use of random and contiguous quadrats in the study of the structure of plant communities. Ann. Bot. N. S. **16:** 293-316.

Griggs, R. F. 1934. The problem of arctic vegetation. Wash. Acad. Sci. Jour. **24:** 153-175.

———. 1936. The vegetation of the Katmai district. Ecology **17**: 380-417.

Grisebach, A. 1838. Ueber den Einfluss des Climas auf die Begränzung der natürlichen Floren. Linnaea **12**: 159-200.

———. 1872. Die Vegetation der Erde nach ihrer klimatischen Anordnung. Ein Abriss der vergleichenden Geographie der Pflanzen. Leipzig: Engelmann. 2 vols., 567 & 693 pp.

———. 1877. La végétation du globe d'après sa disposition suivant les climats, esquisse d'une géographie comparée des plantes. Paris: Baillière et Fils. 2 vols., 765 & 905 pp.

Grønlie, A. M. 1948. The ornithocoprophilous vegetation of the bird cliffs of Røst in the Lofoten Islands, northern Norway. Nyt Mag. Naturvidensk. **86**. Reviewed Jour. Ecol. **37**: 424-425, 1949.

Hairston, N. G. 1951. Interspecies competition and its probable influence upon the vertical distribution of Appalachian salamanders of the genus *Plethodon*. Ecology **32**: 266-274.

Halliday, W. E. D. 1935. Report on vegetation and site studies, Clear Lake, Riding Mountain National Park, Manitoba, Summer, 1932. Dept. of Interior, Forest Serv., Ottawa, Res. Note No. **42**: 1-44.

Haman, M. & B. R. Wood. 1928. The forests of British Guiana. Tropical Woods (Yale Univ. School Forestry) **15**: 1-13.

Hamilton, W. J., Jr. 1937. The biology of microtine cycles. Jour. Agr. Res. **54**: 779-790.

Hansen, A. A. 1921. The terminology of the ultimate vegetation. Ecology **2**: 125-126.

Hansen, H. Mølholm. 1928. Livsform og Alder. Bot. Tidsskr. **40**: 193-203.

———. 1930. Studies on the vegetation of Iceland. Copenhagen: Frimodt. 186 pp.

———. 1932. Nørholm hede, en formationsstatistisk vegetationsmonografi. (English summary). K. Danske Vidensk. Selsk. Skrifter, Nat. og. Math., Ser. 9, **3**(3): 96-196.

Hansen, H. P. 1947. Postglacial forest succession, climate, and chronology in the Pacific Northwest. Amer. Phil. Soc. Trans., N. S. **37**: 1-130.

Harshberger, J. W. 1929a. Preliminary notes on American snow patches and their plants. Ecology **10**: 275-281.

———. 1929b. The vegetation of Campos de Jordão, Brazil. Amer. Phil. Soc. Proc. **68**: 83-92.

Haskell, E. F. 1940. Mathematical systematization of "environment," "organism" and "habitat." Ecology **21**: 1-16.

Hausrath, H. 1942. Heide und Wald. Allg. Forst. u. Jagd. Ztg. **118**(1): 2-10.

Hayashi, N. 1935. On a snow-patch association at Mt. Hakkôda. Tôhoku Imp. Univ., Sendai, Sci. Repts. 4th Ser. (Biol.) **9**: 253-278.

Heer, O. 1835. Die Vegetationsverhältnisse des südöstlichen Theils des Cantons Glarus; ein Versuch, die pflanzengeographischen Erscheinungen der Alpen aus climatologischen und Bodenverhältnissen abzuleiten. Zurich: Orell, Füssli und Co. (*fide* Rübel 1929).

Heimburger, C. C. 1934. Forest-type studies in the Adirondack region. N. Y. (Cornell) Agr. Expt. Sta. Mem. **165**: 1-122.

Hess, E. 1910. Über die Wuchsformen der alpinen Geröllpflanzen. Bot. Centbl. Beih., Abt. 2, **27**: 1-170.

Hesse, R., W. C. Allee, & K. P. Schmidt. 1937. Ecological animal geography. New York: Wiley & Sons. 597 pp.

Högbom, B. 1914. Über die geologische Bedeutung des Frostes. Uppsala Univ. Geol. Inst. Bull. **12**: 257-390.

Holdaway, F. G. 1930. Field populations and natural control of *Lucilia sericata*. Nature [London] **126**: 648-649.

Holttum, R. E. 1922. The vegetation of west Greenland. Jour. Ecol. **10**: 87-108.

———. 1924. The vegetation of Gunong Belemut in Johore. Straits Settlements Gardens' Bull. **3**: 245-257.

———. 1941. Plant life in Malaya. Victorian Nat. **58**: 106-107.

Hope-Simpson, J. F. 1940. Studies of the vegetation of the English chalk. VI. Late stages in succession leading to chalk grassland. Jour. Ecol. **28**: 386-402.

Hopkins, D. M. & R. S. Sigafoos. 1951. Frost action and vegetation patterns on Seward Peninsula, Alaska. U. S. Geol. Surv. Bull. **974-C**: 51-101.

Hopkins, H., F. W. Albertson & A. Riegel. 1948. Some effects of burning upon a prairie in west-central Kansas. Kansas Acad. Sci. Trans. **51**: 131-141.

Hopkinson, J. W. 1927. Studies on the vegetation of Nottinghamshire. I. The ecology of the Bunter Sandstone. Jour. Ecol. **15**: 130-171.

Horton, J. S. 1941. The sample plot as a method of quantitative analysis of chaparral vegetation in southern California. Ecology **22**: 457-468.

Hotchkiss, N. & R. E. Stewart. 1947. Vegetation of the Patuxent Research Refuge, Maryland. Amer. Midland Nat. **38**: 1-75.

Hou, H. Y. 1944. The plant communities of acid and calcium soils in southern Kweichow. Natl. Geol. Surv. China, Spec. Soils Bull. **5**: 1-75.

Hough, A. F. & R. D. Forbes. 1943. The ecology and silvics of forests in the high plateaus of Pennsylvania. Ecol. Monog. **13**: 299-320.

Hult, R. 1885. Blekinges vegetation. Ett bidrag till Växtformationernas utvecklingshistoria. Medd. Soc. pro Fauna et Flora Fenn. **12**: 161-251. (Bot. Centbl. **27**: 192-193, 1886).

Humbert, H. 1937. La protection de la nature dans les pays intertropicaux et subtropicaux. Soc. Biogéog., Paris, Mem. **5**: 159-180.

———. 1938. Les aspects biologiques du problème des feux de brousse et la protection de la nature dans les zones intertropicales. Séances Inst. Roy. Col. Belge Bull. **9**: 811-835.

Humboldt, A. von. 1806. Ideen zu einer Physiognomik der Gewächse. Tubingen. (*fide* authors).

Humboldt, A. von & A. Bonpland. 1807. Ideen zu einer Geographie der Pflanzen nebst einem Naturgemälde der Tropenländer. Tubingen: F. G. Cotta. 182 pp.

Hutchinson, G. E. 1941. Ecological aspects of succession in natural populations. Amer. Nat. **75**: 406-418.

———. 1944. Limnological studies in Connecticut. VII. A critical examination of the supposed relationship between phytoplankton periodicity and chemical changes in lake waters. Ecology **25**: 3-26.

———. 1948. Circular causal systems in ecology. In "Teleological mechanisms." N. Y. Acad. Sci. Ann. **50**: 221-246.

———. 1950. Survey of contemporary knowledge of biogeochemistry. 3. The biogeochemistry of vertebrate excretion. Amer. Mus. Nat. Hist. Bull. **96**: 1-554.

Hutchinson, G. E. & A. Wollack. 1940. Studies on Connecticut lake sediments. II. Chemical analyses of a core from Linsley Pond, North Branford. Amer. Jour. Sci. **238**: 493-517.

Huxley, J. S. & N. E. Odell. 1924. Notes on surface markings in Spitsbergen. Geog. Jour. **63**: 207-229.

Iaroshenko, P. D. 1946. On changes in the plant cover (In Russian). Bot. Zhur. S.S.S.R. **31**: 29-40 (*fide* Major 1951: 398).

Iljinski, A. P. 1921. Essai de la définition concrète de l'équilibre mobile dans les groupements des plantes (Russian with French summary). Izv. Glav. Botan. Sad, Leningrad **20**: 151-166.

Ilinskii, A. P. & M. A. Poselskaia. 1929. The sociability of plants. (Russian with English summary.) Appl. Bot., Genet. & Pl.-Breed. Bull., Leningrad **20**: 459-473. (Biol. Abstr. **5**: 3828).

Ilvessalo, Y. 1929. Notes on some forest (site) types in North America. Acta Forest. Fenn. **34**(39): 1-111.

Issler, E 1926. Les associations silvatiques haut-rhinoises. Classification sociologique des Forêts du département du Haut-Rhin a l'exclusion du Sundgau et du Jura alsacien (avec une carte). Soc. Bot. de France Bull. **63**(Fasc. Sess. Extraord.): 62-142.

———. 1944. Die Pflanzenwelt der Hochvogesen mit besonderer Berücksichtigung der Waldverhältnisse. Allg. Forst u. Jagd. Ztg. **120**(7/9): 59-66.

Ives, R. L. 1942a. Atypical subalpine environments. Ecology **23**: 89-96.

———. 1942b. The beaver-meadow complex. Jour. Geomorphol. **5**: 191-203.

Jarret, P. H. & A. H. K. Petrie. 1929. The vegetation of the Blacks' Spur region: A study in the ecology of some Australian mountain *Eucalyptus* forests. II. Pyric succession. Jour. Ecol. **17**: 249-281.

Jenny, H. 1941. Factors of soil formation; a system of quantitative pedology. New York: McGraw-Hill. 281 pp.

Jenny-Lips, H. 1930. Vegetationsbedingungen und Pflanzengesellschaften auf Felsschutt. Phytosoziologische Untersuchungen in den Glarner Alpen. Bot. Centbl. Beih., Abt. 2, **46**: 119-296.

Johnson, D. S. & H. H. York. 1915. The relation of plants to tide-levels. A study of factors affecting the distribution of marine plants. Carnegie Inst. Wash. Publ. **206**: 1-162.

Johnston, I. M. 1941. Gypsophily among Mexican desert plants. Arnold Arboretum Jour. **22**: 145-170.

Jones, E. W. 1945. The structure and reproduction of the virgin forests of the North Temperate Zone. New Phytol. **44**: 130-148.

Jones, G. N. 1936. A botanical survey of the Olympic Peninsula, Washington. Wash. [State] Univ. Publ. Biol. **5**: 1-286.

———. 1938. The flowering plants and ferns of Mount Rainier. Wash. [State] Univ. Publ. Biol. **7**: 1-192.

Juraszek, H. 1928. Pflanzensoziologische Studien über die Dünen bei Warschau. Bull. Internat. Polon. Acad. des Sci. et Lettres, Cl. de Sci. Math. et Nat., Sér., B: Sci. Nat. 1927: 565-610.

Katz, N. J. 1926. Sphagnum bogs of central Russia: phytosociology, ecology and succession. Jour. Ecol. **14**: 177-202.

———. 1930. Zur Kenntnis der Moore Nordosteuropas. Bot. Centbl. Beih., Abt. 2, **46**: 297-394.

Kell, Lucille L. 1938. The effect of the moisture-retaining capacity of soils on forest succession in Itasca Park, Minnesota. Amer. Midland Nat. **20**: 682-694.

Keller, B. A. 1925-6. Die Vegetation auf den Salzböden der russischen Halbwüsten und Wüsten (Versuch einer ökologischen Präliminaranalyse). Ztschr. f. Bot. **18**: 113-137.

———. 1927. Distribution of vegetation on the plains of European Russia. Jour. Ecol. **15**: 189-233.

———. 1930. Die Methoden zur Erforschung der Ökologie der Steppen- und Wüstenpflanzen. Handb. Biol. Arbeitsmeth. ed. E. Abderhalden XI, **6**: 1-128.

Kelley, W. P. 1922. Variability of alkali soil. Soil Sci. **14**: 177-189.

Kendeigh, S. C. 1948. Bird populations and biotic communities in northern lower Michigan. Ecology **29**: 101-114.

Kerner von Marilaun, A. 1863. Das Pflanzenleben der Donauländer. Innsbruck. Transl. Conard 1951.

Kielhauser, G. E. 1939. Zur Oekologie des Quercetum galloprovinciale pubescentetosum. Österr. Bot. Ztschr. **88**: 24-42.

Killian, C. & A. Dubuis. 1937. Études des conditions édaphiques qui déterminent la rèpartition des végétaux de rochers dans l'Atlas Mitidjien. Rev. Gén. de Bot. **49**: 405-423.

Killian, C. & H. Moussu. 1948. Dégradation de la végétation et des sols dans le massif de la Bouzarea, ses caractères phytosociologiques, pédologiques et microbiologiques. Acad. des Sci. Compt. Rend. **226**: 118-120.

King, W. 1685. On the bogs and loughs of Ireland. Roy. Soc. London Phil. Trans. **15**: 948-960 (*fide* Clements 1928:8).

Kirstein, K. 1929. Lettlands Waldtypen. Acta Forest. Fenn. **34**(33): 1-20.

Kleist, C. de. 1929. Recherches phytosociologiques sur les tourbières de la région des dunes de la rive droite de la Vistule aux environs de Varsovie. Bull. Intern. Polon. Acad. des Sci. et Lettres, Cl. de Sci. Math. et. Nat., Sér. B: Sci. Nat. 1929(1): 41-104.

Klika, J. 1927. Příspěvek ke geobotanickému výzkumu Velké Fatry. 1. O lesních společenstvech (French summary). Preslia 1927 (5): 6-35.

———. 1932. Der Seslerion coeruleae-Verband in den Westkarpathen (Eine vergleichende soziologische Studie). Bot. Centbl. Beih., Abt. 2, **49**: 133-175.

———. 1936. Das Klimax-Gebiet der Buchenwälder in den Westkarpathen. Bot. Centbl. Beih., Abt. B, **55**: 373-418.

———. 1939. Zur Kenntnis der Waldgesellschaften im Böhmischen Mittelgebirge (Wälder des Milleschauer Mittelgebirges). Bot. Centbl. Beih., Abt. B, **60**: 249-286.

Knuchel, H., Jr. 1947. Über die Nebelvegetation an der peruanischen Küste. Schweiz. Ztschr. f. Forstw. **98**: 81-84.

Køie, M. 1950. Relations of vegetation, soil and subsoil in Denmark. Dansk Bot. Arkiv. **14**(5): 1-64.

Konovalov, A. N. 1928. Retrogression of forest associations. All-Russ. Congr. Bot., Leningrad, Jour. 1928: 234-236. (Biol. Abstr. **5**: 13488).

Krajina, V. 1933. Die Pflanzengesellschaften des Mlynica-Tales in den Vysoké Tatry (Hohe Tatra). Mit besonderer Berücksichtigung der ökologischen Verhältnisse. Bot. Centbl. Beih., Abt. 2, **50**: 774-957, **51**: 1-224.

Krist, V. 1940. Halofytní vegetace jihozápadního Slovenska a severní části Malé uherské nížiny. (German summary.) Práce Moravska Přírod. Společ., Brunn **12**(10): 1-100.

Krogerus, R. 1927. Beobachtungen über die Succession einiger Insektenbiocoenosen in Fichtenstümpfen. Notulae Ent. **7**: 121-126.

Krogh, A. & K. Berg. 1931. Über die chemische Zusammensetzung des phytoplanktons aus dem Frederiksborg-Schlosssee und ihre Bedeutung für die Maxima der Cladoceren. Internat. Rev. Gesam. Hydrobiol. Hydrogr. **25**: 204-218.

Küchler, A. W. 1946. The broadleaf deciduous forests of the Pacific Northwest. Assoc. Amer. Geog. Ann. **36**: 122-147.

———. 1949. A physiognomic classification of vegetation. Assoc. Amer. Geog. Ann. **39**: 201-210.

———. 1951. The relation between classifying and mapping vegetation. Ecology **32**: 275-283.

Lack, D. 1944. Ecological aspects of species-formation in passerene birds. Ibis **86**: 260-286.

———. 1945. The ecology of closely related species with special reference to cormorant (*Phalacrocorax carbo*) and shag (*P. aristotelis*). Jour. Anim. Ecol. **14**: 12-16.

———. 1946. Competition for food by birds of prey. Jour. Anim. Ecol. **15**: 123-129.

Lakari, O. J. 1915. Studien über die Samenjahre und Altersklassenverhältnisse der Kiefernwälder auf dem nordfinnischen Heideboden. Acta Forest. Fenn. **5**(1): 1-211.

Lämmermayr, L. 1926. Materialien zur Systamatik und Ökologie der Serpentinflora. I. Neue Beiträge zur Kenntnis der Flora steirischer Serpentine. Sitzber. Akad. Wiss. Wien, Math.-naturw. Kl., Abt. I, **135**: 369-407.

———. 1927. Materialien zur Systematik und Ökologie der Serpentinflora. II. Das Problem der "Serpentinpflanzen."—Eine kritische ökologische Studie. Sitzber. Akad. Wiss. Wien, Math.-naturw. Kl., Abt. I, **136**:

———. 1928. Weitere Beiträge zur Flora der Magnesit- und Serpentinböden. Sitzber. Akad. Wiss. Wien, Math.-naturw. Kl., Abt. I, **137**: 55-99.

Lane-Poole, C. E. 1925. The forests of Papua and New Guinea. Empire Forestry Jour. **4**: 200-234.

Larson, F. 1940. The role of the bison in maintaining the short grass plains. Ecology **21**: 113-121.

Leach, W. 1930. A preliminary account of the vegetation of some non-calcareous British screes (Gerölle). Jour. Ecol. **18**: 321-332.

Lemée, G. 1937-9. Recherches écologiques sur la végétation du Perche. Rev. Gén. de Bot. **49**: 731-751, **50**: 22-46 etc., **51**: 53-64 etc. (19 sections).

Lemon, P. C. 1949. Successional responses of herbs in the longleaf-slash pine forest after fire. Ecology **30**: 135-145.

Lenoble, F. 1926. À propos des associations végétales Soc. Bot. de France Bull. **73**: 873-893.

———. 1928. Associations végétales et espèces. Arch. Bot. [Caen] **2**(Bull. Mens. 1): 1-14.

Leopold, A. 1947. Game management. New York: Scribners. 481 pp.

Lewis, F. J. & E. S. Dowding. 1926. The vegetation and retrogressive changes of peat areas ("muskegs") in central Alberta. Jour. Ecol. **14**: 317-341.

Lewis, F. J., E. S. Dowding, & E. H. Moss. 1928. The vegetation of Alberta. II. The swamp, moor and bog forest vegetation of central Alberta. Jour. Ecol. **16**: 19-70.

Lindeman, R. L. 1942. The trophic-dynamic aspect of ecology. Ecology **23**: 399-418.

Linkola, K. 1924. Waldtypenstudien in den schweizer Alpen. Veröff. Geobot. Inst. Rübel Zürich **1**: 139-224.

———. 1929. Zur Kenntnis der Waldtypen Eestis. Acta Forest. Fenn. **34**(40): 1-73.

Lippmaa, T. 1929. Pflanzenökologische Untersuchungen aus Norwegisch- und Finnisch-Lappland unter besonderer Berücksichtigung der Lichtfrage. Acta Inst. et Horti Bot. Univ. Tartu **2**(1/2): 1-165.

———. 1933a. Taimeühingute uurimise metoodika ja eesti taimeühingute klassifikatsiooni põhijooni. (German summary). Acta Inst. et Horti Bot. Univ. Tartu **3**(4): 1-169.

———. 1933b. Aperçu général sur la végétation autochtone du Lautaret (Hautes-Alpes) avec des remarques critiques sur quelques notions phytosociologiques. (Estonian summary). Acta Inst. et Horti Bot. Univ. Tartu **3**(3): 1-108.

———. 1935a. La méthode des associations unistrates et la systeme écologique des associations. Acta Inst. et Horti Bot. Univ. Tartu **4**(1/2): 1-7.

———. 1935b. Une analyse des forêts de l'île estonienne d'Abruka (Abro) sur la base des associations unistrates. Acta Inst. et. Horti Bot. Univ. Tartu **4**(1/2): 1-97.

Little, E. L., Jr. 1938. The vegetation of Muskogee County, Oklahoma. Amer. Midland Nat. **19**: 559-572.

Livingston, R. B. 1949. An ecological study of the Black Forest, Colorado. Ecol. Monog. **19**: 123-144.

Ljungqvist, J. E. 1927-9. Vegetationsbilder fr˘n Mästermyr. II., III. Svensk Bot. Tidskr. **21**: 314-343, **23**: 219-232.

Lotka, A. J. 1925. Elements of physical biology. Baltimore: Williams & Wilkins. 460 pp.

———. 1932. The growth of mixed populations: Two species competing for a common food supply. Wash. Acad. Sci. Jour. **22**: 461-469.

Lüdi, W. 1920. Die Sukzession der Pflanzenvereine. Allgemeine Betrachtungen über die dynamisch-genetischen Verhältnisse der vegetation in einem Gebiet des Berner Oberlandes. Naturf. Gesell. in Bern, Mitt. 1919: 9-87.

———. 1921. Die Pflanzengesellschaften des Lauterbrunnentales und ihre Sukzession: Versuch zur Gliederung der Vegetation eines Alpentales nach genetisch-dynamischen Gesichtspunkten. Beitr. Geobot. Landesaufn. Schweiz **9**: 1-364.

———. 1923. Die Untersuchung und Gliederung der Sukzessionsvorgänge in unsere Vegetation. Naturf. Gessell. in Basel, Verhandl. **35**: 277-302.

———. 1929. Sukzession der Pflanzengesellschaften und Bodenreifung in der alpinen Stufe der schweizer

Alpen. Repert. Spec. Novarum Regni. Veg. Beih. **56:** 81-92.

———. 1930. Die Methoden der Sukzessionsforschung in der Pflanzensoziologie. Handb. Biol. Arbeitsmeth. ed. E. Abderhalden. XI, **5:** 527-728.

———. 1940. Die Veränderungen von Dauerflächen in der Vegetation des Alpengartens Schinigeplatte innerhalb des Jahrzehnts 1928/29-1938/39. Ber. Geobot. Forschungsinst. Rübel, Zürich 1939: 93-148.

———. 1941. Die Kastanienwälder von Tesserete. Beitrag zur Soziologie der Kastanienwälder am Südhang der Alpen. Ber. Geobot. Forschungsinst. Rübel, Zürich 1940: 52-84.

———. 1943. Über Rasengesellschaften und alpine Zwergstrauchheide in den Gebirgen des Apennin. Ber. Geobot. Forschungsinst. Rübel, Zürich 1942: 23-68.

———. 1945. Besiedlung und Vegetationsentwicklung auf den jungen Seitenmoränen des Grossen Aletschgletschers, mit einem Vergleich der Besiedlung im Vorfeld des Rhonegletschers und des Oberen Grindelwaldgletschers. Ber. Geobot. Forschungsinst. Rübel, Zürich 1944: 35-112.

———. 1948. Die Pflanzengesellschaften der Schinigeplatte bei Interlaken und ihre Beziehungen zur Umwelt. Eine vergleichend ökologische Untersuchung. Veröff. Geobot. Inst. Rübel, Zürich **23:** 1-400.

McBryde, J. B. 1933. The vegetation and habitat factors of the Carrizo sands. Ecol. Monog. **3:** 247-297.

MacCaughey, V. 1917. The Oahu rain forest. Amer. Forestry **23:** 276-278.

———. 1920. Hawaii's tapestry forests. Bot. Gaz. **70:** 137-147.

McComb, A. L. & W. E. Loomis. 1944. Subclimax prairie. Torrey Bot. Club Bull. **71:** 46-76.

McCrea, R. H. 1926. The salt marsh vegetation of Little Island, Co. Cork. Jour. Ecol. **14:** 342-346.

McDougall, W. B. 1927. Plant ecology. Philadelphia: Lea & Febiger. 326 pp. 4th edn. 1949. 234 pp.

MacFadyen, W. A. 1950. Vegetation patterns in the semidesert plains of British Somaliland. Geog. Jour. **116:** 199-211.

MacGinitie, G. E. 1939. Littoral marine communities. Amer. Midland Nat. **21:** 28-55.

McLuckie, J. & A. H. K. Petrie. 1926. An ecological study of the flora of Mount Wilson. III. The vegetation of the valleys. Linn. Soc. N. S. Wales Proc. **51:** 94-113.

———. 1927. An ecological study of the flora of Mount Wilson. IV. Habitat factors and plant response. Linn. Soc. N. S. Wales Proc. **52:** 161-184.

MacLulich, D. A. 1937. Fluctuations in the numbers of the varying hare (*Lepus americanus*). Toronto Univ. Studies. Biol. Ser. **43:** 1-136.

McVaugh, R. 1943. The vegetation of the granitic flatrocks of the southeastern United States. Ecol. Monog. **13:** 119-165.

Maissurow, D. K. 1941. The role of fire in the perpetuation of virgin forests of northern Wisconsin. Jour. Forestry **39:** 201-207.

Major, J. 1951. A functional, factorial approach to plant ecology. Ecology **32:** 392-412.

Mallner, F. 1944. Vorgeschichte und Werdegang des lettländischen Waldtypensystems. Ztschr. f. Forst. u. Jagdw. **76**(7/9): 105-136.

Marks, J. B. 1950. Vegetation and soil relations in the lower Colorado desert. Ecology **31:** 176-193.

Marler, P. & D. J. Boatman. 1952. An analysis of the vegetation of the northern slopes of Pico—the Azores. Jour. Ecol. **40:** 143-155.

Marshall, R. C. 1934. The physiography and vegetation of Trinidad and Tobago, a study in plant ecology. Oxford Forestry Mem. **17:** 1-56.

Martyn, E. B. 1931. A botanical survey of the Rupunini Development Company's ranch at Waranama, Berbice River. Brit. Guiana Agric. Jour. **4:** 18-25. (*fide* Myers 1936).

Mason, H. L. 1947. Evolution of certain floristic associations in western North America. Ecol. Monog. **17:** 201-210.

Meigen, F. 1896. Die Besiedelung der Reblausherde in der Provinz Sachsen. Bot. Jahrb. **21:** 212-257.

Merkle, J. 1951. An analysis of the plant communities of Mary's Peak, western Oregon. Ecology **32:** 618-640.

Metcalfe, G. 1950. The ecology of the Cairngorms. II. The mountain Callunetum. Jour. Ecol. **38:** 46-74.

Meusel, H. 1935. Die Waldtypen des Grabfelds und ihre Stellung innerhalb der Wälder zwischen Main und Werra. Bot. Centbl. Beih., Abt. B, **53:** 175-251.

Meyer, H. A. & D. D. Stevenson. 1943. The structure and growth of virgin beech-birch-maple-hemlock forests in northern Pennsylvania. Jour. Agr. Res. **67:** 465-484.

Michelmore, A. P. G. 1934. Vegetation succession and regional surveys, with special reference to tropical Africa. Jour. Ecol. **22:** 313-317.

Middleton, A. D. 1934. Periodic fluctuations in British game populations. Jour. Anim. Ecol. **3:** 231-249.

Mikyška, R. 1932. O smilkových pastvinách ve Štiavnickém středohoří. (German summary.) Českoslov. Acad. Zeměděl. Sborn. **7:** A189-216.

———. 1937. Přehled přirozených lesních společenstev ve Slovenskem středohoří. (French & German summaries.) Lesnická Práce **16:** 259-266.

———. 1939. Studie über die natürlichen Waldbestände im Slowakischen Mittelgebirge. (Slovenské stredohorie). Ein Beitrag zur Soziologie der Karpatenwälder. Bot. Centbl. Beih., Abt. B, **59:** 169-244.

Miller, E. H., Jr. 1947. Growth and environmental conditions in southern California chaparral. Amer. Midland Nat. **37:** 379-420.

Milne, G. 1935. Some suggested units of classification and mapping, particularly for East African soils. Soil Res. **4:** 183-198.

———. 1937. Note on soil conditions and two East African vegetation types. Jour. Ecol. **25:** 254-258.

Mohr, C. O. 1943. Cattle droppings as ecological units. Ecol. Monog. **13:** 275-298.

Mood, A. M. 1950. Introduction to the theory of statistics. New York: McGraw-Hill. 433 pp.

Morison, C. G. T., A. C. Hoyle & J. F. Hope-Simpson. 1948. Tropical soil-vegetation catenas and mosaics: A study in the southwestern part of the Anglo-Egyptian Sudan. Jour. Ecol. **36:** 1-84.

Moss, C. E. 1907. Succession of plant formations in Britain. Brit. Assoc. Adv. Sci. Rept. 1906: 742-743. (Bot. Centbl. **107:** 255-256, 1908).

———. 1910. The fundamental units of vegetation: Historical development of the concepts of the plant association and the plant formation. New Phytol. **9:** 18-53.

———. 1913. Vegetation of the Peak District. London: Cambridge Univ. Press. 235 pp. (Reviewed Tansley 1913).

Moss, E. H. 1932. The vegetation of Alberta. IV. The poplar association and related vegetation of central Alberta. Jour. Ecol. **20:** 380-415.

Mowrer, O. H. 1939. An experimental analogue of "regression" with incidental observations on "reaction-formation." Jour. Abnorm. & Soc. Psychol. **35:** 56-87.

Mrugowsky, J. 1931. Die Formation der Gipspflanzen. Beiträge zu ihrer Soziologie und Oekologie. Bot. Arch. **32:** 245-341.

Muller, C. H. 1940. Plant succession in the Larrea-Flourensia climax. Ecology **21:** 206-212.

Müller, J. 1936. Lesní typy Jizerskych hor. (English summary). Lesnická Práce **15:** 477-523.

———. 1938. Jurské smrčiny (Srovnávací typologická studie). (French summary). Lesnická Práce **17:** 418-430.

Munger, T. T. 1940. The cycle from Douglas fir to hemlock. Ecology **21:** 451-459.

Myers, J. G. 1936. Savannah and forest vegetation of the interior Guiana Plateau. Jour. Ecol. **24:** 162-184.

Naumov, S. P. 1939. Fluctuations of number in hares. (Russian with English summary). Voprosy Ekologii i Biotsenologii, Leningrad Univ. **5/6:** 40-82.

Navalkar, B. S. 1951. Succession of the mangrove vegetation of Bombay and Salsette Islands. Bombay Nat. Hist. Soc. Jour. **50:** 157-160.

Nichols, G. E. 1917. The interpretation and application of certain terms and concepts in the ecological classification of plant communities. Plant World **20:** 305-319, 341-353.

———. 1923. A working basis for the ecological classification of plant communities. Ecology **4:** 11-23, 154-179.

———. 1929. Plant associations and their classification. Internat. Congr. Plant Sci., Ithaca, Proc. 1926, **1:** 629-641.

———. 1935. The hemlock-white pine-northern hardwood region of eastern North America. Ecology **16:** 403-422.

Nicholson, A. J. 1933. The balance of animal populations. Jour. Anim. Ecol. **2:** 132-178.

Nicholson, A. J. & V. A. Bailey. 1935. The balance of animal populations. I. Zool. Soc. London Proc. 1935: 551-598.

Niedziałkowski, W. 1935. Réserves de sapin dans la Forêt Dominiale de Tuków. Étude phytogéograph que et forestière. (Polish with French Summary). Inst. Badawczy Lasów Państwow. Rozpr. i Sprawozd. Ser. A, **13:** 1-274.

Nienburg, W. & E. Kolumbe. 1931. Zur Ökologie der Flora des Wattenmeeres. II. Das Neufelder Watt im Elbmündungsgebiet. Wiss. Meeresuntersuch., Abt. Kiel **21:** 73-114.

Nilsson, A. 1899. Några drag ur de svenska växtsamhällenas utvecklingshistoria. Bot. Notiser 1899: 89-101, 123-135. (Bot. Centbl. Beih. **9:** 370-372, 1900).

Nordhagen, R. 1923. Vegetationsstudien auf der Insel Utsire im westlichen Norwegen. Bergens Mus. Aarbok. Naturv. rekke 1920-21(1): 1-149.

———. 1928. Die Vegetation und Flora des Sylenegebietes. I. Die Vegetation. Norske Vidensk. Akad. i Oslo, Mat. Nat. Kl. Skr. 1927 (1): 1-612.

———. 1936. Versuch einer neuen Einteilung der subalpinen-alpinen Vegetation Norwegens. Bergens Mus. Aarbok Naturv. rekke 1936 (H. 2, Nr. 7): 1-88.

———. 1940. Studien über die maritime Vegetation Norwegens I. Die Pflanzengesellschaften der Tangwälle. Bergens Mus. Aarbok, Naturv. rekke 1939/40 (2): 1-123.

———. 1943. Sikilsdalen og Norges Fjellbeiter, en plantesosiologisk Monografi. Bergens Mus. Skr. **22:** 1-607.

Nytzenko, A. A. 1948. Boundaries of plant associations in nature. (Russian). Bot. Zhurnal S.S.S.R. **33:** 487-495. (Biol. Abstr. **24:** 17407).

Oberdorfer, E. 1937. Pflanzensoziologische Probleme des Oberrheingebietes. Deut. Bot. Gesell. Ber. **55**(Generalversammlungs-Heft 1): (187)-(194).

Olmsted, C E. 1937 Vegetation of certain sand plains of Connecticut. Bot. Gaz. **99:** 209-300.

Oosting, H. J. 1942. An ecological analysis of the plant communities of Piedmont, North Carolina. Amer. Midland Nat. **28:** 1-126.

———. 1945. Tolerance to salt spray by plants of coastal dunes. Ecology **26:** 85-89.

———. 1948. The study of plant communities: An introduction to plant ecology. San Francisco: Freeman. 389 pp.

Oosting, H. J. & W. D. Billings. 1942. Factors effecting vegetational zonation on coastal dunes. Ecology **23:** 131-142.

Ostenfeld, C. H. 1908. The land-vegetation of the Faeröes, with special reference to the higher plants. Botany of the Faeröes based upon Danish investigations. Copenhagen: Nordisk forlag. Part. III: 867-1026.

Osvald, H. 1923. Die Vegetation des Hochmoores Komosse. Svenka Växtsoc. Sällsk Handl. I. Akad. Abhandl., Uppsala. 436 pp.

———. 1949. Notes on the vegetation of British and Irish mosses. Acta Phytogeog. Suecica **26:** 1-62.

Paczoski, J. K. 1917. Vegetationsbeschreibung des Gouv. Cherson. II. Die Steppen. (Russian, *fide* Alechin 1926). Cherson.

———. 1921. Die Grundzüge der Phytosoziologie. (*fide* Sukatschew 1929). Cherson.

Palmgren, A. 1922. Zur Kenntnis des Florencharakters des Nadelwaldes. Eine Pflanzengeographische Studie aus dem Gebiete Ålands. Acta Forest. Fenn. **22**(2): 1-115.

———. 1929. Chance as an element in plant geography. Internat. Congr. Plant Sci., Ithaca, Proc. 1926, **1:** 591-602.

Palmgren, P. 1949. Some remarks on the short-term fluctuations in the numbers of northern birds and mammals. Oikos **1:** 114-121.

Park, O. 1941. Concerning community symmetry. Ecology **22:** 164-167.

Patton, R. T. 1930. The factors controlling the distribution of trees in Victoria. Roy. Soc. Victoria Proc., N. S. **42:** 154-210.

Paulsen, O. 1915. Some remarks on the desert vegetation of America. Plant World **18:** 155-161.

Pavillard, J. 1928. Le *Crithmion maritimae* autour de Biarritz. Soc. Bot. de France Bull. **75:** 795-799.

Pearsall, W. H. 1932. Phytoplankton in the English Lakes. II. The composition of the phytoplankton in relation to dissolved substances. Jour. Ecol. **20:** 241-262.

Pechanec, J. F., G. D. Pickford & G. Stewart. 1937. Effects of the 1934 drought on native vegetation of the upper Snake River plains, Idaho. Ecology **18:** 490-505.

Pennak, R. W. 1946. The dynamics of fresh-water plankton populations. Ecol. Monog. **16:** 339-356.

———. 1949. Annual limnological cycles in some Colorado reservoir lakes. Ecol. Monog. **19:** 233-267.

———. 1951. Hydrosere versus limnosere. Ecol. Soc. Amer. Bull. **32:** 65 (Abstract).

Pennington, W. 1941. The control of the numbers of fresh-water phytoplankton by small invertebrate animals. Jour. Ecol. **29:** 204-211.

———. 1943. Lake sediments: The bottom deposits of the North Basin of Windermere, with special reference to the diatom succession. New Phytol. **42:** 1-27.

Pessin, L. J. 1933. Forest associations in the uplands of the lower Gulf Coastal plain (longleaf pine belt). Ecology **14:** 1-14.

Petrie, A. H. K., P. H. Jarrett & R. T. Patton. 1929. The vegetation of the Blacks' Spur region: A study in the ecology of some Australian mountain *Eucalyptus* forests. I. The mature plant communities. Jour. Ecol. **17:** 223-248.

Pfaffenberg, K. & W. Hassenkamp. 1934. Ueber die Versumpfungsgefahr des Waldbodens im Syker Flottsandgebiet. Naturw. Ver. zu Bremen, Abhandl. **29:** 89-121.

Pfeiffer, H. 1943. Über örtliche Feinheiten der Assoziationsverteilung. Biol. Gen. **17:** 147-163.

Phillips, J. F. V. 1928. The principal forest types in the Knysna region—an outline. So. African Jour. Sci. **15:** 188-201.

———. 1931. The biotic community. Jour. Ecol. **19:** 1-24.

———. 1934-5. Succession, development, the climax, and the complex organism: An analysis of concepts. Parts I-III. Jour. Ecol. **22:** 554-571, **23:** 210-246, 488-508.

Pichi-Sermolli, R. E. 1948a. Flora e vegetazione delle serpentine et delle autre ofioliti dell'Alta Valle del Tevere (Toscana). (English summary). Webbia **6:** 1-380.

———. 1948b. An index for establishing the degree of maturity in plant communities. Jour. Ecol. **36:** 85-90.

Pidgeon, Ilma M. 1938. The ecology of the central coastal area of New South Wales. II. Plant succession on the Hawkesbury sandstone. Linn. Soc. N. S. Wales Proc. **63:** 1-26.

———. 1941. The ecology of the central coastal area of New South Wales. IV. Forest types on soils from Hawkesbury sandstone and Wianamatta shale. Linn. Soc. N. S. Wales Proc. **66:** 113-137.

Pittier, H. & L. Williams. 1945. A review of the flora of Venezuela. In "Plants and plant science in Latin America," ed. F. Verdoorn. pp. 102-105.

Platt, R. B. 1951. An ecological study of the mid-Appalachian shale barrens and of the plants endemic to them. Ecol. Monog. **21:** 269-300.

Pohle, R. 1908. Vegetationsbilder aus Nordrussland. Vegetationsbilder **5**(3-5).

Polunin, N. 1934-5. The vegetation of Akpatok Island. Parts I & II. Jour. Ecol. **22:** 337-395, **23:** 161-209.

———. 1936. Plant succession in Norwegian Lapland. Jour. Ecol. **24:** 372-391.

Poole, A. L. 1937. A brief ecological survey of the Pukekura State Forest, South Westland. New Zeal. Jour. Forestry **4:** 78-85.

Porsild, A. E. 1938. Earth mounds in unglaciated arctic northwestern America. Geog. Rev. **28:** 46-58.

Potts, G. & C. E. Tidmarsh. 1937. An ecological study of a piece of Karroo-like vegetation near Bloemfontein. Jour. So. African Bot. **3:** 51-92.

Praeger, R. L. 1934. The botanist in Ireland. Dublin: Hodges, Figgis, & Co. 587 pp.

Purer, Edith A. 1942. Plant ecology of the coastal salt marshlands of San Diego County, California. Ecol. Monog. **12:** 81-111.

Quarterman, Elsie. 1950. Major plant communities of Tennessee cedar glades. Ecology **31:** 234-254.

Ramensky, L. G. 1918. Wiesenuntersuchungen im Gouv. Woronesh.—Materialen zur naturgeschichtlichen Erforschung des Gouv. Woronesh. Lief. 1. Moskow (Russian, *fide* Alechin 1926).

———. 1924. Les lois fondamentales du tapis végétal. (Russian, *fide* Roussin 1948). Westnik Opitnogo Dela Woronej.

Randall, W. E. 1951. Interrelations of autecological characteristics of woodland herbs. Ecol. Soc. Amer. Bull. **32:** 57 (Abstract).

Rasmussen, R. 1946. Vegetationen i de faerøske Fugelbjaerge og deres naermeste Omgivelser. Bot. Tidsskr. **48:** 46-70.

Raunkiaer, C. 1910. Statistik der Lebensformen als Grundlage für die biologische Pflanzengeographie. Bot. Centbl. Beih., Abt. 2, **27:** 171-206d.

———. 1934. The life forms of plants and statistical plant geography; being the collected papers of C. Raunkiaer, translated into English by H. Gilbert Carter, A. G. Tansley, and Miss Fausbøll. Oxford: Clarendon Press. 632 pp.

———. 1937. Plant life forms. Translated by H. Gilbert-Carter. Oxford: Clarendon Press. 104 pp.

Raup, H. M. 1938. Botanical studies in the Black Rock Forest. Black Rock Forest Bull. **7:** 1-161.

———. 1941. Botanical problems in boreal America. Bot. Rev. **7:** 147-248.

———. 1942. Trends in the development of geographic botany. Assoc. Amer. Geog. Ann. **32:** 319-354.

———. 1951. Vegetation and cryoplanation. In "The glacial border—climatic, soil, and biotic features." Ohio Jour. Sci. **51:** 105-116.

Rawitscher, F. 1948. The water economy of the vegetation of the 'Campos Cerrados' in southern Brazil. Jour. Ecol. **36:** 237-268.

Rawson, D. S. 1939. Some physical and chemical factors in the metabolism of lakes. In "Problems on lake biology." Amer. Assoc. Adv. Sci. Publ. **10:** 9-26.

———. 1942. A comparison of some large alpine lakes in western Canada. Ecology **23:** 143-161.

Regel, C. 1932. Pflanzensoziologische Studien aus dem nördlichen Russland. I. Die Fleckentundra von Nowaja Semlja. Beitr. z. Biol. der Pflanz. **20**(1): 7-24.

Renvall, A. 1912. Die periodischen Erscheinungen der Reproduktion der Kiefer an der polaren Waldgrenze. Acta Forest. Fenn. **1**(2): 1-154.

———. 1919. Suojametsäkysymyksestä I. Mäntymetsän elinehdot sen pohjoisrajalla sekä tämän rajan alenemisen syyt. Acta Forest. Fenn. **11**(1): 1-143. German summary 1921. Über die Schutzwaldfrage. Transl. by L. Heikinheimo. Acta Forest. Fenn. **11**(Suppl.): 1-12.

Richards, F. J. 1934. The salt marshes of the Dovey Estuary. IV. The rates of vertical accretion, horizontal extension and scarp erosion. Ann. Bot. **48**: 225-259.

Richards, P. W. 1936. Ecological observations on the rain forest of Mount Dulit, Sarawak. Parts I & II. Jour. Ecol. **24**: 1-37, 340-360.

———. 1939. Ecological studies on the rain forest of southern Nigeria. I. The structure and floristic composition of the primary forest. Jour. Ecol. **27**: 1-61.

———. 1945. The floristic composition of primary tropical rain forest. Cambridge Phil. Soc. Biol. Rev. **20**: 1-13.

Riley, G. A. 1940. Limnological studies in Connecticut. III. The plankton of Linsley Pond. Ecol. Monog. **10**: 279-306.

Rivas Goday, S. & M. M. Box. 1947. Intercalaciones esclerófiloedáficas en el montaño del Moncayo. Soc. Españ. Hist. Nat. Bol. **45**: 79-83.

Roberty, G. 1946. Les associations végétales de la valée moyene du Niger. Veröff. Geobot. Inst. Rübel Zürich **22**: 1-168.

Robyns, M. W. 1938. Considérations sur les aspects biologiques du problème des feux de brousse au Congo belge et au Ruanda-Urundi. Séances Inst. Roy. Colon. Belge Bull. **9**: 382-420.

Rohlena, J. 1927. O vegetačních rozdílech mezi severní a jižní exposicí v Čechách. (German summary). Preslia 1927(5): 52-64.

Roussin, N. 1948. L'évolution des theories phytosociologiques en Russie. Vegetatio **1**: 175-183.

Rübel, E. 1913. Ökologische Pflanzengeographie 3. Die Sukzessionen oder der Formationswechsel. Handwörterbuch der Naturw. **4**: 903-906.

———. 1922. Geobotanische Untersuchungsmethoden. Berlin: Borntraeger. 290 pp.

———. 1929. The present state of geobotanical research in Switzerland. Internat. Congr. Plant Sci., Ithaca, Proc. 1926, **1**: 603-621.

———. 1930. Pflanzengesellschaften der Erde. Bern/Berlin: Huber. 464 pp.

———. 1936. Plant communities of the world. In "Essays in geobotany in honor of William Albert Setchell" ed. T. H. Goodspeed. Berkeley: Univ. Calif. Press. pp. 263-290.

Rudolph, K. & F. Firbas. 1927. Paläofloristische und stratigraphische Untersuchungen böhmischen Moore. III. Die Moore des Riesengebirges. Bot. Centbl. Beih., Abt. 2, **43**: 69-144.

Rudolph, K., F. Firbas & H. Sigmond. 1928. Das Koppenplanmoor im Riesengebirge. (Ein Beispiel für den subalpinen Moortypus in Böhmen). Lotos **76**: 173-222.

Russell, R. S. & P. S. Wellington. 1940. Physiological and ecological studies on an arctic vegetation. I. The vegetation of Jan Mayen Island. Jour. Ecol. **28**: 153-179.

Sambuk, F. 1930. Beobachtungen an Kiefernwäldern und Quellmooren im Tale und im Becken der Obla—eines Nebenflusses der Luga. (Russian with German summary). Trudy Bot. Muzeia Akad. Nauk S.S.S.R. **22**: 277-310.

Sampson, A. W. 1944. Plant succession on burned chaparral lands in northern California. Calif. Agr. Expt. Sta. Bull. **685**: 1-144.

Sampson, H. C. 1930a. Succession in the swamp forest formation in northern Ohio. Ohio Jour. Sci. **30**: 340-357.

———. 1930b. The mixed mesophytic forest community of northeastern Ohio. Ohio Jour. Sci. **30**: 358-367.

Savely, H. E. 1939. Ecological relations of certain animals in dead pine and oak tree trunks. Ecol. Monog. **9**: 321-385.

Saxton, W. T. 1922. Mixed formations in time: A new concept in oecology. Jour. Indian Bot. **3**: 30-33.

———. 1924. Phases of vegetation under monsoon conditions. Jour. Ecol. **12**: 1-38.

Scharfetter, R. 1921. Die Vegetation der Turracher Höhe. Österreich. Bot. Ztschr. **70**: 77-91.

Scheffer, T. H. 1937. Study of a small prairie-dog town. Kans. Acad. Sci. Trans. **40**: 391-395.

Schimper, A. F. W. 1898. Pflanzengeographie auf physiologischer Grundlage. Jena: Gustav Fischer. 876 pp.

Schimper, A. F. W., W. R. Fisher, P. Groom & I. B. Balfour. 1903. Plant-geography upon a physiological basis. Oxford: Clarendon Press. 839 pp.

Schimper, A. F. W. & F. C. von Faber. 1935. Pflanzengeographie auf physiologischer Grundlage. Jena: Gustav Fischer. 3rd. edn. 2 vols., 588 & 1023 pp.

Schmid, E. 1922. Biozönologie und Soziologie. Naturw. Wochenschr., N. F. **21**: 518-523.

———. 1935. Die Kartierung der oekogenetischen Vegetationseinheiten. Zesde Internatl. Bot. Congr. Amsterdam Proc. 1935, **2**: 76-78.

———. 1936. Die Reliktföhrenwälder der Alpen. Beitr. Geobot. Landesaufn. Schweiz. **21**: 1-190.

Schrödinger, E. 1946. What is life? The physical aspect of the living cell. New York: Macmillan. 91 pp.

Schwickerath, M. 1933. Die Vegetation der Kalktriften (Bromion-erecti-Verband) des nördlichen Westdeutschlandes. Bot. Jahrb. **65**: 212-250.

Sears, P. B. 1947. The insolation-exposure factor. Ecology **28**: 316-317.

Seifriz, W. 1923. The altitudinal distribution of plants on Mt. Gedeh, Java. Torrey Bot. Club Bull. **50**: 283-305.

———. 1943. The plant life of Cuba. Ecol. Monog. **13**: 375-426.

Sernander, R. 1910. Ausstellung zur Beleuchtung der Entwicklungsgeschichte der schwedischen Torfmoore. Compt. Rend. XI Congr. Geol. Intern. 203 (*fide* Clements 1928: 151).

———. 1936. Granskär och Fiby urskog. En studie över stormluckornas och marbuskarnas betydelse i den svenska granskogens regeneration. (English summary). Acta Phytogeog. Suecica **8**: 1-232.

Shantz, H. L. 1925. Plant communities in Utah and Nevada. U. S. Natl. Mus., Contrib. U. S. Natl. Herbarium **25**: 15-23.

Shantz, H. L. & R. L. Piemeisel. 1924. Indicator significance of the natural vegetation of the southwestern desert region. Jour. Agr. Res. **28**: 721-802.

———. 1940. Types of vegetation in Escalante Valley, Utah, as indicators of soil conditions. U. S. Dept. Agr. Tech. Bull. **713**: 1-46.

Sharp, R. P. 1942. Soil structures in the St. Elias Range, Yukon Territory. Jour. Geomorph. **5**: 274-301.

Shelford, V. E. 1913. Animal communities in temperate America as illustrated in the Chicago region: A study in animal ecology. Geog. Soc. Chicago Bull. 5: Univ. Chicago Press. 362 pp.

———. 1932. Basic principles of the classification of communities and habitats and the use of terms. Ecology **13**: 105-120.

———. 1951. Fluctuation of forest animal populations in east central Illinois. Ecol. Monog. **21**: 183-214.

Shelford, V. E. & S. Olson. 1935. Sere, climax and influent animals with special reference to the transcontinental coniferous forest of North America. Ecology **16**: 375-402.

Show, S. B. & E. I. Kotok. 1924. The rôle of fire in the California pine forests. U. S. Dept. Agr. Bull. **1294**: 1-80.

Shreve, F. 1914a. A mountain rain-forest. A contribution to the physiological plant geography of Jamaica. Carnegie Inst. Wash. Publ. **199**: 1-110.

———. 1914b. The direct effects of rainfall on hygrophilous vegetation. Jour. Ecol. **2**: 82-98.

———. 1915. The vegetation of a desert mountain range as conditioned by climatic factors. Carnegie Inst. Wash. Publ. **217**: 1-112.

———. 1927. The vegetation of a coastal mountain range. Ecology **8**: 27-44.

———. 1942. The desert vegetation of North America. Bot. Rev. **8**: 195-246.

———. 1951. Vegetation of the Sonoran Desert. Carnegie Inst. Wash. Publ. **591**: 1-192.

Shreve, F. & T. D. Mallery. 1933. The relation of caliche to desert plants. Soil Sci. **35**: 99-113.

Sigafoos, R. S. 1949. The effects of frost action and processes of cryoplanation upon the development of tundra vegetation, Seward Peninsula, Alaska. Amer. Jour. Bot. **36**: 832 (Abstract).

Siivonen, L. 1948. Structure of short-cycle fluctuations in numbers of mammals and birds in the northern parts of the northern hemisphere. Riistatieteellisia Julkaisu. **1**: 1-66.

Sjörs, H. 1948. Myr-vegetation i Bergslagen. Acta Phytogeog. Suecia **21**: 1-299.

Smith, H. S. 1935. The rôle of biotic factors in the determination of population densities. Jour. Econ. Ent. **28**: 873-898.

Smith, W. G. 1912. Anthelia: An arctic-alpine plant association. Scottish Bot. Rev. **1**: 81-89.

———. 1913. Raunkiaer's "life-forms" and statistical methods. Jour. Ecol. **1**: 16-26.

Soó, R. von. 1936. Die Vegetation der Alkalisteppe Hortobágy, Ökologie und Soziologie der Pflanzengesellschaften. Repert. Spec. Novarum Regni. Veg. **39**: 352-364.

Sørensen, T. 1935. Bodenformen und Pflanzendecke in Nordostgrönland. (Beiträge zur Theorie der polaren Bodenversetzungen auf Grund von Beobachtungen über deren Einfluss auf die Vegetation in Nordostgrönland). Medd. om Grønland **93**(4): 1-69.

———. 1943. The flora of Melville Bugt. Medd. om Grønland **124**(5): 1-70.

Spilsbury, R. H. & D. S. Smith. 1947. Forest site types of the Pacific Northwest: A preliminary report. Brit. Columbia Dept. of Lands & Forests, Forest Branch Tech. Publ. T.30. 46 pp.

Sprague, F. L. & H. P. Hansen. 1946. Forest succession in the McDonald Forest, Willamette Valley, Oregon. Northwest Sci. **20**: 89-98.

Stamp, L. D. 1925. The vegetation of Burma from an ecological standpoint. Univ. Rangoon Res. Monog. **1**: 1-58.

Stamp, L. D. & L. Lord. 1923. The ecology of part of the riverine tract of Burma. Jour. Ecol. **11**: 129-159.

Stantschinsky, W. W. 1931. Zur Frage der Bedeutung der Masse der Artensubstanz für das dynamische Gleichgewicht der Biocönosen. (Russian with German summary). Zhurnal Ekologii i Biotsenologii **1**: 88-98.

Stearns, F. W. 1949. Ninety years change in a northern hardwood forest in Wisconsin. Ecology **30**: 350-358.

Stefanoff, B. 1930. A parallel classification of climates and vegetation types. Sbornik Balg. Akad. Nauk. **26**. (Bulgarian and German, reviewed and summarized in English by W. B. Turrill, Jour. Ecol. **20**: 211-213, 1932).

Steffen, H. 1928. Beiträge zur Flora und Pflanzengeographie von Nowaja Semlja, Waigatsch und Kolgujew. Bot. Centbl. Beih., Abt. 2, **44**: 283-361.

Stehlé, H. 1945. Les conditions écologiques, la végétation et les ressources agricoles de l'Archipel des Petites Antilles. In "Plants and plant science in Latin America," ed. F. Verdoorn. pp. 85-100.

Steiger, T. L. 1930. Structure of prairie vegetation. Ecology **11**: 170-217.

Steiner, M. 1934. Zur Ökologie der Salzmarschen der nordöstlichen Vereinigten Staaten von Nordamerika. (Die osmotischen Verhältnisse des Bodens als Standortsfaktor. Die Ökologie der osmotischen Werte und der Zellsaftchemie bei Halophyten). Jahrb. f. Wiss. Bot. **81**: 94-202.

Stewart, G. & W. Keller. 1936. A correlation method for ecology as exemplified by studies of native desert vegetation. Ecology **17**: 500-514.

Steyermark, J. A. 1940. Studies of the vegetation of Missouri. I. Natural plant associations and succession in the Ozarks of Missouri. Field Mus. Nat. Hist., Chicago, Bot. Ser. **9**: 347-475.

Sukachev, V. N. (Sukatschew) 1928. Principles of classification of the spruce communities of European Russia. Jour. Ecol. **16**: 1-18.

———. 1929. Über einige Grundbegriffe in der Phytosoziologie. Deut. Bot. Gesell. Ber. **47**: 296-312.

Summerhayes, V. S. & C. S. Elton. 1923. Contributions to the ecology of Spitsbergen and Bear Island. Jour. Ecol. **11**: 214-286.

———. 1928. Further contributions to the ecology of Spitsbergen. Jour. Ecol. **16**: 193-268.

Swanson, C. H. 1929. The ecology of Turkey Run State Park. I. The flood plain. Ind. Acad. Sci. Proc. 1928, **38**: 165-170.

Szafer, W. 1924. Zur soziologischen Auffassung der Schneetälchenassoziationen. Veröff. Geobot. Inst. Rübel Zürich **1**: 300-310.

Szafer, W., B. Pawłowski & S. Kulczyński. 1924. Die Pflanzenassoziationen des Tatra-Gebirges I. Teil: Die Pflanzenassoziationen des Chockołowska-Tales. Bull. Internat. Polon. Acad. des Sci et Lettres, Cl. de Sci. Math. et Nat., Sér. B: Sci. Nat. 1923 (Suppl.): 1-65.

Talbot, M. W., H. H. Biswell & A. L. Hormay. 1939. Fluctuations in the annual vegetation of California. Ecology **20:** 394-402.

Tansley, A. G. 1911. Types of British vegetation, by members of the central committee for the survey and study of British vegetation. ed. A. G. Tansley. Cambridge: Univ. Press. 416 pp.

———. 1913. Primary survey of the Peak District of Derbyshire. Jour. Ecol. **1:** 275-285.

———. 1916. The development of vegetation. A review of Clements' "Plant Succession," 1916. Jour. Ecol. **4:** 198-204.

———. 1920. The classification of vegetation and the concept of development. Jour. Ecol. **8:** 118-149.

———. 1929. Succession: The concept and its values. Internat. Congr. Plant Sci., Ithaca, Proc. 1926, **1:** 677-686.

———. 1935. The use and abuse of vegetational concepts and terms. Ecology **16:** 284-307.

———. 1939. The British Islands and their vegetation. Cambridge: Univ. Press. 930 pp.

———. 1941. Note on the status of salt-marsh vegetation and the concept of 'formation.' Jour. Ecol. **29:** 212-214.

———. 1947-8. The early history of modern plant ecology in Britain. Jour. Ecol. **35:** 130-137, **36:** 180.

———. 1948. The peculiarities of South Australian vegetation. Jour. Ecol. **36:** 181-183.

Tansley, A. G. & T. F. Chipp. 1926. Aims and methods in the study of vegetation. London: Brit. Emp. Veget. Comm. & Crown Agents for Colonies. 383 pp.

Tanttu, A. 1915. Ueber die Entstehung der Bülten und Stränge der Moore. Acta Forest. Fenn. **4**(1): 1-24.

Tate, G. H. H. 1932. Life zones at Mount Roraima. Ecology **13:** 235-257.

Taylor, N. 1938. A preliminary report on the salt marsh vegetation of Long Island, New York. N. Y. State Mus. Bull. **316:** 21-84.

Tchou Yen-Tcheng. 1949. Études écologiques et phytosociologiques sur les forêts riveraines du Bas-Languedoc (Populetum albae). IV. Vegetatio **1:** 347-384.

Thompson, W. R. 1939. Biological control and the theories of the interactions of populations. Parasitology **31:** 299-388.

Thornber, J. T. 1901. The prairie-grass formation in region I. Univ. of Nebr., Bot. Surv. of Nebr. 5, Studies in the vegetation of the state **1:** 29-143. Lincoln, Nebr.: Bot. Seminar.

Tidestrom, I. 1925. Flora of Utah and Nevada. Introduction. U. S. Natl. Mus., Contrib. U. S. Natl. Herbarium **25:** 7-14.

Timmons, F. L. 1941-2. The rise and decline of cactus in Kansas. Kans. State Bd. Agr. Bien. Rpt. **33:** 37-46.

Tischler, W. 1951. Zur Synthese Biozönotischer Forschung. Acta Biotheoretica **9:** 135-161.

Tolstead, W. L. 1942. Vegetation of the northern part of Cherry county, Nebraska. Ecol. Monog. **12:** 255-292.

Trapnell, C. G. 1933. Vegetation types in Godthaab Fjord in relation to those in other parts of west Greenland and with special reference to Isersiutilik. Jour. Ecol. **21:** 294-334.

Tresner, H. D. 1951. Microflora of the upland forest continuum. Ecol. Soc. Amer. Bull. **32:** 57 (Abstract).

Trochain, J. 1940. Contribution a l'étude de la végétation du Sénégal. Paris: Librairie Larose. 433 pp.

Troll, C. 1944. Strukturböden, Solifluktion, und Frostklimate der Erde. In "Diluvial-Geologie und Klima," ed. C. Troll. Geol. Rundschau Klimaheft **34**(7/8): 545-694.

Troup, R. S. 1926. Problems of forest ecology in India. In "Aims and methods in the study of vegetation," ed. Tansley and Chipp, pp. 283-313.

Tutin, T. G. 1941. The hydrosere and current concepts of the climax. Jour. Ecol. **29:** 268-279.

Tüxen, R. 1928. Zur Arbeitsmethode der Pflanzensoziologie (Nach Braun-Blanquet). In "Mitteilungen der floristisch-soziologischen Arbeitsgemeinschaft in Niedersachsen," ed. R. Tüxen. Jahresber. Naturh. Gesell. Hannover Beih. (1): 11-19.

———. 1933. Klimaxprobleme des nw.-europäischen Festlandes. Nederland. Kruidk. Arch., Leyden **43:** 293-308.

———. 1935. Vegetationskartierung Nordwest Deutschlands und ihre wirtschaftliche Auswertung. Zesde Internat. Bot. Congr. Amsterdam Proc. 1935, **2:** 73-74.

Tüxen, R. & H. Diemont. 1937. Klimaxgruppe und Klimaxschwarm. Ein Beitrag zur Klimaxtheorie. Jahresber. Naturh. Gesell. Hannover **88/89:** 73-87. (Reviewed Conard 1938).

Utida, S. 1950. On the equilibrium state of the interacting population of an insect and its parasite. Ecology **31:** 165-175.

Uvarov, B. P. 1931. Insects and climate. Ent. Soc. London Trans. **79:** 1-247.

Varley, G. C. 1947. The natural control of population balance in the knapweed gall-fly (*Urophora jaceana*). Jour. Anim. Ecol. **16:** 139-187.

Vaughan, R. E. & P. O. Wiehe. 1941. Studies on the vegetation of Mauritus. III. The structure and development of the upland climax forest. Jour. Ecol. **29:** 127-160.

Vaughan, T. W. 1910. The geologic work of mangroves in southern Florida. U. S. Natl. Mus., Smithsonian Misc. Coll. **52:** 461-464.

Verduin, J. 1951. A comparison of phytoplankton data obtained by a mobile sampling method with those obtained from a single station. Amer. Jour. Bot. **38:** 5-11.

Vesey-Fitzgerald, D. 1942. Further studies of the vegetation on islands in the Indian Ocean. Jour. Ecol. **30:** 1-16.

Vestergren, T. 1902. Om den olikformiga snöbetäckningens inflytande på vetetationen i Sarjekfjällen. Bot. Notiser 1902: 241-268.

Villar, E. H. Del. 1929a. Geobotánica. Barcelona: Colección Labor. 339 pp.

———. 1929b. Sur la méthode et la nomenclature employées dans mon étude géobotanique de l'Espagne. Internat. Congr. Plant Sci., Ithaca, Proc. 1926, **1:** 541-564.

Volterra, V. 1926. Variazioni e fluttuazioni del numero

d'individui in specie animali conviventi. R. Accad. Naz. Lincei Cl. Fis. Mat. e Nat. Mem. **2**(3): 31-112.

———. 1931a. Variation and fluctuations of the number of individuals in animal species living together. In "Animal ecology," by R. N. Chapman. New York: McGraw-Hill. pp. 409-448.

———. 1931b. Leçons sur la théorie mathématique de la lutte pour la vie. Paris: Gauthier-Villars. 214 pp.

Wager, H. G. 1938. Growth and survival of plants in the arctic. Jour. Ecol. **26**: 390-410.

Walter, H. 1937. Pflanzensoziologie und Sukzessionslehre. Ztschr. f. Bot. **31**: 545-559.

———. 1943. Die Vegetation Osteuropas unter Berücksichtigung von Klima, Boden, und wirtschaftlicher Nutzung. Deutsche Forscherarbeit in Kolonie und Ausland, ed. K. Meyer. Heft 9. Berlin: Parey. 180 pp.

Walton, J. 1922. A Spitsbergen salt marsh: with observations on the ecological phenomena attendant on the emergence of land from the sea. Jour. Ecol. **10**: 109-121.

Warming, E. 1891. De psammophile Formationer i Danmark. Vidensk. Medd. Naturh. For. Copenhagen (Dansk Naturhist. Forening) 1891: 153-202.

———. 1895. Plantesamfund. Grundträk af den ökologiska Plantegeografi. Copenhagen: Philipsens. 335 pp.

———. 1896. Lehrbuch der ökologischen Pflanzengeographie. Eine Einführung in die Kenntnis der Pflanzenverein. Berlin: Borntraeger. 412 pp.

———. 1909. Oecology of plants; an introduction to the study of plant-communities. Assisted by M. Vahl, prepared for publication in English by P. Groom & I. B. Balfour. Oxford: Clarendon Press. 422 pp.

Warming, E. & P. Graebner. 1933. Lehrbuch der ökologischen Pflanzengeographie. Berlin: Borntraeger. 1157 pp.

Washburn, A. L. 1947. Reconnaissance geology of portions of Victoria Island and adjacent regions, arctic Canada. Geol. Soc. Amer. Mem. **22**: 1-142.

Watson, W. 1925. The bryophytes and lichens of arctic-alpine vegetation. Jour. Ecol. **13**: 1-26.

Watt, A. S. 1923. On the ecology of British beechwoods with special reference to their regeneration. I. The causes of failure of natural regeneration of the beech (*Fagus silvatica* L.). Jour. Ecol. **11**: 1-48.

———. 1924-5. On the ecology of British beechwoods with special reference to their regeneration. II. The development and structure of beech communities on the Sussex Downs. Jour. Ecol. **12**: 145-204, **13**: 27-73.

———. 1936. Studies in the ecology of Breckland. I. Climate, soil and vegetation. Jour. Ecol. **24**: 117-138.

———. 1945. Contributions to the ecology of bracken (*Pteridium aquilinum*). III. Frond types and the make-up of the population. New Phytol. **44**: 156-178.

——— 1947a. Contributions to the ecology of bracken (*Pteridium aquilinum*). IV. The structure of the community. New Phytol. **46**: 97-121.

———. 1947b. Pattern and process in the plant community. Jour. Ecol. **35**: 1-22.

Watt, A. S. & E. W. Jones. 1948. The ecology of the Cairngorms. I. The environment and the altitudinal zonation of the vegetation. Jour. Ecol. **36**: 283-304.

Weaver, H. 1951a. Fire as an ecological factor in the southwestern ponderosa pine forests. Jour. Forestry **49**: 93-98.

———. 1951b. Observed effects of prescribed burning on perennial grasses in the ponderosa pine forests. Jour. Forestry **49**: 267-271.

Weaver, J. E. 1943. Replacement of true prairie by mixed prairie in eastern Nebraska and Kansas. Ecology **24**: 421-434.

———. 1944. Recovery of midwestern prairies from drought. Amer. Phil. Soc. Proc. **88**: 125-131.

Weaver, J. E. & F. W. Albertson. 1936. Effects of the great drought on the prairies of Iowa, Nebraska, and Kansas. Ecology **17**: 567-639.

———. 1939. Major changes in grassland as a result of continued drought. Bot. Gaz. **100**: 576-591.

Weaver, J. E. & W. E. Bruner. 1945. A seven-year quantitative study of succession in grassland. Ecol. Monog. **15**: 297-319.

Weaver, J. E. & F. E. Clements. 1929. Plant Ecology. New York: McGraw-Hill. 520 pp. 2nd. edn. 1938. 601 pp.

Weaver, J. E & R. W. Darland. 1944. Grassland patterns in 1940. Ecology **25**: 202-215.

Weaver, J. E., J. H. Robertson & R. L. Fowler. 1940. Changes in true-prairie vegetation during drought as determined by list quadrats. Ecology **21**: 357-362.

Weaver, J. E., L. A. Stoddart & W. Noll. 1935. Response of the prairie to the great drought of 1934. Ecology **16**: 612-629.

Weiss, R. F. 1923. Die Gipsflora des Südharzes. Eine Vegetationsskizze. Bot. Centbl. Beih., Abt. 2, **40**: 223-252.

Wells, B. W. 1939. A new forest climax: the salt-spray climax of Smith Island, N. C. Torrey Bot. Club Bull. **66**: 629-634.

———. 1942. Ecological problems of the southeastern United States Coastal Plain. Bot. Rev. **8**: 533-561.

———. 1946. Archeological disclimaxes. Elisha Mitchell Sci. Soc. Jour. **62**: 51-53.

Wells, B. W. & I. V. Shunk. 1931. The vegetation and habitat factors of the coarser sands of the North Carolina Coastal Plain; An ecological study. Ecol. Monog. **1**: 465-520.

———. 1938a. Salt spray: An important factor in coastal ecology. Torrey Bot. Club Bull. **65**: 485-492.

———. 1938b. The important rôle of salt spray in coastal ecology. Elisha Mitchell Sci. Soc. Jour. **54**: 185-186.

Went, F. W. 1948. Ecology of desert plants. I. Observations on germination in the Joshua Tree National Monument, California. Ecology **29**: 242-253.

———. 1949. Ecology of desert plants. II. The effect of rain and temperature on germination and growth. Ecology **30**: 1-13.

Went, F. W. & M. Westergaard. 1949. Ecology of desert plants. III. Development of plants in the Death Valley National Monument, California. Ecology **30**: 26-38.

Whitford, H. N. 1906. The vegetation of the Lamao Forest Reserve. Philippine Jour. Sci. **1**: 373-431.

———. 1909. Studies in the vegetation of the Philippines. I. The composition and volume of the dipterocarp forests of the Philippines. Philippine Jour. Sci., C. Bot. **4**: 699-725.

Whitford, P. B. 1949. Distribution of woodland plants

in relation to succession and clonal growth. Ecology **30**: 199-208.

Whittaker, R. H. 1951. A criticism of the plant association and climatic climax concepts. Northwest Sci. **25**: 17-31.

———. 1952. A study of summer foliage insect communities in the Great Smoky Mountains. Ecol. Monog. **22**: 1-44.

Wiener, N. 1948. Cybernetics: Or control and communication in the animal and the machine. New York: Wiley. 194 pp.

Wilde, S. A. 1932. On the relation of soil and forest vegetation in the podsol region of the U.S.A. (Russian with English summary). Pochvovedenie **1**: 5-23.

———. 1933. The relation of soils and forest vegetation of the Lake States region. Ecology **14**: 94-105.

Wilde, S. A., P. B. Whitford & C. T. Youngberg. 1948. Relation of soils and forest growth in the driftless area of southwestern Wisconsin. Ecology **29**: 173-180.

Williams, C. B. 1947. The generic relations of species in small ecological communities. Jour. Anim. Ecol. **16**: 11-18.

———. 1951. Intra-generic competition as illustrated by Moreau's records of East African bird communities. Jour. Anim. Ecol. **20**: 246-253.

Williams, R. M. & H. J. Oosting. 1944. The vegetation of Pilot Mountain, North Carolina: A community analysis. Torrey Bot. Club Bull. **71**: 23-45.

Wood, J. G. 1937. The vegetation of South Australia. Adelaide: Government Printer. 164 pp.

———. 1939. Ecological concepts and nomenclature. Roy. Soc. So. Austral. Trans. **63**: 215-223.

Woodbury, A. M. 1933. Biotic relationships of Zion Canyon, Utah, with special reference to succession. A survey of the geological, botanical, and zoological interrelationships within a part of Zion National Park, Utah. Ecol. Monog. **3**: 147-245.

Woodruff, L. L. 1912. Observations on the origin and sequence of the protozoan fauna of hay infusions. Jour. Exper. Zool. **12**: 205-264.

———. 1913. The effect of excretion products of infusoria on the same and on different species, with special reference to the protozoan sequence in infusions. Jour. Exper. Zool. **14**: 575-582.

Wright, J. C. & E. A. Wright. 1948. Grassland types of south central Montana. Ecology **29**: 449-460.

Yapp, R. H., D. Johns & O. T. Jones. 1917. The salt marshes of the Dovey Estuary. Jour. Ecol. **5**: 65-103.

Yoshii, Y. & N. Hayasi. 1931. Botanische Studien subalpiner Moore auf vulkanischer Asche. Tôhoku Imp. Univ., Sendai, Sci. Repts. 4th Ser. (Biol.) **6**: 307-346.

Zach, L. W. 1950. A northern climax, forest or muskeg? Ecology **31**: 304-306.

Zangheri, P. 1950. Flora e vegetazione dei terreni "ferrettizzati" del Preappennino Romagnolo. Webbia **7**: 1-308.

Zimmermann, M. 1912. Sols fluides et sols polygonaux. Ann. de Géog. **21**: 452-455.

Zinserling, G. D. 1929. Uebersicht der Moorvegetation längs des Mittelstroms der Petschora. Isvestiia Glavnyi Bot. Sad, Leningrad **28**: 95-128.

Zlatník, A. 1928. Aperçu de la végétation des Krkonoše (Riesengebirge). D'après les études du feu M. František Schustler et ses propres par Alois Zlatník. Preslia **7**: 94-152.

Zohary, M. 1942. The vegetational aspect of Palestine soils: A preliminary account of the relation between vegetation and soils in Palestine. Palestine Jour. Bot., J. Ser. **2**: 200-246.

———. 1944. Vegetational transects through the desert of Sinai. Palestine Jour. Bot., J. Ser. **3**: 57-78.

———. 1945. Outline of the vegetation in Wadi Arabia. Jour. Ecol. **32**: 204-213.

———. 1947. A geobotanical soil map of western Palestine. Palestine Jour. Bot., J. Ser. **4**: 24-35.

Zohary, M. & G. Orshansky. 1947. The vegetation of the Huleh Plain. Palestine Jour. Bot., J. Ser. **4**: 90-104.

———. 1949. Structure and ecology of the vegetation in the Dead Sea region of Palestine. Palestine Jour. Bot., J. Ser. **4**: 177-206.

Zotov, V. D. 1938. Survey of the tussock-grasslands of the South Island, New Zealand. Preliminary Report. New Zeal. Jour. Sci. & Technol. **20**(4A): 212A-244A.

15

Reprinted from *Science* **164:**262–270 (Apr. 18, 1969)

The Strategy of Ecosystem Development

An understanding of ecological succession provides a basis for resolving man's conflict with nature.

Eugene P. Odum

The principles of ecological succession bear importantly on the relationships between man and nature. The framework of successional theory needs to be examined as a basis for resolving man's present environmental crisis. Most ideas pertaining to the development of ecological systems are based on descriptive data obtained by observing changes in biotic communities over long periods, or on highly theoretical assumptions; very few of the generally accepted hypotheses have been tested experimentally. Some of the confusion, vagueness, and lack of experimental work in this area stems from the tendency of ecologists to regard "succession" as a single straightforward idea; in actual fact, it entails an interacting complex of processes, some of which counteract one another.

As viewed here, ecological succession involves the development of ecosystems; it has many parallels in the developmental biology of organisms, and also in the development of human society. The ecosystem, or ecological system, is considered to be a unit of biological organization made up of all of the organisms in a given area (that is, "community") interacting with the physical environment so that a flow of energy leads to characteristic trophic structure and material cycles within the system. It is the purpose of this article to summarize, in the form of a tabular model, components and stages of development at the ecosystem level as a means of emphasizing those aspects of ecological succession that can be accepted on the basis of present knowledge, those that require more study, and those that have special relevance to human ecology.

Definition of Succession

Ecological succession may be defined in terms of the following three parameters (*1*). (i) It is an orderly process of community development that is reasonably directional and, therefore, predictable. (ii) It results from modification of the physical environment by the community; that is, succession is community-controlled even though the physical environment determines the pattern, the rate of change, and often sets limits as to how far development can go. (iii) It culminates in a stabilized ecosystem in which maximum biomass (or high information content) and symbiotic function between organisms are maintained per unit of available energy flow. In a word, the "strategy" of succession as a short-term process is basically the same as the "strategy" of long-term evolutionary development of the biosphere—namely, increased control of, or homeostasis with, the physical environment in the sense of achieving maximum protection from its perturbations. As I illustrate below, the strategy of "maximum protection" (that is, trying to achieve maximum support of complex biomass structure) often conflicts with man's goal of "maximum

The author is director of the Institute of Ecology, and Alumni Foundation Professor, at the University of Georgia, Athens. This article is based on a presidential address presented before the annual meeting of the Ecological Society of America at the University of Maryland, August 1966.

production" (trying to obtain the highest possible yield). Recognition of the ecological basis for this conflict is, I believe, a first step in establishing rational land-use policies.

The earlier descriptive studies of succession on sand dunes, grasslands, forests, marine shores, or other sites, and more recent functional considerations, have led to the basic theory contained in the definition given above. H. T. Odum and Pinkerton (*2*), building on Lotka's (*3*) "law of maximum energy in biological systems," were the first to point out that succession involves a fundamental shift in energy flows as increasing energy is relegated to maintenance. Margalef (*4*) has recently documented this bioenergetic basis for succession and has extended the concept.

Changes that occur in major structural and functional characteristics of a developing ecosystem are listed in Table 1. Twenty-four attributes of ecological systems are grouped, for convenience of discussion, under six headings. Trends are emphasized by contrasting the situation in early and late development. The degree of absolute change, the rate of change, and the time required to reach a steady state may vary not only with different climatic and physiographic situations but also with different ecosystem attributes in the same physical environment. Where good data are available, rate-of-change curves are usually convex, with changes occurring most rapidly at the beginning, but bimodal or cyclic patterns may also occur.

Bioenergetics of Ecosystem Development

Attributes 1 through 5 in Table 1 represent the bioenergetics of the ecosystem. In the early stages of ecological succession, or in "young nature," so to speak, the rate of primary production or total (gross) photosynthesis (P) exceeds the rate of community respiration (R), so that the P/R ratio is greater than 1. In the special case of organic pollution, the P/R ratio is typically less than 1. In both cases, however, the theory is that P/R approaches 1 as succession occurs. In other words, energy fixed tends to be balanced by the energy cost of maintenance (that is, total community respiration) in the mature or "climax" ecosystem. The P/R ratio, therefore, should be an excellent functional index of the relative maturity of the system.

So long as P exceeds R, organic matter and biomass (B) will accumulate in the system (Table 1, item 6), with the result that ratio P/B will tend to decrease or, conversely, the B/P, B/R, or B/E ratios (where $E = P + R$) will increase (Table 1, items 2 and 3). Theoretically, then, the amount of standing-crop biomass supported by the available energy flow (E) increases to a maximum in the mature or climax stages (Table 1, item 3). As a consequence, the net community production, or yield, in an annual cycle is large in young nature and small or zero in mature nature (Table 1, item 4).

Comparison of Succession in a Laboratory Microcosm and a Forest

One can readily observe bioenergetic changes by initiating succession in experimental laboratory microecosystems. Aquatic microecosystems, derived from various types of outdoor systems, such as ponds, have been cultured by Beyers (*5*), and certain of these mixed cultures are easily replicated and maintain themselves in the climax state indefinitely on defined media in a flask with only light input (*6*). If samples from the climax system are inoculated into fresh media, succession occurs, the mature system developing in less than 100 days. In Fig. 1 the general pattern of a 100-day autotrophic succession in a microcosm based on data of Cooke (*7*) is compared with a hypothetical model of a 100-year forest succession as presented by Kira and Shidei (*8*).

During the first 40 to 60 days in a typical microcosm experiment, daytime net production (P) exceeds nighttime respiration (R), so that biomass (B) accumulates in the system (*9*). After an early "bloom" at about 30 days, both rates decline, and they become approximately equal at 60 to 80 days. the B/P ratio, in terms of grams of carbon supported per gram of daily carbon production, increases from less than 20 to more than 100 as the steady state is reached. Not only are autotrophic and heterotrophic metabolism balanced in the climax, but a large organic structure is supported by small daily production and respiratory rates.

While direct projection from the small laboratory microecosystem to open nature may not be entirely valid, there is evidence that the same basic trends that are seen in the laboratory are characteristic of succession on land and in large bodies of water. Seasonal successions also often follow the same pattern, an early seasonal bloom characterized by rapid growth of a few dominant species being followed by the development later in the season of high B/P ratios, increased diversity, and a relatively steady, if temporary, state in terms of P and R (*4*). Open systems may not experience a decline, at maturity, in total or gross productivity, as the space-limited microcosms do, but the general pattern of bioenergetic change in the latter seems to mimic nature quite well.

These trends are not, as might at first seem to be the case, contrary to the classical limnological teaching which describes lakes as progressing in time from the less productive (oligotrophic) to the more productive (eutrophic) state. Table 1, as already emphasized, refers to changes which are brought about by biological processes *within* the ecosystem in question. Eutrophication, whether natural or cultural, results when nutrients are imported into the lake from *outside* the lake—that is, from the watershed. This is equivalent to adding nutrients to the laboratory microecosystem or fertilizing a field; the system is pushed back, in successional terms, to a younger or "bloom" state. Recent studies on lake sediments (*10*), as well as theoretical considerations (*11*), have indicated that lakes can and do progress to a more oligotrophic condition when the nutrient input from the watershed slows or ceases. Thus, there is hope that the troublesome cultural eutrophication of our waters can be reversed if the inflow of nutrients from the watershed can be greatly reduced. Most of all, however, this situation emphasizes that it is the entire drainage or catchment basin, not just the lake or stream, that must be considered the ecosystem unit if we are to deal successfully with our water pollution problems. Ecosystematic study of entire landscape catchment units is a major goal of the American plan for the proposed International Biological Program. Despite the obvious logic of such a proposal, it is proving surprisingly difficult to get tradition-bound scientists and granting agencies to look beyond their specialties toward the support of functional studies of large units of the landscape.

Food Chains and Food Webs

As the ecosystem develops, subtle changes in the net ›rk pattern of food chains may be expected. The manner in which organisms are linked together through food tends to be relatively sim-

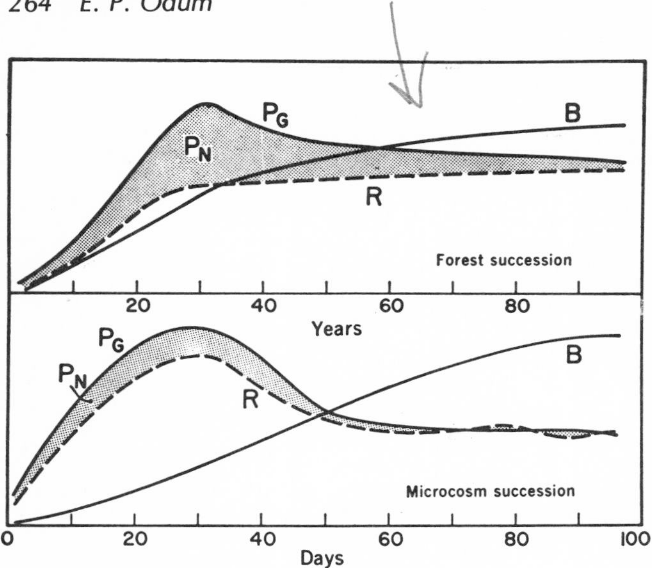

Fig. 1. Comparison of the energetics of succession in a forest and a laboratory microcosm. P_G, gross production; P_N, net production; R, total community respiration; B, total biomass.

ple and linear in the very early stages of succession, as a consequence of low diversity. Furthermore, heterotrophic utilization of net production occurs predominantly by way of grazing food chains—that is, plant-herbivore-carnivore sequences. In contrast, food chains become complex webs in mature stages, with the bulk of biological energy flow following detritus pathways (Table 1, item 5). In a mature forest, for example, less than 10 percent of annual net production is consumed (that is, grazed) in the living state (*12*); most is utilized as dead matter (detritus) through delayed and complex pathways involving as yet little understood animal-microorganism interactions. The time involved in an uninterrupted succession allows for increasingly intimate associations and reciprocal adaptations between plants and animals, which lead to the development of many mechanisms that reduce grazing—such as the development of indigestible supporting tissues (cellulose, lignin, and so on), feedback control between plants and herbivores (*13*), and increasing predatory pressure on herbivores (*14*). Such mechanisms enable the biological community to maintain the large and complex organic structure that mitigates perturbations of the physical environment. Severe stress or rapid changes brought about by outside forces can, of course, rob the system of these protective mechanisms and allow irruptive, cancerous growths of certain species to occur, as man too often finds to his sorrow. An example of a stress-induced pest irruption occurred at Brookhaven National Laboratory, where oaks became vulnerable to aphids when translocation of sugars and amino acids was impaired by continuing gamma irradiation (*15*).

Radionuclide tracers are providing a means of charting food chains in the intact outdoor ecosystem to a degree that will permit analysis within the concepts of network or matrix algebra. For example, we have recently been able to map, by use of a radiophosphorus tracer, the open, relatively linear food linkage between plants and insects in an early old-field successional stage (*16*).

Diversity and Succession

Perhaps the most controversial of the successional trends pertain to the complex and much discussed subject of diversity (*17*). It is important to distinguish between different kinds of diversity indices, since they may not follow parallel trends in the same gradient or developmental series. Four components of diversity are listed in Table 1, items 8 through 11.

The variety of species, expressed as **a species-number ratio or a species-area ratio, tends to increase during the early stages of community development. A second component of species diversity is what has been called equitability, or evenness (*18*), in the apportionment of individuals among the species. For example, two systems each containing 10 species and 100 individuals have the same diversity in terms of species-number ratio but could have widely different equitabilities depending on the apportionment of the 100 individuals among the 10 species—for example, 91-1-1-1-1-1-1-1-1-1 at one extreme or 10 individuals per species at the other. The Shannon formula,**

$$-\Sigma \frac{ni}{N} \log_2 \frac{ni}{N}$$

where ni is the number of individuals in each species and N is the total number of individuals, is widely used as a diversity index because it combines the variety and equitability components in one approximation. But, like all such lumping parameters, Shannon's formula may obscure the behavior of these two rather different aspects of diversity. For example, in our most recent field experiments, an acute stress from insecticide reduced the number of species of insects relative to the number of individuals but increased the evenness in the relative abundances of the surviving species (*19*). Thus, in this case the "variety" and "evenness" components would tend to cancel each other in Shannon's formula.

While an increase in the variety of species together with reduced dominance by any one species or small group of species (that is, increased evenness) can be accepted as a general probability during succession (*20*), there are other community changes that may work against these trends. An increase in the size of organisms, an increase in the length and complexity of life histories, and an increase in interspecific competition that may result in competitive exclusion of species (Table 1, items 12–14) are trends that may reduce the number of species that can live in a given area. In the bloom stage of succession organisms tend to be small and to have simple life histories and rapid rates of reproduction. Changes in size appear to be a consequence of, or an adaptation to, a shift in nutrients from inorganic to organic (Table 1, item 7). In a mineral nutrient-rich environment, small size is of selective advantage, especi y to autotrophs, because of the greater surface-to-volume ratio. As the ecosystem develops, however, inorganic nutrients tend to become more and more tied up in the biomass (that is, to become intrabiotic), so that the selective advantage shifts to larger organisms (either larger individuals of the same species or larger species, or both) which have greater storage capacities and more complex life histories, thus are adapted to exploiting seasonal or periodic releases of nutrients or other resources. The question of whether the seemingly direct relationship between organism size and stability is the result of positive feedback or is merely fortuitous remains unanswered (*21*).

Thus, whether or not species diversity continues to increase during succession will depend on whether the increase in potential niches resulting from increased biomass, stratification (Table 1, item 9), and other consequences of biological organization exceeds the countereffects of increasing size and competition. No one has yet been able to catalogue all the species in any sizable area, much less follow total species diversity in a successional series. Data are so far available only for segments of the community (trees, birds, and so on). Margalef (*4*) postulates that diversity will tend to peak during the early or middle stages of succession and then decline in the climax. In a study of bird populations along a successional gradient we found a bimodal pattern (*22*); the number of species increased during the early stages of old-field succession, declined during the early forest stages, and then increased again in the mature forest.

Species variety, equitability, and stratification are only three aspects of diversity which change during succession. Perhaps an even more important trend is an increase in the diversity of organic compounds, not only of those within the biomass but also of those excreted and secreted into the media (air, soil, water) as by-products of the increasing community metabolism. An increase in such "biochemical diversity" (Table 1, item 10) is illustrated by the increase in the variety of plant pigments along a successional gradient in aquatic situations, as described by Margalef (*4, 23*). Biochemical diversity within populations, or within systems as a whole, has not yet been systematically studied to the degree the subject of species diversity has been. Consequently, few generalizations can be made, except that it seems safe to say that, as succession progresses, organic extrametabolites probably serve increasingly important functions as regulators which stabilize the growth and composition of the ecosystem. Such metabolites may, in fact, be extremely important in preventing populations from overshooting the equilibrial density, thus in reducing oscillations as the system develops stability.

The cause-and-effect relationship between diversity and stability is not clear and needs to be investigated from many angles. If it can be shown that biotic diversity does indeed enhance physical stability in the ecosystem, or is the result of it, then we would have an important guide for conservation practice. Preservation of hedgerows, woodlots, noneconomic species, noneutrophicated waters, and other biotic variety in man's landscape could then be justified on scientific as well as esthetic grounds, even though such preservation often must result in some reduction in the production of food or other immediate consumer needs. In other words, is variety only the spice of life, or is it a necessity for the long life of the total ecosystem comprising man and nature?

Nutrient Cycling

An important trend in successional development is the closing or "tightening" of the biogeochemical cycling of major nutrients, such as nitrogen, phosphorus, and calcium (Table 1, items 15–17). Mature systems, as compared to developing ones, have a greater capacity to entrap and hold nutrients for cycling within the system. For example, Bormann and Likens (*24*) have estimated that only 8 kilograms per hectare out of a total pool of exchangeable calcium of 365 kilograms per hectare is lost per year in stream outflow from a North Temperate watershed covered with a mature forest. Of this, about 3 kilograms per hectare is replaced by rainfall, leaving only 5 kilograms to be obtained from weathering of the underlying rocks in order for the system to maintain mineral balance. Reducing the volume of the vegetation, or otherwise setting the succession back to a younger state, results in increased water yield by way of stream outflow (*25*), but this

Table 1. A tabular model of ecological succession: trends to be expected in the development of ecosystems.

Ecosystem attributes	Developmental stages	Mature stages
	Community energetics	
1. Gross production/community respiration (*P/R* ratio)	Greater or less than 1	Approaches 1
2. Gross production/standing crop biomass (*P/B* ratio)	High	Low
3. Biomass supported/unit energy flow (*B/E* ratio)	Low	High
4. Net community production (yield)	High	Low
5. Food chains	Linear, predominantly grazing	Weblike, predominantly detritus
	Community structure	
6. Total organic matter	Small	Large
7. Inorganic nutrients	Extrabiotic	Intrabiotic
8. Species diversity—variety component	Low	High
9. Species diversity—equitability component	Low	High
10. Biochemical diversity	Low	High
11. Stratification and spatial heterogeneity (pattern diversity)	Poorly organized	Well-organized
	Life history	
12. Niche specialization	Broad	Narrow
13. Size of organism	Small	Large
14. Life cycles	Short, simple	Long, complex
	Nutrient cycling	
15. Mineral cycles	Open	Closed
16. Nutrient exchange rate, between organisms and environment	Rapid	Slow
17. Role of detritus in nutrient regeneration	Unimportant	Important
	Selection pressure	
18. Growth form	For rapid growth ("*r*-selection")	For feedback control ("*K*-selection")
19. Production	Quantity	Quality
	Overall homeostasis	
20. Internal symbiosis	Undeveloped	Developed
21. Nutrient conservation	Poor	Good
22. Stability (resistance to external perturbations)	Poor	Good
23. Entropy	High	Low
24. Information	Low	High

greater outflow is accompanied by greater losses of nutrients, which may also produce downstream eutrophication. Unless there is a compensating increase in the rate of weathering, the exchangeable pool of nutrients suffers gradual depletion (not to mention possible effects on soil structure resulting from erosion). High fertility in "young systems" which have open nutrient cycles cannot be maintained without compensating inputs of new nutrients; examples of such practice are the continuous-flow culture of algae, or intensive agriculture where large amounts of fertilizer are imported into the system each year.

Because rates of leaching increase in a latitudinal gradient from the poles to the equator, the role of the biotic community in nutrient retention is especially important in the high-rainfall areas of the subtropical and tropical latitudes, including not only land areas but also estuaries. Theoretically, as one goes equatorward, a larger percentage of the available nutrient pool is tied up in the biomass and a correspondingly lower percentage is in the soil or sediment. This theory, however, needs testing, since data to show such a geographical trend are incomplete. It is perhaps significant that conventional North Temperate row-type agriculture, which represents a very youthful type of ecosystem, is successful in the humid tropics only if carried out in a system of "shifting agriculture" in which the crops alternate with periods of natural vegetative redevelopment. Tree culture and the semiaquatic culture of rice provide much better nutrient retention and consequently have a longer life expectancy on a given site in these warmer latitudes.

Selection Pressure: Quantity versus Quality

MacArthur and Wilson (*26*) have reviewed stages of colonization of islands which provide direct parallels with stages in ecological succession on continents. Species with high rates of reproduction and growth, they find, are more likely to survive in the early uncrowded stages of island colonization. In contrast, selection pressure favors species with lower growth potential but better capabilities for competitive survival under the equilibrium density of late stages. Using the terminology of growth equations, where r is the intrinsic rate of increase and K is the upper asymptote or equilibrium population size, we may say that "r selection" predominates in early colonization, with "K selection" prevailing as more and more species and individuals attempt to colonize (Table 1, item 18). The same sort of thing is even seen within the species in certain "cyclic" northern insects in which "active" genetic strains found at low densities are replaced at high densities by "sluggish" strains that are adapted to crowding (*27*).

Genetic changes involving the whole biota may be presumed to accompany the successional gradient, since, as described above, quantity production characterizes the young ecosystem while quality production and feedback control are the trademarks of the mature system (Table 1, item 19). Selection at the ecosystem level may be primarily interspecific, since species replacement is a characteristic of successional series or seres. However, in most well-studied seres there seem to be a few early successional species that are able to persist through to late stages. Whether genetic changes contribute to adaptation in such species has not been determined, so far as I know, but studies on population genetics of *Drosophila* suggest that changes in genetic composition could be important in population regulation (*28*). Certainly, the human population, if it survives beyond its present rapid growth stage, is destined to be more and more affected by such selection pressures as adaptation to crowding becomes essential.

Overall Homeostasis

This brief review of ecosystem development emphasizes the complex nature of processes that interact. While one may well question whether all the trends described are characteristic of all types of ecosystems, there can be little doubt that the net result of community actions is symbiosis, nutrient conservation, stability, a decrease in entropy, and an increase in information (Table 1, items 20–24). The overall strategy is, as I stated at the beginning of this article, directed toward achieving as large and diverse an organic structure as is possible within the limits set by the available energy input and the prevailing physical conditions of existence (soil, water, climate, and so on). As studies of biotic communities become more functional and sophisticated, one is impressed with the importance of mutualism, parasitism, predation, commensalism, and other forms of symbiosis. Partnership between unrelated species is often noteworthy (for example, that between coral coelenterates and algae, or between mycorrhizae and trees). In many cases, at least, biotic control of grazing, population density, and nutrient cycling provide the chief positive-feedback mechanisms that contribute to stability in the mature system by preventing overshoots and destructive oscillations. The intriguing question is, Do mature ecosystems age, as organisms do? In other words, after a long period of relative stability or "adulthood," do ecosystems again develop unbalanced metabolism and become more vulnerable to diseases and other perturbations?

Relevance of Ecosystem Development Theory to Human Ecology

Figure 1 depicts a basic conflict between the strategies of man and of nature. The "bloom-type" relationships, as exhibited by the 30-day microcosm or the 30-year forest, illustrate man's present idea of how nature should be directed. For example, the goal of agriculture or intensive forestry, as now generally practiced, is to achieve high rates of production of readily harvestable products with little standing crop left to accumulate on the landscape—in other words, a high P/B efficiency. Nature's strategy, on the other hand, as seen in the outcome of the successional process, is directed toward the reverse efficiency—a high B/P ratio, as is depicted by the relationship at the right in Fig. 1. Man has generally been preoccupied with obtaining as much "production" from the landscape as possible, by developing and maintaining early successional types of ecosystems, usually monocultures. But, of course, man does not live by food and fiber alone; he also needs a balanced CO_2–O_2 atmosphere, the climatic buffer provided by oceans and masses of vegetation, and clean (that is, unproductive) water for cultural and industrial uses. Many essential life-cycle resources, not to mention recreational and esthetic needs, are best provided man by the less "productive" landscapes. In other words, the landscape is not just a supply depot but is also tl *oikos*—the home—in which we must live. Until recently mankind has more or less taken for granted the gas-exchange, water-purification, nutrient-cycling, and other protective functions of self-maintaining ecosystems, chiefly because neither his numbers nor his environmental manipulations have been great enough to affect regional and global balances. Now, of course, it is painfully evident that such balances are being affected, often detrimentally. The "one problem, one solution approach" is no longer adequate and must be replaced by some form of ecosystem analysis that considers man as a part of, not apart from, the environment.

The most pleasant and certainly the safest landscape to live in is one containing a variety of crops, forests, lakes, streams, roadsides, marshes, seashores, and "waste places"—in other words, a mixture of communities of different ecological ages. As individuals we more or less instinctively surround our houses with protective, nonedible cover (trees, shrubs, grass) at the same time that we strive to coax extra bushels from our cornfield. We all consider the cornfield a "good thing," of course, but most of us would not want to live there, and it would certainly be suicidal to cover the whole land area of the biosphere with cornfields, since the boom and bust oscillation in such a situation would be severe.

The basic problem facing organized society today boils down to determining in some objective manner when we are getting "too much of a good thing." This is a completely new challenge to mankind because, up until now, he has had to be concerned largely with too little rather than too much. Thus, concrete is a "good thing," but not if half the world is covered with it. Insecticides are "good things," but not when used, as they now are, in an indiscriminate and wholesale manner. Likewise, water impoundments have proved to be very useful man-made additions to the landscape, but obviously we don't want the whole country inundated! Vast man-made lakes solve some problems, at least temporarily, but yield comparative little food or fiber, and, because of high evaporative losses, they may not even be the best device for storing water; it might better be stored in the watershed, or underground in aquafers. Also, the cost of building large dams is a drain on already overtaxed revenues. Although as individuals we readily recognize that we can have too many dams or other large-scale environmental changes, governments are so fragmented and lacking in systems-analysis capabilities that there is no effective mechanism whereby negative feedback signals can be received and acted on before there has been a serious overshoot. Thus, today there are governmental agencies, spurred on by popular and political enthusiasm for dams, that are putting on the drawing boards plans for damming every river and stream in North America!

Table 2. Contrasting characteristics of young and mature-type ecosystems.

Young	Mature
Production	Protection
Growth	Stability
Quantity	Quality

Society needs, and must find as quickly as possible, a way to deal with the landscape as a whole, so that manipulative skills (that is, technology) will not run too far ahead of our understanding of the impact of change. Recently a national ecological center outside of government and a coalition of governmental agencies have been proposed as two possible steps in the establishment of a political control mechanism for dealing with major environmental questions. The soil conservation movement in America is an excellent example of a program dedicated to the consideration of the whole farm or the whole watershed as an ecological unit. Soil conservation is well understood and supported by the public. However, soil conservation organizations have remained too exclusively farm-oriented, and have not yet risen to the challenge of the urban-rural landscape, where lie today's most serious problems. We do, then, have potential mechanisms in American society that could speak for the ecosystem as a whole, but none of them are really operational (*29*).

The general relevance of ecosystem development theory to landscape planning can, perhaps, be emphasized by the "mini-model" of Table 2, which contrasts the characteristics of young and mature-type ecosystems in more general terms than those provided by Table 1. It is mathematically impossible to obtain a maximum for more than one thing at a time, so one cannot have both extremes at the same time and place. Since all six characteristics listed in Table 2 are desirable in the aggregate, two possible solutions to the dilemma immediately suggest themselves. We can compromise so as to provide moderate quality and moderate yield on all the landscape, or we can deliberately plan to compartmentalize the landscape so as to simultaneously maintain highly productive and predominantly protective types as separate units subject to different management strategies (strategies ranging, for example, from intensive cropping on the one hand to wilderness management on the other). If ecosystem development theory is valid and applicable to planning, then the so-called multiple-use strategy, about which we hear so much, will work only through one or both of these approaches, because, in most cases, the projected multiple uses conflict with one another. It is appropriate, then, to examine some examples of the compromise and the compartmental strategies.

Pulse Stability

A more or less regular but acute physical perturbation imposed from without can maintain an ecosystem at some intermediate point in the developmental sequence, resulting in, so to speak, a compromise between youth and maturity. What I would term "fluctuating water level ecosystems" are good examples. Estuaries, and intertidal zones in general, are maintained in an early, relatively fertile stage by the tides, which provide the energy for rapid nutrient cycling. Likewise, freshwater marshes, such as the Florida Everglades, are held at an early successional stage by the seasonal fluctuations in water levels. The dry-season drawdown speeds up aerobic decomposition of accumulated organic matter, releasing nutrients that, on reflooding, support a wet-season bloom in productivity. The life histories of many organisms are intimately coupled to this periodicity. The wood stork, for example, breeds when the water levels are falling and the small fish on which it feeds become concentrated and easy to catch in the drying pools. If the water level remains high during the usual dry season or fails to rise in the wet season, the stork will not nest (*30*). Stabilizing water levels in the Everglades by means of dikes, locks, and impoundments, as is now advocated by some, would, in my opinion, destroy rather than preserve the Everglades as we now know them just as surely as complete drainage would. Without periodic drawdowns and fires, the shallow basins would fill up with organic matter and succession would proceed from the present pond-and-prairie condition toward a scrub or swamp forest.

It is strange that man does not readily recognize the importance of recurrent changes in water level in a natural situation such as the Everglades when similar pulses are the basis for some of his most enduring food culture systems (*31*). Alternate filling and draining of ponds has been a standard procedure in fish culture for centuries in Europe and the Orient. The flooding, draining, and soil-aeration procedure in rice culture is another example. The rice paddy is thus the cultivated analogue of the natural marsh or the intertidal ecosystem.

Fire is another physical factor whose periodicity has been of vital importance to man and nature over the centuries. Whole biotas, such as those of the African grasslands and the California chaparral, have become adapted to periodic fires producing what ecologists often call "fire climaxes" (*32*). Man uses fire deliberately to maintain such climaxes or to set back succession to some desired point. In the southeastern coastal plain, for example, light fires of moderate frequency can maintain a pine forest against the encroachment of older successional stages which, at the present time at least, are considered economically less desirable. The fire-controlled forest yields less wood than a tree farm does (that is, young trees, all of about the same age, planted in rows and harvested on a short rotation schedule), but it provides a greater protective cover for the landscape, wood of higher quality, and a home for game birds (quail, wild turkey, and so on) which could not survive in a tree farm. The fire climax, then, is an example of a compromise between production simplicity and protection diversity.

It should be emphasized that pulse stability works only if there is a complete community (including not only plants but animals and microorganisms) adapted to the particular intensity and frequency of the perturbation. Adaptation—operation of the selection process —requires times measurable on the evolutionary scale. Most physical stresses introduced by man are too sudden, too violent, or too arrhythmic for adaptation to occur at the ecosystem level, so severe oscillation rather than stability results. In many cases, at least, modification of naturally adapted ecosystems for cultural purposes would seem preferable to complete redesign.

Prospects for a Detritus Agriculture

As indicated above, heterotrophic utilization of primary production in mature ecosystems involves largely a delayed consumption of detritus. There is no reason why man cannot make greater use of detritus and thus obtain food or other products from the more protective type of ecosystem. Again, this would represent a compromise, since the short-term yield could not be as great as the yield obtained by direct exploitation of the grazing food chain. A detritus agriculture, however, would have some compensating advantages. Present agricultural strategy is based on selection for rapid growth and edibility in food plants, which, of course, make them vulnerable to attack by insects and disease. Consequently, the more we select for succulence and growth, the more effort we must invest in the chemical control of pests; this effort, in turn, increases the likelihood of our poisoning useful organisms, not to mention ourselves. Why not also practice the reverse strategy—that is, select plants which are essentially unpalatable, or which produce their own systemic insecticides while they are growing, and then convert the net production into edible products by microbial and chemical enrichment in food factories? We could then devote our biochemical genius to the enrichment process instead of fouling up our living space with chemical poisons! The production of silage by fermentation of low-grade fodder is an example of such a procedure already in widespread use. The cultivation of detritus-eating fishes in the Orient is another example.

By tapping the detritus food chain man can also obtain an appreciable harvest from many natural systems without greatly modifying them or destroying their protective and esthetic value. Oyster culture in estuaries is a good example. In Japan, raft and long-line culture of oysters has proved to be a very practical way to harvest the natural microbial products of estuaries and shallow bays. Furukawa (*33*) reports that the yield of cultured oysters in the Hiroshima Prefecture has increased tenfold since 1950, and that the yield of oysters (some 240,000 tons of meat) from this one district alone in 1965 was ten times the yield of natural oysters from the entire country. Such oyster culture is feasible along the entire Atlantic and Gulf coasts of the United States. A large investment in the culture of oysters and other seafoods would also provide the best possible deterrent against pollution, since the first threat of damage to the pollution-sensitive oyster industry would be immediately translated into political action!

The Compartment Model

Successful though they often are, compromise systems are not suitable nor desirable for the whole landscape. More emphasis needs to be placed on compartmentalization, so that growth-type, steady-state, and intermediate-type ecosystems can be linked with urban and industrial areas for mutual benefit. Knowing the transfer coefficients that define the flow of energy and the movement of materials and organisms (including man) between compartments, it should be possible to determine, through analog-computer manipulation, rational limits for the size and capacity of each compartment. We might start, for example, with a simplified model, shown in Fig. 2, consisting of four compartments of equal area, partitioned according to the basic biotic-function criterion —that is, according to whether the area is (i) productive, (ii) protective, (iii) a compromise between (i) and (ii) or (iv), urban-industrial. By continually refining the transfer coefficients on the basis of real world situations, and by increasing and decreasing the size and capacity of each compartment through computer simulation, it would be possible to determine objectively the limits that must eventually be imposed on each compartment in order to maintain regional and global balances in the exchange of vital energy and of materials. A systems-analysis procedure provides at least one approach to the solution of the basic dilemma posed by the question "How do we determine when we are getting too much of a good thing?" Also it provides a means of evaluating the energy drains imposed on ecosystems by pollution, radiation, harvest, and other stresses (*34*).

Implementing any kind of compartmentalization plan, of course, would require procedures for zoning the landscape and restricting the use of some land and water areas. While the principle of zoning in cities is universally accepted, the procedures now followed do not work very well because zoning restrictions are too easily overturned by short-term economic and population pressures. Zoning the landscape would require a whole new order of thinking. Greater use of legal measures providing for tax relief, restrictions on use, scenic easements, and public ownership will be required if appreciable land and water areas are to be held in the "protective" categories. Several states (for example, New Jersey and California), where pollution and population pressure are beginning to hurt, have made a start in this direction by enacting "open space" legislation designed to get as much unoccupied land as possible into a "protective" status so that future uses can be planned on a rational and scientific basis. The United States as a whole is fortunate in that large areas of the country are in national forests, parks, wildlife refuges, and so on. The fact that such areas, as well as the bordering oceans, are not quickly exploitable gives us time for the accelerated ecological study and programming needed to determine what proportions of different types of landscape provide a safe balance between man and nature. The open oceans, for example, should forever be allowed to remain protective rather than productive territory, if Alfred Redfield's (*35*) assumptions are correct. Redfield views the oceans, the major part of the hydrosphere, as the biosphere's governor, which slows down and controls the rate of decomposition and nutrient regeneration, thereby creating and maintaining the highly aerobic terrestrial environment to which the higher forms of life, such as man, are adapted. Eutrophication of the ocean in a last-ditch effort to feed the populations of the land could well have an adverse effect on the oxygen reservoir in the atmosphere.

Until we can determine more precisely how far we may safely go in expanding intensive agriculture and urban sprawl at the expense of the protective landscape, it will be good insurance to hold inviolate as much of the latter as possible. Thus, the preservation of natural areas is not a peripheral luxury for society but a capital investment from which we expect to draw interest. Also, it may well be that restrictions in the use of land and water are our only practical means of avoiding overpopulation or too great an exploitation of resources, or both. Interestingly enough, restriction of land use is the analogue of a natural behavioral control mechanism known as "territoriality" by which

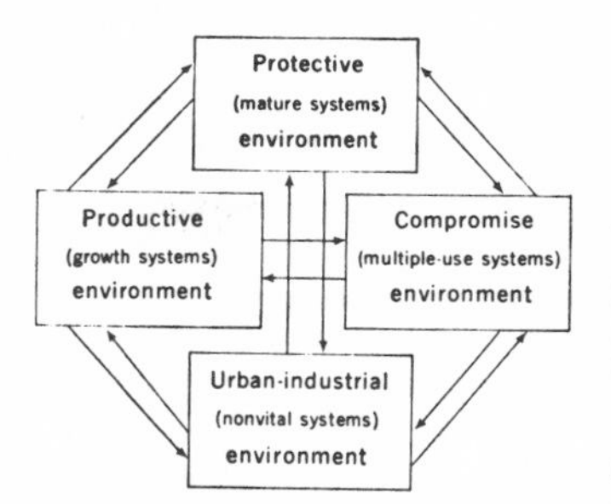

Fig. 2. Compartment model of the basic kinds of environment required by man, partitioned according to ecosystem development and life-cycle resource criteria.

many species of animals avoid crowding and social stress (*36*).

Since the legal and economic problems pertaining to zoning and compartmentalization are likely to be thorny, I urge law schools to establish departments, or institutes, of "landscape law" and to start training "landscape lawyers" who will be capable not only of clarifying existing procedures but also of drawing up new enabling legislation for consideration by state and national governing bodies. At present, society is concerned—and rightly so—with human rights, but environmental rights are equally vital. The "one man one vote" idea is important, but so also is a "one man one hectare" proposition.

Education, as always, must play a role in increasing man's awareness of his dependence on the natural environment. Perhaps we need to start teaching the principles of ecosystem in the third grade. A grammar school primer on man and his environment could logically consist of four chapters, one for each of the four essential kinds of environment, shown diagrammatically in Fig. 2.

Of the many books and articles that are being written these days about man's environmental crisis, I would like to cite two that go beyond "crying out in alarm" to suggestions for bringing about a reorientation of the goals of society. Garrett Hardin, in a recent article in *Science* (*37*), points out that, since the optimum population density is less than the maximum, there is no strictly technical solution to the problem of pollution caused by overpopulation; a solution, he suggests, can only be achieved through moral and legal means of "mutual coercion, mutually agreed upon by the majority of people."

Earl F. Murphy, in a book entitled *Governing Nature* (*38*), emphasizes that the regulatory approach alone is not enough to protect life-cycle resources, such as air and water, that cannot be allowed to deteriorate. He discusses permit systems, effluent charges, receptor levies, assessment, and cost-internalizing procedures as economic incentives for achieving Hardin's "mutually agreed upon coercion."

It goes without saying that the tabular model for ecosystem development which I have presented here has many parallels in the development of human society itself. In the pioneer society, as in the pioneer ecosystem, high birth rates, rapid growth, high economic profits, and exploitation of accessible and unused resources are advantageous, but, as the saturation level is approached, these drives must be shifted to considerations of symbiosis (that is, "civil rights," "law and order," "education," and "culture"), birth control, and the recycling of resources. A balance between youth and maturity in the socio-environmental system is, therefore, the really basic goal that must be achieved if man as a species is to successfully pass through the present rapid-growth stage, to which he is clearly well adapted, to the ultimate equilibrium-density stage, of which he as yet shows little understanding and to which he now shows little tendency to adapt.

References and Notes

1. E. P. Odum, *Ecology* (Holt, Rinehart & Winston, New York, 1963), chap. 6.
2. H. T. Odum and R. C. Pinkerton, *Amer. Scientist* **43**, 331 (1955).
3. A. J. Lotka, *Elements of Physical Biology* (Williams and Wilkins, Baltimore, 1925).
4. R. Margalef, *Advan. Frontiers Plant Sci.* **2**, 137 (1963); *Amer. Naturalist* **97**, 357 (1963).
5. R. J. Beyers, *Ecol. Monographs* **33**, 281 (1963).
6. The systems so far used to test ecological principles have been derived from sewage and farm ponds and are cultured in half-strength No. 36 Taub and Dollar medium [*Limnol. Oceanog.* **9**, 61 (1964)]. They are closed to organic imput or output but are open to the atmosphere through the cotton plug in the neck of the flask. Typically, liter-sized microecosystems contain two or three species of nonflagellated algae and one to three species each of flagellated protozoans, ciliated protozoans, rotifers, nematodes, and ostracods; a system derived from a sewage pond contained at least three species of fungi and 13 bacterial isolates [R. Gordon, thesis, University of Georgia (1967)]. These cultures are thus a kind of minimum ecosystem containing those small species originally found in the ancestral pond that are able to function together as a self-contained unit under the restricted conditions of the laboratory flask and the controlled environment of a growth chamber [temperature, 65° to 75°F (18° to °C); photoperiod, 12 hours; illumination, 100 to 1000 footcandles].
7. G. D. Cooke, *BioScience* **17**, 717 (1967).
8. T. Kira and T. Shidei, *Japan. J. Ecol.* **17**, 70 (1967).
9. The metabolism of the microcosms was

monitored by measuring diurnal pH changes, and the biomass (in terms of total organic matter and total carbon) was determined by periodic harvesting of replicate systems.

10. F. J. H. Mackereth, *Proc. Roy. Soc. London Ser. B* **161**, 295 (1965); U. M. Cowgill and G. E. Hutchinson, *Proc. Intern. Limnol. Ass.* **15**, 644 (1964); A. D. Harrison, *Trans. Roy. Soc. S. Africa* **36**, 213 (1962).
11. R. Margalef, *Proc. Intern. Limnol. Ass.* **15**, 169 (1964).
12. J. R. Bray, *Oikos* **12**, 70 (1961).
13. D. Pimentel, *Amer. Naturalist* **95**, 65 (1961).
14. R. T. Paine, *ibid.* **100**, 65 (1966).
15. G. M. Woodwell, *Brookhaven Nat. Lab. Pub. 924(T-381)* (1965), pp. 1–15.
16. R. G. Wiegert, E. P. Odum, J. H. Schnell, *Ecology* **48**, 75 (1967).
17. For selected general discussions of patterns of species diversity, see E. H. Simpson, *Nature* **163**, 688 (1949): C. B. Williams, *J. Animal Ecol.* **22**, 14 (1953); G. E. Hutchinson, *Amer. Naturalist* **93**, 145 (1959); R. Margalef, *Gen. Systems* **3**, 36 (1958); R. MacArthur and J. MacArthur, *Ecology* **42**, 594 (1961); N. G. Hairston, *ibid.* **40**, 404 (1959); B. C. Patten, *J. Marine Res. (Sears Found. Marine Res.)* **20**, 57 (1960); E. G. Leigh, *Proc. Nat. Acad. Sci. U.S.* **55**, 777 (1965); E. R. Pianka, *Amer. Naturalist* **100**, 33 (1966); E. C. Pielou, *J. Theoret. Biol.* **10**, 370 (1966).
18. M. Lloyd and R. J. Ghelardi, *J. Animal Ecol.* **33**, 217 (1964); E. C. Pielou, *J. Theoret. Biol.* **13**, 131 (1966).
19. G. W. Barrett, *Ecology* **49**, 1019 (1969).
20. In our studies of natural succession following grain culture, both the species-to-numbers and the equitability indices increased for all trophic levels but especially for predators and parasites. Only 44 percent of the species in the natural ecosystem were phytophagous, as compared to 77 percent in the grain field.
21. J. T. Bonner, *Size and Cycle* (Princeton Univ. Press, Princeton, N.J., 1963); P. Frank, *Ecology* **49**, 355 (1968).
22. D. W. Johnston and E. P. Odum, *Ecology* **37**, 50 (1956).
23. R. Margalef, *Oceanog. Marine Biol. Annu. Rev.* **5**, 257 (1967).
24. F. H. Bormann and G. E. Likens, *Science* **155**, 424 (1967).
25. Increased water yield following reduction of vegetative cover has been frequently demonstrated in experimental watersheds throughout the world [see A. R. Hibbert, in *International Symposium on Forest Hydrology* (Pergamon Press, New York, 1967), pp. 527–543]. Data on the long-term hydrologic budget (rainfall input relative to stream outflow) are available at many of these sites, but mineral budgets have yet to be systematically studied. Again, this is a prime objective in the "ecosystem analysis" phase of the International Biological Program.
26. R. H. MacArthur and E. O. Wilson, *Theory of Island Biogeography* (Princeton Univ. Press, Princeton, N.J., 1967).
27. Examples are the tent caterpillar [see W. G. Wellington, *Can. J. Zool.* **35**, 293 (1957)] and the larch budworm [see W. Baltensweiler, *Can. Entomologist* **96**, 792 (1964)].
28. F. J. Ayala, *Science* **162**, 1453 (1968).
29. Ira Rubinoff, in discussing the proposed sea-level canal joining the Atlantic and Pacific oceans [*Science* **161**, 857 (1968)], calls for a "control commission for environmental manipulation" with "broad powers of approving, disapproving, or modifying all major alterations of the marine or terrestrial environments. . . ."
30. See M. P. Kahl, *Ecol. Monographs* **34**, 97 (1964).
31. The late Aldo Leopold remarked long ago [*Symposium on Hydrobiology* (Univ. of Wisconsin Press, Madison, 1941), p. 17] that man does not perceive organic behavior in systems unless he has built them himself. Let us hope it will not be necessary to rebuild the entire biosphere before we recognize the worth of natural systems!
32. See C. F. Cooper, *Sci. Amer.* **204**, 150 (April 1961).
33. See "Proceedings Oyster Culture Workshop, Marine Fisheries Division, Georgia Game and Fish Commission, Brunswick" (1968), pp. 49–61.
34. See H. T. Odum, in *Symposium on Primary Productivity and Mineral Cycling in Natural Ecosystems*, H. E. Young, Ed. (Univ. of Maine Press, Orono, 1967), p. 81; ———, in *Pollution and Marine Ecology* (Wiley, New York, 1967), p. 99; K. E. F. Watt, *Ecology and Resource Management* (McGraw-Hill, New York, 1968).
35. A. C. Redfield, *Amer. Scientist* **46**, 205 (1958).
36. R. Ardrey, *The Territorial Imperative* (Atheneum, New York, 1967).
37. G. Hardin, *Science* **162**, 1243 (1968).
38. E. F. Murphy, *Governing Nature* (Quadrangle Books, Chicago, 1967).

16

Reprinted from *The Arnold Arbor. J.* **54**(3):331–368 (1973)

SUCCESSION *

WILLIAM H. DRURY AND IAN C. T. NISBET

IN ITS WIDEST SENSE the term "succession" refers to observed sequences of vegetation associations or animal groups. Some occur in space, such as a sequence of zones of grasses, shrubs, and trees on the side of many ponds; or a sequence of zones of vegetation on the side of a mountain. Other sequences occur in time, as, for example, a sequence of vegetation types occurring after an "old field" or a gravel pit is abandoned.

DEFINITION

In ecological literature, the term succession is usually used to imply sequences in time. However, only short-term changes can be observed directly, and most descriptions of long-term changes are based on observation of spatial sequences. In order to provide a unified description on which a general theory can be based, it is necessary to assume a homology between a spatial sequence of zones of vegetation visible at one time in a landscape and a long-term sequence of vegetation types on a single site (Gleason 1927: 320 and 324). This assumption is consistent with and reinforced by the classical geomorphological theory (Davis 1909) of landscape development (for discussion see Hack & Goodlett 1960, McCormick 1968, Drury & Nisbet 1971).

Functional effects have been ascribed to the observed sequences. For example, MacArthur and Connell (1966) describe the changes in organisms on an abandoned field, in stored grain, and in a jar of water. Then they comment: "This gives us a clue to all of the true replacements of succession: *each species alters the environment in such a way that it can no longer grow so successfully as others.*" (Italics theirs.)

Another property ascribed to succession is that the changes are progressive, that is, directional, and it is possible to predict which species will replace others in the course of a succession. MacArthur and Connell, in the same chapter (1966) say:

> In the case of forest succession, each species is able to stand deeper shade than the previous one, and as the forest grows the canopy becomes thicker

* Contribution No. 90 from the Scientific Staff, Massachusetts Audubon Society.

> and casts an even deeper shade. In this new, deeper shade other species are more successful. . . . Foresters have tables of 'tolerance' of different tree species; tolerant species are those that are successful in shade. As expected, the climax forests are composed of the most tolerant species. . . .

Hence the theory of succession allows one to predict the characteristics of those species which will replace others, and to recognize unstable or successional species. Thus, species and associations of organisms have been assigned to places in a successional order.

Succession is supposed to continue until the species combination best suited to the regional climate and the site are established. Many authors believe this reflects the development of a particularly well-adjusted configuration of species (Braun 1950, Clements 1916, 1936, Dansereau 1957, Daubenmire 1968, Odum 1959, Oosting 1958); other authors believe it to be simply running out of available species (Gleason 1926, 1927, Cooper 1926). But obviously the concept of succession is inseparable from that of climax (Whittaker 1965), even if it is possible to equivocate by suggesting that change continues at a very slow rate in the "climax."

The concept of climax has been used in a highly inconsistent way (Whittaker 1953), and there is a rich literature challenging, defending, and redefining climax and the nature of plant communities (see reviews by Whittaker 1953, MacIntosh 1967 and the replies in *The Botanical Review* 1968, and Langford & Buell 1969). The concept of succession has seldom been critically reviewed (but see Egler 1954 and McCormick 1968).

There are several vegetational types conventionally referred to as successional. Cowles (1911) classified successions into regional, topographic, and biotic types. These include (a) the "major vegetation zones" (biomes or latitudinal and altitudinal sequences); (b) pond margins, peat bogs, and marshes, or sand dunes; (c) old fields, or changes following fire. Among other successions are included patterns of plants on floodplains, around the snouts of retreating glaciers, on landslides, and the sequences of species in the course of the history of colonization of islands and the microcosms of laboratory infusions.

The primary bases for assigning plant associations to a position in a succession are size of the dominant plants, complexity of the vegetation structure and the number of species present. Successional series are generally recognized as extending from associations of low stature, few species and simple structure to associations of tall plants, many species and complex structure: (a) encrusting, prostrate, decumbent or emergent forms; (b) grasses and sedges; (c) perennial wildflowers and low thickets; (d) tall shrubs and scattered, taller trees, or stands of uniformly aged trees with little underbrush; (e) a canopy of trees with understory of saplings and ground cover of several levels.

On the longest time-scale, some successions are credited to sites not previously occupied by vegetation; these are called Primary Successions. If the site has been previously occupied, the succession is called Secondary. Tansley (1935) distinguished Autogenic Successions, "in which successive changes are brought about by the action of the plants themselves on the habitat," from Allogenic Successions, "in which the changes are brought about by external factors." Primary successions were divided by Clements (1916) into hydroseres, xeroseres, etc., according to the moisture conditions of the site.

Ponds and lakes have been put into a quasi-successional sequence from Oligotrophic (deep lakes, poor in nutrients, with little visible floating vegetation) to Eutrophic (shallow lakes, rich in nutrients, often covered with floating plants). Zones of vegetation occur in the sea below high tide and have been described on the coasts of the Pacific (Ricketts & Calvin 1968) and the Atlantic, of Europe (Yonge 1949) and the United States (Dexter 1947, Lamb & Zimmermann 1964, Stephenson & Stephenson 1954). However, we know of no assignment of such algal zones to successional stages, although the zones of emergent vegetation immediately above high tide in salt marshes are usually classified in successional terms.

Historical Review

Descriptions of zones of vegetation and of changes in vegetation on one site are available from Theophrastus (300 B.C.) onwards. The idea that one community of plants might alter its site [1] and prepare the way for another is also old (Buffon 1742). Several authors during the 19th century (e.g. Kerner 1863) described zones and sequences of vegetation, but did not formulate a model for the ecological events involved. In 1863 Thoreau recognized that pine stands on upland soils in central New England were followed after logging by even-aged hardwood stands which today are the main forest type of the region. He named this trend forest succession (Spurr 1952). Hult (1885) recognized "developmental changes" in vegetation, yet it was the work of Cowles (1899, 1901, 1911) which formulated the concept of succession. Clements's brilliant and exhaustive studies (1905, 1916, 1920, 1928, 1934, 1936) appear to be responsible for its general acceptance, because he made of it the mechanism of progress in his visionary system for plant ecology. His developmental model, in which climax was regarded as a superorganism and succession its embryonic development (Clements 1916: 161) was consistent with deterministic, closed system models current in several branches of earth science of that time.

Apparently it was Clements's experience in the grasslands which convinced him of the validity and integrity of grassland communities (in contrast to the previously held opinion that they were unnatural) and led him to the concept of the vital integrity of plant associations. Following Clements, ecologists were for several decades preoccupied with mapping and monitoring the natural communities of North America and discovering their structures, consistencies and inconsistencies. In this

[1] In botanical usage, site refers to a place and a set of habitat conditions suitable for the growth of a particular species or vegetation association. For example, in the course of succession, "each species alters the site."

period, strongly divergent schools of thought developed with regard to the nature of plant and animal communities (MacIntosh 1967, Langford & Buell 1969).

During the last decades there has been a shift to attempts to develop a general theory of structural and functional characteristics of communities (Hutchinson 1957, 1965, MacArthur *et al.* 1966, Levins 1966, 1968). Yet there has been little or no resolution of the contradictory assumptions about the nature of plant and animal communities. Most theoretical formulations retain the deterministic assumptions to a greater or lesser degree; hence, implicitly or explicitly, they relegate the majority of species and communities to "successional" status, with less than maximum fitness. It is no wonder then, that there is a lack of clarity in current theoretical formulations of community interactions or of successional trends (Johnson & Valentine 1971).

Contemporary Concepts of Succession

In spite of little uniformity in detail there is considerable agreement on the general trends of "community development," as quotations from recent reviews by Odum (1969) and Whittaker (1970) illustrate. Our intent in quoting these passages is to identify what is generally accepted by contemporary ecologists,[2] not to assign opinions.

Whittaker (1970: 73):

> In any particular habitat in the landscape, however, the climax community may have been destroyed or may not yet have developed. In this habitat the communities go through progressive development of parallel and interacting changes in environments and communities, a succession. Through the course of succession community production, height, and mass, species-diversity, relative stability, and soil depth and differentiation all tend to increase (though there are exceptions). The end point of succession is a climax community of relatively stable species composition and steady-state function, adapted to its habitat and essentially permanent in its habitat if undisturbed.

Odum (1969: 262):

> Ecological succession may be defined in terms of the following three parameters: (i). It is an orderly process of community development that is reasonably directional and, therefore, predictable. (ii) It results from modification of the physical environment by the community; that is succession is community-controlled even though the physical environment determines the pattern, the rate of change, and often sets limits as to how far development can go. (iii) It culminates in a stabilized ecosystem in which maximum biomass (or high information content) and symbiotic function between organisms are maintained per unit of available energy flow. In a word, the "strategy" of succession as a short-term process is basically

[2] In what follows, we refer to this as the contemporary view of succession. Perhaps it should already be called the ***traditional*** view, because some younger ecologists (e.g., Johnson & Valentine 1971, Krebs 1972) have recently started to question the logical and factual basis of some of its assertions.

> the same as the "strategy" of long-term evolutionary development of the biosphere — namely, increased control of, or homeostasis with, the physical environment in the sense of achieving maximum protection from its perturbations.

Generally agreed upon successional trends are listed in TABLE 1 (after Odum 1969).

REQUIREMENTS OF A GENERALIZED THEORY

It will be apparent from the foregoing survey that a number of different types of phenomena have been drawn together under the term succession. For the purposes of this paper, it is useful to classify successions into three broad categories.

a) Temporal sequences on one site with climate and physiography substantially stable. This category includes most of the classical types of secondary succession and perhaps some primary successions.

b) Temporal sequences on one site with the local environment changing under the influence of extrinsic factors (e.g., climatic change, erosion, deposition, changes in drainage, input of nutrients, etc.). This category (allogenic succession of Tansley) includes many primary successions (especially successions in lakes and ponds) and some secondary successions.

c) Spatial sequences on adjacent sites. These are used to infer temporal changes, at least on the time-scale appropriate to vegetational change. Changes on a longer (geological) time-scale are not now usually included in the successional scheme.

Theories of succession are usually formulated explicitly to describe and explain temporal sequences of type (a). However, as we pointed out at the beginning of this paper, only short-term observations of such changes are available: hence most of the observational basis for the theories is derived from sequences of type (c). Accordingly the primary task of a review is to examine how well observations of type (c) and of short-term changes in type (a) conform to the generalizations of the contemporary theory.

The extent to which observations of type (b) successions are relevant to contemporary theory is less clear. They were explicitly included in the classical theories, as developmental sequences leading to the climatically determined end-point (Clements 1916, 1936). Contemporary theories, however, although recognizing control by the physical environment (see quotation from Odum 1969 above), place primary emphasis on community-controlled changes and tend to treat cases where external factors are dominant (e.g. agricultural management, or eutrophication of lakes) as exceptions to or temporary reversals of the successional process (Odum 1959, 1969). On the other hand one of the most thorough studies of changes in species diversity within developing lakes (Goulden 1969) was interpreted entirely within the framework of the contemporary theory. In this paper we take the view that external factors make at least some contribution to change in all real systems (i.e., that "pure" type (a)

TABLE 1. A tabular model of ecological succession: trends to be expected in the development of ecosystems. (After Odum's *Table 1.* 1969)

	ECOSYSTEM ATTRIBUTES	DEVELOPMENTAL STAGES	MATURE STAGES
		COMMUNITY ENERGETICS	
1.	Gross production/community respiration (P/R ratio)	Greater or less than 1	Approaches 1
2.	Gross production/standing crop biomass (P/B ratio)	High	Low
3.	Biomass supported/unit energy flow (B/E ratio)	Low	High
4.	Net community production (yield)	High	Low
5.	Food chains	Linear, predominantly grazing	Weblike, predominantly detritus
		COMMUNITY STRUCTURE	
6.	Total organic matter	Small	Large
7.	Inorganic nutrients	Extrabiotic	Intrabiotric
8.	Species diversity — variety component	Low	High
9.	Species diversity — equitability component	Low	High
10.	Biochemical diversity	Low	High
11.	Stratification and spatial heterogeneity (pattern diversity)	Poorly organized	Well organized
		LIFE HISTORY	
12.	Niche specialization	Broad	Narrow
13.	Size of organism	Small	Large
14.	Life cycles	Short, simple	Long, complex
		NUTRIENT CYCLING	
15.	Mineral cycles	Open	Closed
16.	Nutrient exchange rate, between organisms and equitability component	Rapid	Slow
17.	Role of detritus in nutrient regeneration	Unimportant	Important
		SELECTION PRESSURE	
18.	Growth form	For rapid growth ("r-selection")	For feedback control ("K-selection")
19.	Production	Quantity	Quality
		OVERALL HOMEOSTASIS	
20.	Internal symbiosis	Undeveloped	Developed
21.	Nutrient conservation	Poor	Good
22.	Stability (resistance to external perturbations)	Poor	Good
23.	Entropy	High	Low
24.	Information	Low	High

successions are an abstraction), and that any acceptable generalization about succession must be able to describe adequately a reasonable number of real systems. However, for a fair test of the contemporary theory, it is necessary to consider primarily cases where the effects of external factors are relatively modest. Extreme cases of type (c) successions will be considered separately at the end of this review.

REVIEW OF AVAILABLE FIELD EVIDENCE

In this section we will examine the Odum-Whittaker criteria by reviewing observational studies of successional systems, and we will show that many of the observed structural or physiological characteristics of the communities do not conform to their generalization. That is, changes in species or life form are not necessarily or consistently accompanied by the other functional changes that are supposed to accompany them. We do not assume that every succession should show all the characteristics ascribed by the ideal sequence. On the other hand, we think that the counter-instances we present are sufficient to show that the Odum-Whittaker formulation of succession is not acceptable as a generalization on the basis of the informal and anecdotal arguments presented to date.

The following general comments may be a desirable introduction to this review.

We have not attempted to cover all the vast literature on vegetational sequences. For each topic under discussion, we have selected the two or three best-documented studies known to us. It is not often realized how few detailed quantitative studies have been made of even short-term changes in vegetation. While there are doubtless important studies that we have not quoted, we have not selectively omitted studies supporting the contemporary theory. Indeed, many of the studies we quote appear in Odum and Whittaker's bibliographies.

Our review is largely restricted to studies of forest succession in areas of temperate climate (including studies of lakes and bogs in forested regions). Succession in other habitats has not been studied in comparable detail. We know of no adequate quantitative evidence for extending contemporary theories of succession to such habitats as grassland, deserts, or tundra.

We have placed emphasis on studies of old-field succession, as this is the best documented example of secondary succession. It might be argued that old fields, having been previously forested, are already "prepared" for the regrowth of the climax forest. However, the contemporary theory of succession is itself based very heavily on studies of old fields (Odum 1969, Whittaker 1970, MacArthur & Connell 1966, etc.), so it is appropriately tested by observations in them.

Although we present evidence opposing contemporary generalizations about succession we do not claim that generalizations about vegetational change are impossible. We ourselves, in concluding, present a very broad generalization as an alternative explanation. More limited generalizations may be possible about certain groups of species or certain floristic regions; comparison of these generalizations may give more biological insight than attempts to develop comprehensive theories.

In evaluating evidence and in our alternative explanation, we have placed some emphasis on natural selection. Natural selection has been neglected in theories of succession because most of them have been primarily concerned with adducing community properties. However, natural selection must act on every species, including "successional" species, to adjust them to their environment, including the communities in which they live. In our view, generalizations about the behavior of communities should be viewed with caution unless they can be reconciled with the action of natural selection on the individual organism.

BIOENERGETICS

STATEMENT OF HYPOTHESIS. Margalef (1968: 30): "Biomass increases during succession as, almost always, does primary production; however, the ratio of primary production to total biomass drops." FIGURE 1 (from

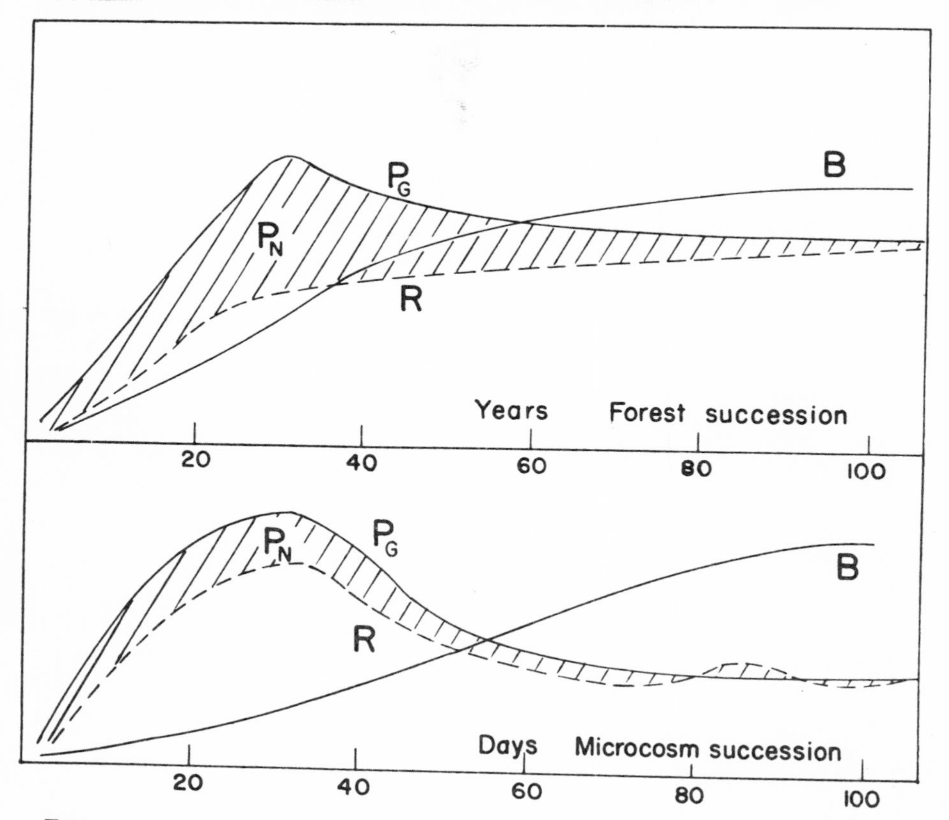

FIGURE 1 (after Odum, 1969). Comparison of the energetics of succession in a forest and a laboratory microcosm. P_G, gross production; P_N, net production; R, total community respiration; B, total biomass.

Odum 1969: 264) illustrates the energetic changes during succession, comparing a forest with a laboratory microcosm. Both production and diversity were described as increasing toward the climax, but both reached a maximum just before the climax stage was reached. They continued at a high, but not maximum, value in the climax. At that stage community respiration is approximately equated to photosynthesis (Bormann 1969:163). Therefore production and diversity should be correlated.

FIELD EVIDENCE. a) Old field successions have been studied in detail as illustrative of the developmental changes involved in more general cases.

Odum (1960) made a detailed study over seven years of changes in species composition and organic production in an old field in Georgia (FIGURE 2, from Odum 1969) about which he wrote the following paragraphs (p. 48):

> If we consider that species composition and species diversity represent "structural" features of the community and that productivity is a "functional" attribute, then it is clear that structurally the "old-field" community changed gradually and continuously, but that functionally a temporary steady-state was established during the period of forb dominance. The study clearly showed that productivity does not necessarily change with change in species, nor necessarily increase with succession as has often been assumed.
>
> The trends so far observed suggest a tentative hypothesis regarding the relationship of energy to succession. From the functional standpoint succession may involve a series of steady-states each associated with a major life form rather than a continuous change associated with species change as is usually postulated. It may be further suggested that an increase in productivity, even if only a temporary "bloom," would be most likely to occur in transition from one life form to another since a new life form may be able to utilize accumulated limiting materials not available to the previous life form.

b) Whittaker (1965:251) cast doubts on the hypothesized correlation between increasing diversity and increasing production in the course of a successional development. In his extensive studies of plant communities, plant production, foliage insects, and birds in the Great Smoky Mountains, he has given measures of diversity (the number of vascular plant species) in quadrats and measured net production.

> Variations in species-diversity do not simply parallel variations in community production. In the Great Smoky Mountains, production and diversity are not significantly correlated either in vegetation samples or in samples of foliage insects. The magnificent redwood forests of the California and Oregon coasts, probably among the most productive of temperate-zone climax forests, have low species diversity.

TABLE 2, plotted in FIGURE 3, was made by combining *Table 1* from Whittaker (1965) with *Table VI* from Whittaker (1966). It supports the conclusion that there is no demonstrated increase in production with increase in diversity.

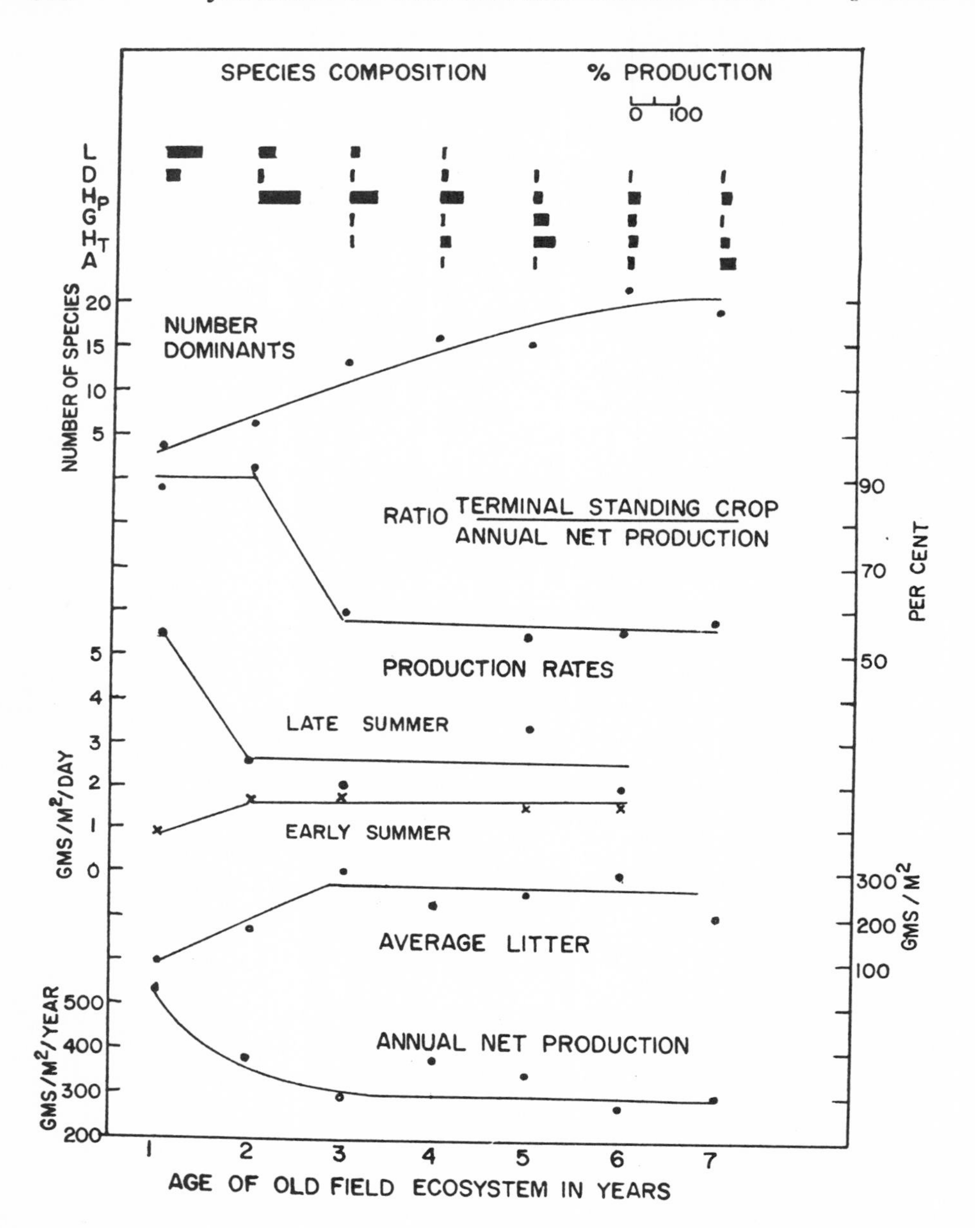

FIGURE 2 (after Odum, 1960). An old-field ecosystem during the first seven years of natural succession, comparing the major structural attributes (species composition and diversity) and functional attributes (productivity). The relative importance in terms of the per cent of total annual production of six major taxa are shown in the top bar diagram. The six taxa are: L — *Leptilon*; D — *Digitaria*; H_P — *Haplopappus*; G — *Gnaphalium obtusifolium*; H_T — *Heterotheca*; A — *Andropogon*. The number of dominants includes all species contributing more than 1 gm/m²/year to annual production. The relative stability of productivity features contrasts with marked changes in species composition and increase in species diversity.

Table 2. Species diversity and above ground net production, Great Smoky Mountains. (After Whittaker, 1965, 1966)

Sample No.		Diversity *	Production †
		Heath Balds	
9	High mixed heath	8	983
		Forest Heaths	
10	Pine forest, Cade's Cove	27	875
11	Pine forest, Pittman Center	32	991
12	Pine heath, Brushy Mountain	23	578
13	Pine heath, Greenbrier Pinnacle	20	419
14	Chestnut oak heath	23	539
15	Hemlock-beech cove forest	40	1333
16	Hemlock-rhododendron forest	5	1022
17	Spruce-rhododendron forest	7	812
		Forests	
18	Deciduous cove forest	43	1221
19	Cove forest transition	44	1911
20	Oak-hickory forest	26	1203
21	Chestnut oak forest	27	1465
22	Successional tulip forest	35	2408
23	Upper deciduous cove forest	39	1097
24	Hemlock mixed cove forest	31	1183
25	Gray beech forest	29	668
26	Gray beech forest	21	906
27	Northern red oak forest	41	828
28	Red oak, white oak forest	32	568
29	Spruce-fir forest	17	1024
30	Spruce-fir forest	14	944
31	Spruce-fir forest	14	1402
32	Fraser fir forest	14	566
33	Fraser fir forest	6	653

* Column 1 — Diversity indicates the number of species in quadrat, from *Table 1* in Whittaker 1965.

† Column 2 — Production equals total net above ground production, from *Table 6* in Whittaker 1966.

Soil Development

Statement of hypothesis. Whittaker (1970:69) stated: "A number of trends or progressive developments underlie most successional processes. There is usually progressive development of the soil, with increasing depth, increasing organic content, and increasing differentiation of layers or horizons toward the mature soil of the final community. . . ."

Two points in this statement call for examination. First, the length of time involved in development, and second, the differentiation of soils into layers as they mature. Central to the argument is the idea that successive generations of plants leave their remains which are incorporated

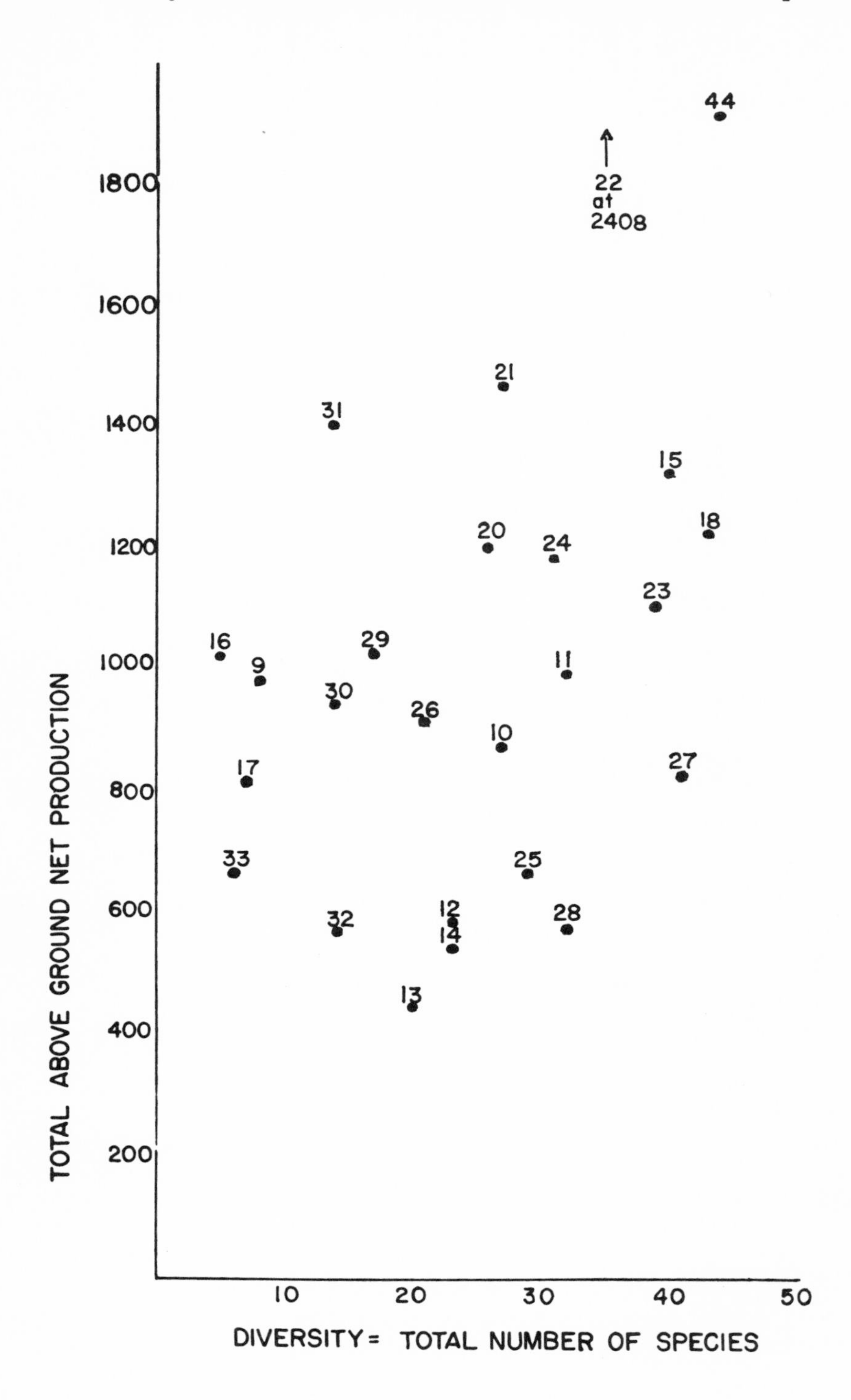
44
22
at
2408
1800
1600
21
31
1400
15
18
20
24
1200
23
16
29
9
11
1000
30
26
10
27
17
800
33
25
12
28
600
32
14
13
400
200
10
20
30
40
50
TOTAL ABOVE GROUND NET PRODUCTION
DIVERSITY = TOTAL NUMBER OF SPECIES

into the mineral soil and humified while nutrients and minerals are leached and deposited at specific depths in the process of soil formation.

a) The time scale within which progressive soil development takes place has seldom been specified explicitly. But it is clearly implied that development of a mature soil profile is coterminous with succession. and is therefore a process which approaches completion and equilibrium in the climax. Because successions are generally believed to take a very long time (Clements 1916, Braun 1950) it is clearly implied that soil development requires a very long time for its consummation. "Succession in general appears to be very slow, and the vegetational history of any particular habitat is more apt to be measured in thousands of years. or even tens of thousands, than in centuries or decades" (Gleason 1927:301). "Cowles and others cherished the hope that the old black oak dunes might *eventually* succeed to typical mesophytic forest even though they had not done so in all of post-glacial time — that it was merely a question of too slow a rate and too little time" (Olson 1958:155). Cowles (1901:177) blamed "the slowness of humus accumulation . . . . perhaps more than all else."

b) Soil development involves increasing differentiation of soil horizons, so that processes which stir up and homogenize the soil into a uniform layer must be considered to reverse the process of soil development and to run counter to successional development.

FIELD EVIDENCE. a) Time scale of soil development. Olson (1958) reexamined chronological details of soil development in the vegetation of the Indiana sand dunes studied a half-century earlier by Cowles.

The ages of the dunes and hence of the period available for development of soils upon them were derived from ages of trees or dates of subsequently buried fences on younger dunes and from radiocarbon dating of older ridges and raised shore lines. Age estimates are of course subject to error (up to 20%) but as Olson said: "Present Great Lakes chronology is far superior to any available only a decade ago and covers a much longer period than most previous quantitative studies of succession."

Our FIGURE 4 (Olson 1958, *Fig. 19*) graphs cation exchange relations as function of time age in the process of soil development. "The cation exchange capacity of dune soils is another important variable that increases along with humus content. The exchange complex is initially saturated with calcium and other metallic ions that are valuable in plant nutrition, but gradually these basic ions are partially replaced by hydrogen ions (exchangeable acidity) as soil development proceeds under the influence of vegetation" (Olson 1958:160). FIGURE 4 shows rapid changes

---

FIGURE 3 (after Whittaker 1965, 1966). Community diversity is compared to community productivity in the Great Smoky Mountains vegetation. Diversity (the number of vascular plant species) in quadrat samples numbered 9–33 in *Table I* of Whittaker (1965) is compared to productivity (total above ground net production) in the same numbered quadrat samples taken from *Table VI* of Whittaker (1966).

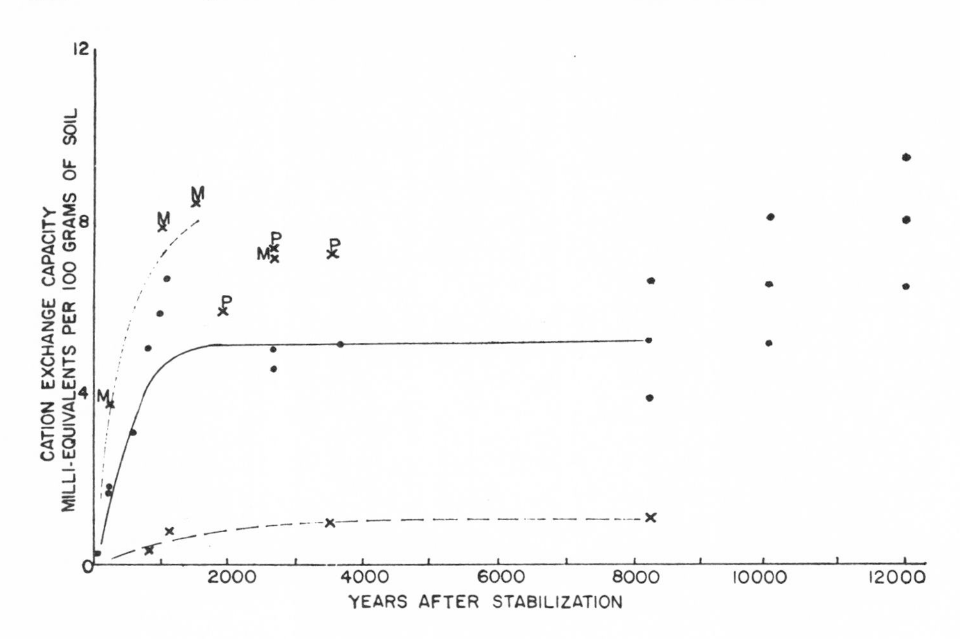

FIGURE 4. Cation exchange capacity (soil nitrogen or humus content) as a function of dune age (Olson 1958). "Cation exchange capacity of dune soils . . . increases along with humus content. . . . The exchange capacity of both surface soils and lower horizons increases in proportion to soil nitrogen. Heavy line for 0–1 dm. layer, main black oak dune series. Closely spaced light dashed line is a possible alternative trend for mesic soils. Open dashed line, 1–4 dm. layer. Samples for three areas with more than the usual prairie cover are marked by "P." Samples from mesophytic pockets or lee slopes are marked by "M." These pockets show basswood, red oak, or sugar maple which were absent on more "normal" topographic situations. The two oldest dune systems (Glenwood and Calumet) do not seem to fit the ideal scheme: instead of scattering around the same trend, thin levels of nitrogen and other properties related to humus content seem to be higher. . . . One may recall the higher silt and clay contents of these soils and might suspect that the resulting textural and moisture differences would help maintain higher nitrogen levels here." (From Olson 1958.)

in the first 1000 years, followed by many thousands of years without further development.

Olson in his summary (p. 168) concluded that:

> 3. Soil analyses of carbon, nitrogen, moisture equivalent, carbonates, acidity, and cation exchange relations show how most soil improvement of the original barren dune sand occurs within about a thousand years after stabilization. The pattern of change on older dunes promises little further improvement and perhaps even deterioration of fertility.

> 4. Low fertility favors vegetation with low nutrient requirements. But such vegetation probably is relatively ineffective in returning nutrient to the dune surface in its litter and thus aggravates low fertility. Leaching of nutrients out of the sand-dune ecosystem and the low moisture reserve of

most dune surfaces help account for the poor prospects for successional replacement of the black oak-blueberry community by the more exacting species of the mesophytic forest. . . .

Olson's studies indicated that soil development has been rapid in terms of the ages of the present dunes. Soils on a relatively young dune surface have developed to a steady state, but those on surfaces many times older have not "progressed." The growth of trees appeared to be in response to physical stabilization of the dune-hollow rather than soil preparation.

b) Stratification in soil development. Goodlett (1954) studied the distribution of forest tree species related to the depths and structure of soil. He found wind-throw to be an active force in determining the structure of soils in the deciduous forests of Pennsylvania:

> The unplowed areas of Potter County are characterized by pronounced microrelief of mounds and pits. . . . Most of the mounds are old, and have resulted from the fall of trees that grew in the presettlement forest. . . . The evidence indicates that wind-throw has disturbed the soil materials to a depth of 2 to 3 feet throughout the past 200 to 300 years, and probably since the reoccupation of the region by forest trees. Lutz and Griswold (1939) made a study of the influence of tree wind-throw in southern New Hampshire, and suggested that "all soils which bear, or in the past have borne, forest stands have been more or less disturbed."
> Disturbance of the upper parts of the surficial deposits by wind-throw over long periods of time probably produces the high degree of variability observed in soil profiles over short horizontal distances. Existing soil profiles are disturbed, inverted, or destroyed within the depth of upheaval. . . .

Lutz (1940) described the effect of uprooting trees upon the structure of the soil in southern New Hampshire, a forest vegetation which would be generally accepted as climax:

> . . . as a result of disturbance, horizons may be very irregular, occasionally with long tongues forming the upper layers penetrating deeply into the layers below. Further, horizons may be discontinuous and masses of soil material may be translocated to positions above or below those normally occupied. Frequently, material from upper and lower horizons is rather intimately mixed. The vertical and horizontal movements of rocks two or more feet in diameter is evidence of the tremendous forces involved. . . .

Farmers plowing the soil have leveled the mound-and-pit topography characteristic of the forest floor in old-growth stands. Because most of the soils of the northeastern United States have been plowed, and most studies of forests have until recently been made in that region, the real condition of the soils of the pre-Colonial "climax" forests has not been examined.

These studies of soil development and modification in early and late forest successional stages indicate that soils develop rapidly in response to the vegetation types on the site. No evidence indicated that development of a profile is associated with the replacement of the vegetation type responsible by another one. In the northeastern forest region there is evidence that the mature old-growth forest grew on soils which were immature by theoretical standards.

SEQUENCE OF VEGETATION TYPES

STATEMENT OF HYPOTHESIS. Whittaker (1970) describes the process: "One dominant species modified the soil and microclimate in ways that made possible the entry of a second species, which became dominant and modified [its] environment in ways that suppressed the first and made possible the entry of a third dominant, which in turn altered its environment."

In order to understand why species compositions change, we need to know if species associations replace each other as groups; whether later communities replace early stages, or if many of the final and climax species are present on the site from the start. Also, do the species of the early stages facilitate or inhibit the establishment of the species of later stages? If the early stages are removed, is the appearance of later communities delayed or accelerated? And finally, do successions always go in one direction? Do lateral and retrogressive successions occur and under what conditions?

The question — Are the early successional stages made up exclusively of species of lower life form or is it possible that most species are present as seedlings from the start and the visible successional sequences are the overt expression of sequential conspicuousness reflecting different growth rates and sizes at maturity?

In a review of old field succession Egler (1954) presented two opposed hypotheses. In his "relay floristics" one floristic group relays the site to another until some relatively stable stage is reached. This sequence would illustrate simple or classical succession.

In his "initial floristic composition," "Up to the year of abandonment, the land is receiving many species as seeds and living roots. . . . After abandonment, development unfolds from this initial flora, without additional increments by further invasion (for the purpose of this discussion)."

Typological thinking (Mayr 1963) as applied to plant associations could make this contrast too black and white, but if a moderate number of the species of later stages should prove to be present as small individuals in the early successional stages, the observation would throw doubt on the purported importance of the early successional stages in modifying the environment.

FIELD EVIDENCE from several studies is cited in the paragraphs which follow.

a) Clements (1916:59, 1928:103) observed that cases of "initial floristic composition" occur:

> Secondary areas such as burns, fallow fields, drained areas, etc., contain a large number of germules often representing several successive stages. In

some cases it seems that the seeds and fruits for the dominants of all stages, including the climax, are present at the time of initiation.

b) Niering and Egler (1955: 359), referring to plant succession along power lines in Connecticut, said:

> It has been found that the majority of woody plants of the so-called later "stages" and old field "succession" have not been currently invading. To the contrary, it appears that they had invaded at the same time as or previous to the earliest stages.

c) Hack and Goodlett (1960) studying a forested valley in Virginia damaged by a flood, reported:

> Sycamore and black locust trees rapidly covered the scarred areas on the valley bottoms, but six years after the flood most of the tree species characteristic of the undamaged mature forest were present.

Elsewhere these same authors reported that the tree seedlings first colonizing the landslides on the slopes were identical to the species of trees found on older, pre-Columbian forested landslides. They said:

> Many of the landslides terminated in large piles of rock, soil and uprooted trees. Old debris fans, mantled by mature trees of large diameter, indicate that similar floods have occurred in this area in the recent past, creating slides similar to those formed by the 1949 cloud-burst. The vegetation of these areas of older slides shows an adjustment of composition to topographic form.

d) Marquis (1967) reported on the vegetation composition on a site after clear cutting of an old growth stand of northern hardwoods. See TABLE 3 (from Marquis 1967). Tree species are identified as shade tolerant (i.e. climax), intermediate, and intolerant (i.e. sun requiring or successional). He said (p. 6):

> Five years after cutting, four tolerant species comprised over half of the total numbers of stems. This proportion has increased gradually as the stand matured so that tolerant species account for 63 per cent of the stems at 30 years of age. . . . If left to mature naturally, this stand will become increasingly similar to an uneven-aged stand. . . . Red maple and white ash should survive past 100 years, but eventually they too will mature and the stand will then be dominated by the original, long-lived climax species — beech and sugar maple with small amounts of yellow birch. . . .

Detailed studies are needed to establish at what stages "climax" species colonize a site. For example, for a critical resolution of this in Marquis's (1967) study, it would be necessary to have marked trees in the understory at the time of clear cutting. It would also of course be necessary to eliminate from consideration stump sprouts and coppice which came up after cutting. Nevertheless, this and the previous studies suggest that individual trees which make up the "climax" forest were present from the start. Hence the studies throw doubt upon the functional aspects of alteration of the environment by successional species.

TABLE 3. **Species composition in northern hardwoods stand.** (From Marquis 1967)

	At time of cutting		5 years after cutting	30 years after cutting
	Old-growth stand	Sapling understory		
	Trees 4.6 inches d.b.h. and larger	Trees 1 foot high to 1.5 inches d.b.h.	Trees 1 foot high or higher	Trees 0.6 inches d.b.h. or larger
	Percent of basal area	Percent of stems per acre	Percent of stems per acre	Percent of stems per acre
TOLERANT				
Beech	61	26	19	30
Sugar maple	15	56	29	29
Striped maple		16	4	2
Conifers		1	<.5	2
INTERMEDIATE				
Yellow birch	15	1	8	12
White ash	tr.	<.5	1	2
Red maple	tr.	—	1	4
INTOLERANT				
Paper birch	tr.	—	10	11
Aspen		<.5	3	2

Another question: Do the species of the early stages facilitate or inhibit the establishment of species of later stages?

A phenomenon called reaction, "the effect which a plant or a community exerts upon its habitat" (Clements 1916) is generally acknowledged to be central to autogenic replacement of communities. Clements (1916:79) said: "Reaction is thus the keynote to all succession, for it furnishes the explanation of the orderly progression by stages and the increasing stabilization which produces a final climax."

FIELD EVIDENCE from several studies is cited here.

a) McCormick (1968:9) said:

> There is no question that reactions do occur. A single plant, whether lichen or redwood, casts shade, changes the pattern of air movements, produces organic material that may become humus, and affects the site in other ways. The importance of such effects in succession, however, has not been tested adequately and certainly the generalization that these effects constitute the driving force of succession is not substantiated.

He then went on to examine the effects at an early or pioneer stage of succession on an old field and reports:

> In a series of experiments initiated at the Waterloo Mills Research Station, Devon, Pennsylvania, in 1965 and repeated in 1966 and 1967, annual plants were removed as seedlings from some sections of a recently plowed field, but were allowed to grow elsewhere. According to reaction theory, an annual vegetation is necessary to "prepare the way" for perennial plants on such a site. By the end of each summer, however, perennial plants were several times as abundant on areas kept free of annuals. The biomass (dry weights) of individual perennial plants on the annual-free areas were many (15 to 82) times as great as those on areas with annuals. Many goldenrods, asters, black-eyed susans, and other perennial plants flowered on annual-free sections, but were sterile on plots covered with annuals (McCormick *et al.*, Mss.). This experiment does not refute the general theory of the reaction mechanism. However, it does seriously question the reality of the theory and indicates that the theory was not valid for the early old field situation in which it was tested.

b) Pound and Egler (1953) found that on a fire line cleared and spike-harrowed in 1934–1936 ". . . the accidental absence of suitable conditions for germination and development of tree seeds and the absence of root-suckering trees allowed the formation of a dense low vegetation that has since prevented the mass invasion of trees."

c) Niering and Egler (1955) documented the exclusion of trees of the "next successional" stage by a closed canopy of the shrub nannyberry over a period of 25 years. They suggested that minor silvicultural steps could lead to more or less indefinite exclusion of trees by the closed shrub canopy once it is established.

> The *Viburnum* thicket has apparently been in existence for a quarter of a century as a self-perpetuating shrub type. Some of the tallest, oldest *Vi-*

burnum stems are dead or dying, and will be replaced by younger stems, with no change in the overall height or appearance of the community. . . . The transgressives, barring accidents, would presumably grow to maturity and change at least a portion of the area into forest. . . . It seems reasonably certain, however, that the *Viburnum* thicket is not "permanent," and that eventually, in 30, 60 or 90 years, an occasional tree will break through and convert the area to forest. On the other hand, the number of such maturing trees of the past 25 years has been so small that economic control of them, as by artificial means, would appear to be an extremely low-cost procedure requiring attention only at intervals much greater than a decade. . . .

In most instances, it has been found that shrub-dominated vegetation tends to resist invasion by tree seedlings, whereas upland eastern grasslands are relatively open to invasion by pines, ashes, maples, elms, birches and tulip poplar.

d) Both "early" and "late" stage trees may be suppressed if overtopped. In the same way that red cedar and gray birch can be overtopped and suppressed in oak-hickory forests of southeastern New England, so saplings of red oak, white oak, elm, white ash, sugar maple, red maple, pignut hickory, basswood, and beech can be found in the understory of tall forests of old field pine in central New England (Raup & Carlson 1941, Lutz & Cline 1947). While the pines are on average 90 to 100 years old (aged by counts of annual rings on increment borer samples), the hardwoods, though slender, are on average 50 to 70 years old. With the cutting of the overtopping pine, the suppressed hardwoods flourish.

e) J. A. Kadlec (in litt.) has observed in experimental game management areas in Michigan that removal of "pioneer" trees such as aspens speeded the appearance of species of later successional stages.

f) Olson (1958) observed that on the sand dunes he studied the first tree species to be established inhibited the successful colonization of the site by competitors. A closed stand of black oak-blueberry tended to regenerate itself instead of being replaced by the beech-maple "climax," because of the acid leaf litter these species produce and the low rate of return of nutrients to the soil.

g) The production by plants, through the roots or in the litter, of chemicals which inhibit the germination or growth of plants of their own or other species has received wide attention as a physiological phenomenon for many years (review by Whittaker & Feeny 1971).

That plants should release products which facilitate the growth of their own species and inhibit the growth of other species makes better sense according to natural selection than that they should alter their habitat in such a way as to facilitate the growth of other competing species. Several publications indicate that some "early successional" species do produce chemicals which inhibit the growth of later stage species.

Rice (1964) showed inhibition of nitrogen fixing and nitrifying bacteria by old field grasses which apparently gave them competitive advantage over grass of later successional stages. Brown and Roti (1963) found that perennial goldenrods produce a chemical which inhibited the germination of seeds of jack pine, a pioneer tree on old fields in Michigan. Brian (1949) and Harley (1952) showed that the shrub, ling, produced substances which inhibit mycorrhizal fungi — and, therefore, trees which require them.

An additional question: Are successions unidirectional?

Pond margins, marshes, and bogs have been generally cited as illustrations of typical successions (Smith 1966:182–191, BSCS "Green Version" 1963). The classical illustration of bog succession follows the sequence of (a) floating aquatics, (b) sedge-sphagnum floating bog which is followed by progressive compaction and humification, (c) succession of sphagnum species to produce a heath peat bog, and (d) (a hypothetical extension) a bog forest covering a peat-filled lowland.

FIELD EVIDENCE from several studies is cited here.

a) Several detailed studies of large bogs in the coniferous forest regions of North America and Europe have failed to find this standard sequence and led their authors to doubt even that it necessarily occurs. Drury (1956) in Alaska, and Sjörs (1948, 1950, 1955) in Sweden and the Hudson's Bay lowland drew conclusions similar to those of Heinselman in the region of Glacial Lake Agassiz of Minnesota (1963:370):

> Neither the process of bog expansion nor the patterned bogs and fens of the Lake Agassiz region fit the classical picture of succession in the Lake States. Conclusions are that: (1) Few bogs in this region are the result of a single successional sequence. (2) The bog types cannot be regarded as stages in an orderly development toward mesophytism. (3) Raising of bog surfaces by peat accumulation does not necessarily mean progression toward mesophytism. Such rises often cause concurrent rises of the water table and promote site deterioration. (4) The climax concept does not contribute to understanding bog history in this region.

b) Johnson and Raup (1947) found that the peat of a peat island in a salt marsh on the Taunton River in Massachusetts gave no indication of succession of marsh grasses. The peat indicated continuous stream cutting and redeposition and most plant remains were of one type, salt-water cord-grass. Progression of vegetation types in a successional sequence was not found.

c) Walker (1970) summarized his study of the courses and rates of British post-glacial hydroseres using peat and mud stratigraphy, amplified by pollen diagrams by saying:

> . . . Although certain sequences of transitions are 'preferred' in certain site types, variety is the keynote of the hydroseral succession. In spite of this, the data clearly indicated that bog is the natural 'climax' of autogenic hydroseres throughout the British Isles and the transition from fen to oakwood is unsubstantiated.

The term bog as referred to in this quotation is distinguished by a variety of *Sphagnum* species and acid tolerant plants such as Dwarf Cranberry, Sweet Gale, Cotton Grass, and Heath.

To summarize this section on the sequence of vegetation types, those cases in which detailed studies of forest succession have been made do not indicate an orderly replacement of early successional species by members of later communities. They indicate instead that in the usual case almost all species have established before the "succession starts" or that they colonize during the first few years. Hence in many cases the manifest "succession" is a sequential expression of conspicuousness which reflects the maturing of species having different speeds of growth and sizes at maturity.

In fact it appears in those cases where critical examination has been made, that early and middle successional stages suppress climax species and delay the expression of later stages. Detailed studies of events in the hydrosere indicate that successional trends are readily deflected or reversed in response to changes in water level which are externally controlled. In one case at least early successional stages (mosses) invade a forest and stimulate a rise in water table — hence reverse succession by autogenic processes. Their presence leads to replacement of the forest by heath-birch stage (Sjörs 1948, Drury 1956, Heinselman 1963).

DIVERSITY, STABILITY, AND REPRODUCTIVE STRATEGIES

STATEMENT OF HYPOTHESIS from Margalef (1968:31):

> Diversity very often increases. Sometimes diversity increases to a certain value and then decreases again toward the final stage of succession. . . . Fluctuations are damped and rhythms change from reactions directly induced by external agents to indirect responses to stimuli associated with ecologically significant factors; the ultimate trend is to endogenous rhythms. . . .
> In the later states of succession, a relative constancy of numbers is achieved, and populations are not forced to reconstruct themselves rapidly after drastic and extensive destruction. The natural trend is toward a reduction in the number of offspring produced and better protection for the young.

First: DIVERSITY.

STATEMENT OF HYPOTHESIS: Diversity in most contemporary usage refers simply to the total number of species present. The term has in the past, however, referred to all ill-defined, more complex (and more useful) concepts of community diversity including structural variety and relative frequencies of species in addition to taxonomic variety. There have been such "diverse" uses of the term that Hurlbert (1971) said: "The recent literature on species diversity contains many semantic, conceptual, and technical problems. It is suggested that, as a result of these problems species diversity has become a meaningless concept. . . ." It is used here in its simplest sense — the number of species present.

In order to clarify the ecological significance of the increase in species diversity in the course of "successional" sequences, we need to establish which of the following statements is valid:

a) The successional sequence is the result of a process of replacement of less successful by more successful communities in which the more diverse communities are more successful.

b) The increase of species diversity is a reflection of later emergence into prominence of species which were present at the start as inconspicuous seeds and seedlings.

c) The process of succession combines (b) with progressive species enrichment (independent of modification) as a result of continuing colonization of the site. Then, species will be added first which are numerous, close at hand, and/or have highly developed means of dispersal. With the passing of time, less numerous species — those occurring at a greater distance and/or those having less effective means of dispersal — will arrive "by chance."

d) Diversity is a result of microtopographic and other special influences such as herbivore predation, and thus only partially a result of "intrinsic" processes.

e) "Diversity" is uniformly expressed in all parts of a community. Diversity in the herb layer is correlated with diversity in the shrub and tree layers, and with diversity in the insect and bird species.

Although (a) conforms to the classical descriptions of succession, contemporary ecologists consider the process of succession to result primarily from species by species replacements. Evidence of community by community replacements is inadequate.

Evidence for (b), that some successional changes result from sequential growth rates of the members of a largely complete initial floristic composition, was presented in the first part of this section. Very detailed studies which are not available at present are necessary to confirm that this is the case. Similarly, detailed studies would be necessary to eliminate the probability that (c) is a major ancillary process in most successional sequences.

Evidence that (d) plays a decisive part and that there are inconsistencies with regard to (e) would suggest that diversity is an ecological phenomenon independent of succession.

FIELD EVIDENCE from several studies is cited here.

a) Goulden (1969) discussed temporal changes in diversity correlated with what he called development of species associations in fresh water lakes. He based his conclusions on counts of the numbers of species found in cores of lake sediments, including the earliest stages of the lake's history. Interpreting his data from one lake, he described the sequence of events as follows:

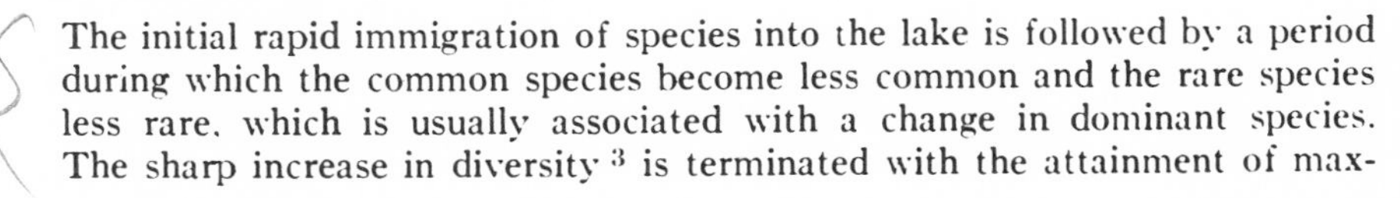
The initial rapid immigration of species into the lake is followed by a period during which the common species become less common and the rare species less rare, which is usually associated with a change in dominant species. The sharp increase in diversity [3] is terminated with the attainment of max-

[3] In this passage, Goulden used the term "diversity" to refer to the information-theoretic measure of weighted species abundance.

> imum diversity for a given number of species. A third phase follows in which additional rare species are established. . . .

The pattern Goulden reports is consistent with ideas presented by Darwin (1859) and Wallace (1876, 1880) and discussed in general by Darlington (1957) and in detail by MacArthur and Wilson (1967). Gleason (1917:474) emphasized that the arrival of additional species simply by migration is a major aspect of "successional" change. At first, common, widespread and well dispersed species rapidly appear. These are followed by less common, well dispersed or widespread species, which have different competitive strategies such as larger size or longer life span. With their arrival we can expect a change in dominance. Last, the widely dispersed, infrequent "rarities" appear at increasing time intervals.

b) Hack and Goodlett (1960) found in the floodplains of main valleys (i.e. higher order valleys) of the Little River in northern Virginia, a northern hardwood forest enriched by many species "extremely rare or absent in the northern hardwood forests of the hollows. . . . The number of non-tree species is generally larger than in the northern hardwood stands of first order valleys." In other words the floodplain, subject to repeated ravages and destruction by high water, and therefore the youngest, most unstable surface, supports the most diverse forests. The hollows, which for geomorphic reasons are subject to a higher rate of creep than most slopes, also support a relatively diverse northern hardwood forest, and the noses, the most stable surfaces of the area, support forests of the lowest diversity.

c) Observed diversity may indeed reflect external circumstances more than the actual number of species present, as has been shown by some experiments carried out in old fields in an originally forested area of Britain (Harper 1969:49):

> Generalizing from the controlled experiments and the results of post Myxomatosis surveys, the effects of rabbit removal appear to be (i) the release and demonstration of hitherto unsuspected floristic richesse in the rabbit-grazed communities: species which in the past had been regularly nibbled and suppressed, flowered and showed their true identity. . . .

When protection from grazing was continued over a long period, in controlled experiments using fertilizers and fenced plots on native hill pastures in central Wales: "The effect of this long period of freedom from grazing in each year was to reduce the number of species present, and the normal result was virtually exclusive development of one or two species of grass. . . ." These results show that "unrestricted stock access was a conservative force on the vegetation and only controlled (heavy) but intermittent grazing allowed the richer flora to develop." They further show that complete removal of grazing was also a conservative force in some cases, while in other cases there was a "later return of a new diversity associated with the entry of species belonging to phases of succession towards scrub. . . ." (Milton 1940, 1947).

According to these field studies, the number of species in a plant association may reflect primarily special circumstances of soil, microtopography, disturbance, or predation by grazing mammals (and release from grazing) which allow certain species a local differential advantage and which suppress the tendency of some species to become dominant and exclude other species.

d) Whittaker (1965:251, 252), referring to his own field experience, said:

> Diversity is as high in the disturbed, immature forests as in mature stable forests of corresponding environments. Tree-stratum diversities increase from high elevations toward low. Diversities of herb and shrub species show no clear relation to elevation below 1400 meters in the Smokies. In both the Siskiyou and Santa Catalina Mountains herb diversities increase from low to middle elevations and decrease from these to high elevations. Along the moisture gradient in the Smokies, the highest tree diversities are in intermediate sites, the highest shrub diversities are in intermediate-to-dry sites, and the highest herb densities are in moist sites. Diversities of the three strata are scarcely correlated, and numbers of insect and bird species are not simply correlated with numbers of plant species.

Hence diversity is inconsistently expressed and increased diversity is by no means an exclusive property of the later stages of vegetation succession.

Second: STABILITY.

STATEMENT OF HYPOTHESIS. No generally accepted definition of stability in ecological systems is available (Brookhaven Symposia in Biology No. 22, 1969). It is not clear whether the term means dampened fluctuations of numbers or long survival until extinction. We know of no comparative study which has confirmed the generally accepted idea that all or the majority of species populations fluctuate less in areas of greater ecological diversity. It is in fact clear that some animal species vary greatly in numbers (squirrels, rabbits, mice, chickadees, foxes) in some complex habitats, while others in the same habitats do not. On the other hand, not all species in the "simple ecosystems" in the north show the periodic spectacular changes in numbers which successional theory predicts.

Third: REPRODUCTIVE STRATEGIES.

STATEMENT OF HYPOTHESIS. Wynne-Edwards (1959, 1962) argued in general that reproductive rates are adjusted to mortality rates in order to avoid overexploitation of food supplies and consequent crashes. MacArthur and Connell (1966) rephrased the argument and applied it specifically to climax habitats. Adjustment of reproductive rate to mortality rate also appears inherent in the concept of "K-selection," at least in its original formulation for closed populations on islands (MacArthur & Wilson 1968:149).

FIELD EVIDENCE.

Adjusted reproduction has been demonstrated only in specially formulated genetic models and there is no evidence that they correspond to natural populations. It seems extremely improbable in plants, which produce vast quantities of seed and in any case do not overexploit their "food supply" or suffer crashes.

In animals, overproduction has been demonstrated even in a climax species by Kluyver and Tinbergen (1953); Kluyver (1966) demonstrated by experiment in the same species that mortality was adjusted to reproduction and not vice versa. However, the subject is still debated (Lack 1966, Skutch 1967). It is indeed likely that animals raise more young in areas where net production is very high (Crook 1965), but this can be understood as a response to variable food supplies within the context of r-selection (Lack 1966, 1968).

ALLOGENIC VS. AUTOGENIC — SAND DUNES, PONDS AND MARSHES, FLOOD PLAINS, AND MORAINES

STATEMENT OF HYPOTHESIS. One of Odum's (1969) hypotheses about succession is that the process is community controlled. Olson (1958:132) put it that change of an ecosystem could proceed as a function of time *even if all other variables remained constant.* If, however, vegetation changes occur in response to site changes produced by extrinsic events, then the "sequences" are not developmental but simply reflect the ecological truism that different species require different habitats.

FIELD EVIDENCE from several studies is cited here.

a) Excavations in Alaskan bogs (Drury 1956) indicated that the "late successional" bog forests were simply perched upon a thick peat composed of species of emergent mosses mixed with alluvial and aeolian silt. Where the water table had been locally lowered by deposition of silt, forest had grown up. Where the water table had risen, the forest had been replaced by bog. Other studies of bogs (Sjörs 1955, Heinselman 1963) also indicate that the observed changes are not developmental. Johnson and Raup's (1947) excavations at Grassy Island gave no indication of succession in the sediment record, but instead showed evidence of a continuing process of stream erosion and deposition, with plants colonizing newly deposited banks and being destroyed as the river's meanders migrated across the marsh.

b) Moraines abandoned below mountain glaciers have been used to illustrate primary succession from bare soil through tundra and willow-alder shrubs to needle-leaved forest. Cooper's work in Glacier Bay (1923) and Sigafoos and Hendricks's on Mt. Rainier (1969) suggest on the contrary that moraine vegetation is generally at equilibrium with the present conditions. In fact, where local conditions are suitable, vegetation of an "advanced successional stage" grows on or next to the ice. In 1948 and 1950 Drury found a spruce forest 50–60 feet tall growing on the moraine, underlain by ice, of the stagnant Matanuska Glacier in Alaska. In Iceland at the edge of Vatnajökull Glacier, birch grows next to the ice and on sites covered by ice as recently as 1934 (Lindroth 1970).

In these extreme habitats, a site's edaphic and microclimate conditions differ according to topography and geomorphic changes beyond the control of (although influenced by) the vegetation. At each site various species whose propagules are already available or are constantly reaching the site are able to become established. Hence each vegetation type or zone appears to be in equilibrium with the present environment, in the same way as the several zones of vegetation on a mountainside.

SKETCH OF AN ALTERNATIVE EXPLANATION

The statements of Odum, Whittaker and Margalef quoted in this paper (including TABLE 1) represent a set of generalizations about succession as now understood. Steps toward a theory of succession have been made by Margalef (1963, 1968), who related succession to an increase in information content, by MacArthur and Connell (1966) in their revival of Clements's (1916, 1936) analogy between succession and embryonic development, and by Odum (1969) in his statement that succession is community-controlled. The common feature of these theories is that they attribute succession to properties of the community.

In this paper we have quoted a number of detailed studies which disagree with the Odum-Whittaker generalization. Whittaker (1953) and McCormick (1968) have previously mentioned a number of similar exceptions. Hence, although there may be other studies which are consistent with it, it is unacceptable as a generalization. Nevertheless, there is a body of interrelated facts to be explained. In the remainder of this paper we will sketch in broad outline an alternative explanation of these facts as we see them.

POINT OF DEPARTURE. The starting-point for such an explanation is three sets of observations on succession in forested regions [4] which we believe to be widely accepted.

1) On sites which are cleared but initially free of seeds, certain kinds of plant species (usually those of depressed life forms) tend to appear first; others (including those of successively larger and more complex life forms) appear later. Sites illustrating this sequence ("primary succession") are rare in nature: examples include exposed sandbanks, volcanic deposits, receding glaciers and mine tailings. Although many textbooks give hypothetical accounts of such successions, very few cases have been actually observed and only the early stages in succession have been described. Evidence about later stages in primary succession is conflicting (e.g., Olson 1958) or lacking.

2) On sites which are cleared but not initially free of seeds, there is usually a similar vegetational sequence, often apparently involving the same kinds of plants, but usually taking place more rapidly ("secondary

[4] In our opinion there are as yet insufficient precise observations on successional sequences in non-forested ecosystems to support any generalizations about them, or to incorporate them into a comprehensive theory.

succession"). Again most published studies are of the early stages in succession and accounts of the later stages are largely hypothetical. The evidence summarized in this paper suggests that the later stages of succession are not consistently unidirectional, and that forest succession is better represented by the lower half of FIGURE 1 than by the upper half.

3) Spatial gradients of several kinds of stress (climatic, chemical or radiological) give rise to similar sequences in space of life forms and vegetation. The congruence between sequences produced by different kinds of stress is reasonably well documented (see review by Woodwell 1970), although few detailed descriptions are available for any but climatic stresses.

The main task in constructing a theory of forest succession is to explain the apparent congruence between the stress-induced sequences and the early and middle stages of primary and secondary successions.

STRESS GRADIENTS. Vegetational sequences on stress gradients are reasonably well understood (Woodwell 1970). Each species appears to be specialized to a certain type of site and is competitively superior within a limited range of conditions, so that different species are dominant at different points along environmental gradients (MacIntosh 1967, Whittaker 1953). Plants of higher life form dominate under low stress conditions, but are unable to withstand higher stresses. The contrasting selection pressures have been considered by Raunkiaer (1934), who classified plants into life forms according to the location of their protected growing points. Those plants, geophytes, whose buds are well protected in the ground must grow their photosynthetic structures each year. They are readily overtopped. Tall, dominant plants, phanerophytes, have competitive advantages but their buds are vulnerable to drying winds, ice, fire, and radiation. Chamaephytes, whose buds are within a couple of meters of the ground, are intermediate in adaptations.

Woodwell (1970) emphasized that plants of higher life form usually have a larger ratio of supportive structures to productive structures: hence they have a smaller margin of reserve productive capacity to repair damage and are more susceptible to stress. Thus much of the observed correlation between stress gradients and vegetational sequences can be related to the problems of organization of a complex organism, together with the competitive pressure for plant species to evolve specializations to a limited range of environments.

SECONDARY SUCCESSIONS. Turning to secondary successions, most of the studies cited in this paper suggest that the early stages in secondary successions can be understood adequately in terms of differential growth. Changes in the later stages are not consistently unidirectional and appear to be influenced primarily by external factors such as grazing, stream cutting, and storm damage, which recur on a time-scale shorter than that required for intrinsic development. Whereas the classical and contemporary theories of succession place primary importance on the role of early successional plants in modifying their environment and permitting later vegetational stages to become established, the studies cited in this paper suggest that modifications of the environment by plants act frequently or generally to *delay* succession. The mechanisms involved include competition for nutrients, shading, pH modification and allelopathy (allotoxicity).[5]

The studies of secondary successions cited in this paper suggest that the plants dominant in the later stages were present either from the beginning, or at least from a very early stage. Their inconspicuousness in the early stages appears to be a consequence of their slow growth and/or their suppression by other species (and, to some extent, to the preoccupation of botanists with the dominant species). That is, most of the detailed observations are consistent with the "initial floristic composition" model of secondary succession (Egler 1954).

Cases in which one species, or a group of species of the same life-form, succeed in dominating the vegetation and maintaining a quasi-stationary state for a time, do not necessarily provide good evidence for the "relay floristics" model. The data of Olson (1958), Niering and Egler (1955), Raup and Carlson (1941), McCormick (1968), and Brown and Roti (1963) suggest that they should be interpreted as delays in the growth sequence, effected by successful suppression of growing subdominant plants by the temporarily dominant species. The replacement of the dominant plants by species of the next higher life form may require external disturbance (Niering & Egler 1955, Olson 1958, Raup & Carlson 1941). In the contemporary theory of succession, this replacement is in part explained as a consequence of continued high net productivity. In fact, however, net productivity is relatively low during the quasi-stationary periods and increases rapidly when a new set of plants becomes dominant (Odum 1960). These flushes of productivity are most easily explained by the hypothesis of release from suppression by the previously dominant plants.

PRIMARY SUCCESSIONS. Genuine observations on primary successions are meager. It seems to be generally accepted that primary succession takes place more slowly than secondary succession, but the relative roles of soil development, external supply of nutrients, exposure, and seed immigration in controlling the rate of development remain largely hypothetical.

[5] Autotoxicity constitutes a limited exception to this generalization. There is at least one well documented case in which autotoxicity in a herbaceous annual appeared to accelerate or at least permit its rapid replacement in a grassland succession (Wilson & Rice 1968). However, even in this case the autotoxic chemicals were also allotoxic (see also Parenti & Rice 1969). The importance of autotoxicity in successions needs to be investigated further, but its explanation should be sought within the framework of natural selection theory. Tentatively, we suggest that autotoxicity might be a by-product of selection for producing allotoxic chemicals whose primary function would be in interspecific competition. Autotoxicity would not be selected against in species whose individuals disperse their seeds beyond the range of their own toxins, and might even serve a function in intraspecific competition in such species. To interpret its function as facilitating the establishment of other species would be teleological.

The observations quoted in this paper of "late successional" trees growing on stagnant glaciers and close behind retreating glaciers suggest that the role of soil development in primary successions has been exaggerated, and that species of later stages can become established within a few years of initial exposure. The rapid growth of forest on Krakatoa and the dominance of phanerophytes there from the very start (Richards 1952) suggests the same conclusion. The data of Olson (1958) suggest that the availability of water and nutrients may be more important limiting factors on the rate of establishment of species than the age, thickness, or organic content of the soil. Tentatively, however, we also ascribe importance to the rate of dispersal of seeds, since "early successional" species tend also to have adaptations for rapid dispersal.

CONGRUENCE OF TEMPORAL AND SPATIAL SEQUENCES. Species dominant in early stages of secondary or primary successions are those characteristic of high-stress sites (Woodwell 1970). According to the present interpretation, they achieve early dominance primarily because they grow faster than species adapted to lower-stress sites. A generalized explanation for the differential growth is given in the arguments of Raunkiaer and Woodwell summarized above. The early dominant species maintain dominance for as long as possible by suppressing competitors, but are replaced because in general they live less long and grow less tall than their successors. In primary successions (but to a much lesser extent in secondary successions) the establishment of the low-stress species may be delayed by delays in immigration and by physical stresses.

According to this highly generalized explanation, therefore, most of the phenomena of succession should be understood as resulting from the differential growth, differential survival, and perhaps differential dispersal of species adapted to grow at different points on stress gradients. The appearance of orderly replacement of "communities" of successively higher life forms results, at least in part, from the temporary dominance of certain species over their successors. Net productivity remains high throughout succession simply because the plants are growing, and it is reduced during the periods at which full-grown plants suppress their potential successors. Net productivity rises when larger plants replace smaller plants, but this is a consequence of the growth of the larger plants and is not a cause of the replacement (as the contemporary theories imply).

Under this interpretation, succession on a single site usually involves a sequence of species (rather than simply the growth of the ultimately dominant species) because no one species can dominate the vegetation throughout the period of growth. In other words the basic cause of the phenomenon of succession is the known correlation between stress tolerance, rapid growth, small size, short life, and wide dispersal of seed. The few exceptions to this general correlation — mangroves, redwoods, coconut palms, douglas firs — are also exceptions to the generalization of succession: they are both "early successional" and "climax" species.

Adaptations to diverse sites. Contemporary theories of geomorphological development (e.g. Chorley 1962) emphasize the continuity of the processes of uplift and erosion (in contrast to the classical concepts of cycles of erosion (Davis 1909) on which the classical theory of plant ecology was based (Clements 1936: for discussion see Drury & Nisbet 1971)). As a consequence, gradients of exposure, soil type, moisture, and other stresses can be assumed to exist continuously in essentially every geographic region. This permits the specialization of different species to different points along the stress gradients, under the pressure of interspecific competition. One extreme evolutionary strategy is to occupy sites at the extreme upper end of the stress gradient, and to grow and reproduce rapidly (during the favorable growing season) before fluctuations in weather make the site intolerable. The other extreme strategy is to occupy the most sheltered sites, to dominate the vegetation by growing as large as possible, and to live as long as possible. Species which adopt these extreme strategies may be equally well adapted to their own environments: there is no need to suppose that the species adapted to the exposed sites are "immature" or "less fit." In geological terms, the sheltered sites are as transitory as the exposed sites.

When new unvegetated sites become exposed, plants adapted to exposed sites are pre-adapted to occupy them and to grow well, provided that they can colonize them before they are occupied by competing species. Thus efficient seed dispersal will give a competitive advantage to individuals of the stress tolerant species. This advantage would contribute to the evolution of efficient dispersal mechanisms in these species.

Essential hypotheses. To place these arguments into a deductive framework, it is necessary to formulate the following hypotheses which differ at least in emphasis from those in contemporary theories of succession.

1) Gradients of soil conditions and exposure to stress exist continuously in essentially every geographic region, as a result of continuous geological processes of uplift and erosion.

2) Different species are specialized to grow under different site conditions, under the pressure of interspecific competition.

3) Individual plants already growing on a site generally have a competitive advantage over seedlings and immigrant individuals, but the advantage is often reduced by disturbance.

4) Dispersal mechanisms and tolerance of physical stress are correlated (together they constitute colonizing ability).

5) Colonizing ability and growth rate tend to be inversely correlated with size at maturity and with longevity.

Hypothesis (1) is geologically trivial, but it needs to be stated explicitly because it conflicts with the classical concept of landscape stability which underlay classical succession theory. Hypotheses (2) and (3) are biologically trivial, but represent the explicit introduction of natural selection into succession theory, in which it is customarily neglected.

Hypotheses (4) and (5) are the key statements underlying our explanation of succession. They amount to the assertion that certain adaptive strategies are mutually exclusive: species whose seeds travel far and grow fast in harsh conditions cannot also grow large and live long. There is a well-known correlation between size, longevity, and low growth-rate. We suggest that this correlation, together with the Raunkiaer-Woodwell argument, comprise a sufficient explanation of the broad features of succession.

An explanation of the correlation might be sought either in theories of senescence, or in theories of the organization of organisms. In any case, we suggest that a complete theory of vegetational succession should be sought at the organismic, physiological or cellular level, and not in emergent properties of populations or communities. The former approach seems more consistent with the theory of natural selection.

SUMMARY

In contemporary usage the term succession refers to a sequence of changes in the species composition of a community, which is supposed to be associated with a sequence of changes in its structural and functional properties. The term succession is generally used for temporal sequences of vegetation on the same site, but only the early stages in the sequence can be observed and the later stages are usually inferred from observations on spatial sequences on adjacent sites. The term is customarily applied to sequences in many types of environment, although few detailed studies have been reported except for temperate forested regions. Contemporary theories of succession in general ascribe the observed phenomena to the action of the community itself in changing the environment.

This paper discusses a number of detailed studies of succession in forested regions, and shows that most of them do not conform to the contemporary generalization. The changes in structural and functional properties are not consistently associated with changes in species composition; the later stages in succession are not consistently unidirectional; and the effects of species already on the site appear frequently to delay rather than facilitate successional replacement.

An alternative explanation is outlined which is based on the observed congruence between temporal sequences of vegetation and spatial sequences along environmental gradients. According to this explanation, most of the phenomena of succession can be understood as consequences of differential growth, differential survival (and perhaps also differential colonizing ability) of species adapted to growth at different points on environmental gradients. The appearance of successive replacement of one "community" or "association" by another results in part from interspecific competition which permits one group of plants temporarily to suppress more slowly growing successors. The structural and functional changes associated with successional change result primarily from the known correlations in plants between size, longevity, and slow growth. A comprehensive theory of succession should be sought at the organismic or cellular level, and not in emergent properties of communities.

ACKNOWLEDGMENTS

Discussions with Madhav Gadgil and Robert Trivers helped greatly in clarifying the ideas in this paper. Rolla Tryon made helpful suggestions for revising an earlier draft. We thank also three anonymous reviewers whose comments made clear to us that the traditional view of succession is alive and well among our peers.

SCIENTIFIC NAMES OF PLANT SPECIES REFERRED TO IN THE TEXT

Common name	Scientific name
Balsam fir	*Abies balsamea* (L.) Mill.
Fraser fir	*Abies Fraseri* (Pursh) Poir.
Hemlock	*Tsuga canadensis* (L.) Carr.
Spruce	*Picea* spp.
White pine	*Pinus Strobus* L.
Jack pine	*Pinus Banksiana* Lamb.
Red cedar	*Juniperus virginiana* L.
Redwood	*Sequoia sempervirens* (Lamb.) Endl.
Douglas fir	*Pseudotsuga taxifolia* (Poir.) Britt.
Salt-water cord-grass	*Spartina alterniflora* Loisel.
Cotton grass	*Eriophorum* sp.
Coconut palm	*Cocos nucifera* L.
Greenbrier	*Smilax* spp.
Quaking aspen (Aspen)	*Populus tremuloides* Michx.
Sweet Gale	*Myrica Gale* L.
Pignut hickory	*Carya glabra* (Mill.) Sweet
Yellow birch	*Betula lutea* Michx. f.
Gray birch	*Betula populifolia* Marsh.
Paper birch	*Betula papyrifera* Marsh.
Beech	*Fagus grandifolia* Ehrh.
White oak	*Quercus alba* L.
Chestnut oak	*Quercus Prinus* L.
Red oak	*Quercus rubra* L.
Black oak	*Quercus velutina* Lam.
American elm	*Ulmus americana* L.
Tulip poplar	*Liriodendron Tulipifera* L.
Sycamore	*Platanus occidentalis* L.
Pin cherry	*Prunus pensylvanica* L. f.
Black locust	*Robinia Pseudo-Acacia* L.
Striped maple	*Acer pensylvanicum* L.
Sugar maple	*Acer saccharum* Marsh.
Red maple	*Acer rubrum* L.
Basswood	*Tilia americana* L.
Dwarf cranberry	*Vaccinium Oxycoccos* L.
Blueberry	*Vaccinium* spp.
Heath	*Erica Tetralix* L.

Ling	*Calluna-vulgaris* (L.) Hull
White ash	*Fraxinus americana* L.
Nannyberry	*Viburnum Lentago* L.
Goldenrod	*Solidago* spp.
	Solidago uliginosa Nutt.
	Solidago juncea Ait.
Aster	*Aster* sp.
Black-eyed Susan	*Rudbeckia serotina* Nutt.

LITERATURE CITED

BIOLOGICAL SCIENCES CURRICULUM STUDY. MOORE, J. A., *et al.* 1963. Biological science: an inquiry into life. Harcourt Brace and World, New York.

BORMANN, F. H. 1969. A holistic approach to nutrient cycling problems in plant communities, pp. 149–165. In K. N. H. GREENIDGE (*ed.*) Essays in plant geography and ecology. Nova Scotia Museum, Halifax.

BOTANICAL REVIEW, several authors. 1968. The continuum concept of vegetation: responses. Bot. Rev. **34:** 253–332.

BRAUN, E. L. 1950. Deciduous forests of eastern North America. Blackiston, Philadelphia. 596 pp.

BRIAN, P. W. 1949. The production of antibiotics by microörganisms in relation to biological equilibria in the soil. Symp. Soc. Exptl. Biol. **3:** 357–372.

BROOKHAVEN SYMPOSIA IN BIOLOGY, No. 22. 1969. Diversity and stability in ecological systems. Brookhaven National Laboratory, May 26–28, 1969. U.S. Dept. of Commerce, Springfield, Va. 264 pp.

BROWN, R. T., & J. R. ROTI. 1963. The "Solidago factor" in jack pine seed germination. Bull. Ecol. Soc. **44:** 113.

BUFFON, G. L. L. 1742. Memoire sur la culture des forêts. Hist. Acad. Roy. Sci. (1742): 233–246.

CHORLEY, R. J. 1962. Geomorphology and general systems theory. U.S. Geol. Survey Prof. Paper 500-B.

CLEMENTS, F. E. 1916. Plant succession. Carnegie Inst. Wash. Publ. 242. 512 pp.

———. 1920. Plant indicators. Carnegie Inst. Wash. Publ. 290.

———. 1928. Plant succession and indicators. H. W. Wilson, New York. 453 pp.

———. 1934. The relict method in dynamic ecology. Jour. Ecology **22:** 39–68.

———. 1936. Nature and structure of the climax. *Ibid.* **24:** 252–284.

COOPER, W. S. 1923. The recent ecological history of Glacier Bay, Alaska. Ecology **4:** 93–128, 223–246, 355–365.

———. 1926. The fundamentals of vegetational change. *Ibid.* **7:** 391–413.

COWLES, H. C. 1899. The ecological relations of the vegetation on the sand dunes of Lake Michigan. Bot. Gaz. **27:** 95–117, 167–202, 281–308, 361–391.

———. 1901. The physiographic ecology of Chicago and vicinity. *Ibid.* **31:** 73–108, 145–182.

———. 1911. The causes of vegetative cycles. *Ibid.* **51:** 161–183.

CROOK, J. H. 1965. The adaptive significance of avian social organizations. Symp. Zool. Soc. London **14:** 181–218.

DANSEREAU, P. 1957. Biogeography — an ecological perspective. Ronald Press Co., New York. 392 pp.

DARLINGTON, P. J. 1957. Zoogeography: the geographical distribution of animals. John Wiley & Sons, Inc., New York. 675 pp.

DARWIN, C. 1859. On the origin of species by means of natural selection. John Murray, London. 490 pp.

DAUBENMIRE, R. 1968. Plant communities. Harper & Row Publ., New York, Evanston, and London. 300 pp.

DAVIS, W. M. 1909. Geographical essays. Ginn & Co., Boston. 777 pp.

DEXTER, R. W. 1947. The marine communities of a tidal inlet at Cape Ann, Massachusetts: a study in bio-ecology. Ecol. Monogr. **17:** 261–294.

DRURY, W. H. 1956. Bog flats and physiographic processes in the Upper Kuskokwim River Region, Alaska. Contr. Gray Herbarium, Harvard Univ. No. 178, Cambridge. 130 pp.

——— & I. C. T. NISBET. 1971. Inter-relations between developmental models in geomorphology, plant ecology and animal ecology. General Systems **16:** 57–68.

EGLER, F. E. 1954. Vegetation science concepts. 1. Initial floristic composition — a factor in old-field vegetation development. Vegetatio **4:** 412–417.

GLEASON, H. A. 1917. The structure and development of the plant association. Bull. Torrey Bot. Club **44:** 463–481.

———. 1926. The individualistic concept of the plant association. *Ibid.* **53:** 7–26.

———. 1927. Further views on the succession concept. Ecology **8:** 299–326.

GOODLETT, J. C. 1954. Vegetation adjacent to the border of the Wisconsin drift in Potter County, Pennsylvania. Harvard Forest Bull. 25. 93 pp.

GOULDEN, C. E. 1969. Temporal changes in diversity, pp. 96–100. In Brookhaven Symposia in Biology 22.

HACK, J. T., & J. C. GOODLETT. 1960. Geomorphology and forest ecology of a mountain region in the central Appalachians. U.S. Geol. Surv. Prof. Paper 347. 64 pp.

HARLEY, J. L. 1952. Association between microörganisms and higher plants (*mycorrhiza*). Ann. Review Microbiol. **6:** 367–386.

HARPER, J. L. 1969. The role of predation in vegetational diversity, pp. 48–61. In Brookhaven Symposia in Biology 22.

HEINSELMAN, M. L. 1963. Forest sites, bog processes, and peatland types in the Glacial Lake Agassiz Region, Minnesota. Ecol. Monogr. **33:** 327–374.

HULT, R. 1885. Blekinges vegetation. Ett bidrag till växformationernas utvecklingshistorie. Medd. Soc. Fenn. XII. 1885. 161 pp.

HURLBERT, S. H. 1971. The nonconcept of species diversity: a critique and alternative parameters. Ecology **52:** 577–586.

HUTCHINSON, G. E. 1957. Concluding remarks, pp. 415–427. In Cold Spring Harbor Symposia on Quantitative Biology 22.

———. 1965. The ecological theater and the evolutionary play. Yale Univ. Press, New Haven. 139 pp.

JOHNSON, E., & W. VALENTINE. 1971. Two recent textbooks reviewed. Ecology **52:** 733, 734.

JOHNSON, F., & H. M. RAUP. 1947. Grassy Island — archaeological and botanical investigations of an Indian site on the Taunton River, Mass. Papers of the Robert S. Peabody Foundation for Archaeology **1:** 1–68.

KERNER, A. VON M. 1863. Das Pflanzenleben des Danaulandes. The plantlife of the Danube Basin. Trans. by H. S. CONARD in "The background of plant ecology." Iowa State College Press, Ames. 238 pp.

KLUYVER, H. N. 1966. Regulation of a bird population. Ostrich, Suppl. **6:** 389–396.

——— & L. TINBERGEN. 1953. Territory and the regulation of density in titmice. Arch. Neerl. Zool. **10:** 265–289.

LACK, D. 1966. Population studies of birds. Clarendon Press, Oxford. 341 pp. (Appendix Section 3, Animal dispersion, pp. 299–312.)

———. 1968. Ecological adaptations for breeding in birds. Methuen & Co., Ltd., London. 409 pp.

LAMB, I. M., & M. H. ZIMMERMANN. 1964. Marine vegetation of Cape Ann, Essex County, Mass. Rhodora **66:** 217–254.

LANGFORD, A. N., & M. F. BUELL. 1969. Integration, identity and stability in the plant association, pp. 83–135. In J. B. CRAGG (*ed.*) Advances in ecological research, Vol. 6.

LEVINS, R. 1966. The strategy of model building in ecology. Am. Sci. **54:** 421–431.

———. 1968. Evolution in changing environments. Princeton Univ. Press, Princeton. 120 pp.

LINDROTH, C. H. 1970. Survival of animals and plants on ice-free refugia. Endeavour **29:** 129–134.

LUTZ, H. J. 1940. Disturbance of forest soil resulting from the uprooting of trees. Yale School of Forestry Bull. 45.

——— & A. C. CLINE. 1947. Results of the first thirty years of experimentation in silviculture in the Harvard Forest, 1908–1938. Part 1. The conversion of stands of old field origin by various methods of cutting and subsequent treatments. Harvard Forest Bull. 23. 182 pp.

——— & F. S. GRISWOLD. 1939. The influence of tree roots on soil morphology. Am. Jour. Sci. **237:** 389–400.

MACARTHUR, R. H., & J. H. CONNELL. 1966. The biology of populations. J. Wiley & Sons, New York. 200 pp.

———, H. RECHER, & M. CODY. 1966. On the relation between habitat selection and species diversity. Am. Nat. **100:** 319–327.

——— & E. O. WILSON. 1967. The theory of island biogeography. Monographs in population biology. Princeton Univ. Press, Princeton. 203 pp.

MACINTOSH, R. P. 1967. The continuum concept of vegetation. Bot. Review **33:** 130–187.

MARGALEF, R. 1963. On certain unifying principles in ecology. Am. Nat. **97:** 357–374.

———. 1968. Perspectives in ecological theory. Univ. Chicago Press, Chicago. 111 pp.

MARQUIS, D. A. 1967. Clearcutting in northern hardwoods. U.S. Forest Service Research Paper NE-85. N.E. Forest Expt. Sta., Upper Darby, Pa. 13 pp.

MCCORMICK, J. 1968. Succession. VIA 1, p. 22–35, 131, 132. Student Publication, Graduate School of Fine Arts, Univ. Pennsylvania.

MILTON, W. E. J. 1940. The effect of manuring, grazing and cutting on the yield, botanical and chemical composition of natural hill pastures. I. Yield and botanical section. Jour. Ecology **28:** 326–356.

——— & R. O. DAVIES. 1947. The yield, botanical and chemical composition of natural hill herbage under manuring, controlled grazing and hay conditions. I. Yield and botanical section. II. Chemical section. *Ibid.* **35:** 65–95.

NIERING, W. A., & F. E. EGLER. 1955. A shrub community of *Viburnum lentago*, stable for twenty-five years. Ecology **36**: 356–360.

ODUM, E. P. 1959. Fundamentals of ecology (2nd edition). W. B. Saunders Co., Philadelphia. 546 pp.

———. 1960. Organic production and turnover in old field succession. Ecology **41**: 34–48.

———. 1969. The strategy of ecosystem development. Science **164**: 262–270.

OLSON, J. S. 1958. Rates of succession and soil changes on southern Lake Michigan sand dunes. Bot. Gaz. **119**: 125–170.

OOSTING, H. J. 1956. The study of plant communities (2nd edition). W. H. Freeman & Co., San Francisco. 440 pp.

PARENTI, R. L., & E. L. RICE. 1969. Inhibitional effects of *Digitaria sanguinalis* and possible role in old-field succession. Bull. Torrey Bot. Club **96**: 70–78.

POUND, C. E., & F. E. EGLER. 1953. Brush control in southeastern New York: fifteen years of stable tree-less communities. Ecology **34**: 63–73.

RAUNKIAER, C. 1934. The life forms of plants and statistical plant geography. Clarendon Press, Oxford. 632 pp.

RAUP, H. M., & R. E. CARLSON. 1941. The history of land use in the Harvard Forest. Harvard Forest Bull. 20. 64 pp.

RICHARDS, P. W. 1952. The tropical rain forest. Cambridge University Press, Cambridge. 450 pp.

RICKETTS, E. F., & J. CALVIN. 1968. Between Pacific tides. (Revised by J. H. HEDGPETH.) Stanford, Calif. 614 pp.

RICE, E. L. 1964. Inhibition of nitrogen-fixing and nitrifying bacteria by seed plants (1). Ecology **45**: 824–837.

SIGAFOOS, R. S., & E. L. HENDRICKS. 1969. The time interval between stabilization of alpine glacial deposits and establishment of tree seedlings. U.S. Geol. Survey Prof. Paper 650-B, pp. B89–B93.

SJÖRS, H. 1948. Myrvegetation i Bergslagen. Acta Phytogeogr. Suec. 21. Uppsala. 340 pp.

———. 1950. Regional studies in north Swedish mire vegetation. Bot. Not. **1950**: 175–222.

———. 1955. Remarks on ecosystems. Svensk Bot. Tidskr. **49**(1–2): 155–169.

SKUTCH, A. F. 1967. Adaptive limitation of the reproductive rate of birds. Ibis **109**: 579–599.

SMITH, R. L. 1966. Ecology and field biology. Harper & Row Publ., New York, Evanston, & London. 686 pp.

SPURR, S. H. 1952. Origin of the concept of forest succession. Ecology **33**: 426, 427.

STEPHENSON, T. A., & A. STEPHENSON. 1954. Life between the tide-marks in North America. IIIA. Nova Scotia and Prince Edward Island: description of the region. IIIB. Nova Scotia and Prince Edward Island: the geographical features of the region. Jour. Ecology **42**: 14–45, 46–70.

——— & ———. 1961. Life between tide-marks in North America. IVA, IVB. Vancouver Island I, II. *Ibid.* **49**: 1–29, 229–243.

TANSLEY, A. G. 1935. The use and abuse of vegetational terms and concepts. Ecology **16**: 284–307.

THEOPHRASTUS. ca. 300 B.C. An enquiry into plants. Book IV: "Of the trees and plants special to particular districts and positions." Sir Arthur Hort Edition 1916. Heinemann, London.

THOREAU, H. D. 1860. Succession of forest trees. Mass. Board Agriculture Report VIII.

WALKER, D. 1970. Direction and rate in some British post glacial hydroseres. Studies in the Vegetational History of the British Isles: 117–139. D. WALKER & R. G. WEST, *eds.* Cambridge.

WALLACE, A. R. 1876. The geographical distribution of animals. 2 Vols.

———. 1880. Island life. London.

WHITTAKER, R. H. 1953. A consideration of climax theory: the climax as a population and pattern. Ecol. Monogr. **23:** 41–78.

———. 1965. Dominance and diversity in land plant communities. Science **147:** 250–260.

———. 1966. Forest dimensions and production in the Great Smoky Mountains. Ecology **47:** 103–121.

———. 1967. Gradient analysis of vegetation. Biol. Rev. **42:** 207–264.

———. 1970. Communities and ecosystems. The Macmillan Co., London & Toronto. 162 pp.

——— & P. P. FEENY. 1971. Allelochemics: chemical interactions between species. Science **171:** 757–770.

WILSON, R. E., & E. L. RICE. 1968. Allelopathy as expressed by *Helianthus annuus* and its role in old-field succession. Bull. Torrey Bot. Club **95:** 432–448.

WOODWELL, G. M. 1970. Effects of pollution on the structure and physiology of ecosystems. Science **168:** 429–433.

WYNNE-EDWARDS, V. C. 1959. The control of population-density through social behaviour: a hypothesis. Ibis **101:** 436–441.

———. 1962. Animal dispersion in relation to social behaviour. Hafner, New York. 653 pp.

YONGE, C. M. 1949. The sea shore. Collins, London.

Part V

COMMUNITY STABILITY

Editor's Comments on Papers 17 Through 21

17 LEWONTIN
The Meaning of Stability

18 MacARTHUR
Fluctuations of Animal Populations, and a Measure of Community Stability

19 HURD et al.
Stability and Diversity at Three Trophic Levels in Terrestrial Successional Ecosystems

20 LOUCKS
Evolution of Diversity, Efficiency, and Community Stability

21 MAY
Will a Large Complex System Be Stable?

In this part we will move into the present and consider a subject that has unusual immediacy. This topic is community stability, and it is of special significance to successional interests because ecological succession can be interpreted as the recovery of community equilibrium after a distance has been removed. In this interpretation of succession, climax communities are defined as dynamic equilibrium states that are persistent through time. A severe disturbance upsets this equilibrium or destroys the community, and recovery of an equilibrium condition eventually occurs through ecological succession.

Stability is defined in Webster's dictionary as firmness, permanence, resistance to change, or the capacity to return to equilibrium after having been displaced. The variety of technical definitions of stability to be discussed share the sense of these common English usages. The problems with the definition of stability, however, are the same as those considered in part four for community—what are the temporal and spatial scales used to evaluate persistence? And, what are the criteria used to determine equilibrium states of the community?

In my opinion, there is no universally acceptable answer to the first question. The scales chosen in an investigation depend upon its purpose and objectives. If we are managing portions of the landscape, we want to know if it will require tens of years or hundreds of years for the community to achieve a given condition. If we are concerned with theory, then we may consider time as a variable and examine the possible states of the community at a variety of different time periods under varying environmental and biological conditions. The second question is more manageable and will be considered further.

Richard C. Lewontin of Harvard University has examined the concept of stability in an ecological sense and describes several different concepts of stability in Paper 17. Neighborhood stability is a yes-no concept and essentially concerns the ability of a system to return to the equilibrium point after disturbance. While it is "mathematically most tractable," most natural communities probably possess neighborhood stability since they exist. Global stability is more interesting ecologically—this concept concerns the number of stable endpoints possible. If the system converges on one point from all other points, then it is globally stable. Lewontin asks, "Can there be more than one stable community composition in a given habitat?" This is the same question we asked in several previous chapters presented in a new context.

Of further theoretical interest is the concept of relative stability. The magnitude of the perturbation required to disturb the system, the distance the system moves from equilibrium, and the time required to return to equilibrium are all measures of relative stability. And further, these measures can be used to compare communities in regard to their potential stability. Two features of relative stability are of special interest. The first is the *resistance* of a system to displacement—a system that is difficult to displace is highly resistant and, in this sense, very stable. The second is *resilience*. If a system returns to its original condition rapidly and directly after disturbance, it is more resilient than another system that responds more slowly. There may be an inverse relationship between resistance and resilience, as suggested by systems ecologists Webster, Waide, and Patten, with systems having high resistance with low resilience and vice versa. These ecologists hypothesize, in an unpublished paper, that evolution would seem to involve a compromise or balance between resistance and resilience.

In an evaluation of community stability, we often have a difficult job choosing the criteria used to determine if the system is stable. Numbers of species, species composition and species diversity have been most often used in these determinations, but so have energy flux or metabolism, nutrient cycling and flow of information. In Odum's paper on ecosystem development (Paper 15 in Part Four), a variety of system characteristics of stability were presented. Here we will consider several of these characteristics which are especially pertinent to stability theory.

Robert MacArthur [1930–1972], a Professor at Princeton and one of the most creative theorists in modern ecology, emphasized the relation of species composition to stability in his 1955 paper (Paper 18). MacArthur argued that the greater the number of potential pathways for energy flow within a community, the more stable that community should be. MacArthur's index of stability, derived form information theory, relates the proportion of energy flowing along a given pathway and the number of potential pathways. Pathways are represented by species populations linked by feeding activity.

The dictum that diversity of populations on energy pathways and stability were directly related dominated ecological thinking for several years. It was the motivation for the 1969 Symposium at Brookhaven National Laboratory, from which Richard Lewontin's paper is reprinted. Evidence, such as that of E. P. Odum (Paper 6 in Part Two), for example, seemed to support the concept, yet by 1970 contradictory observations began to appear. One piece of such evidence was reported by a group of ecologists at Syracuse. Hurd, Mellinger, Wolf, and McNaughton (Paper 19) reported in *Science* experiments where they artifically perturbed old-field systems. They showed that stability defined as resistance to perturbations, declined with succession time and with species diversity of both herbivores and carnivores.

Orie Loucks, a Professor at the University of Wisconsin, also examined evidence on diversity and stability in a large data set on forest communities in Wisconsin. Loucks in (Paper 20) develops a model showing the response of a forest to random disturbance and concludes that the diversity of plants and animals is understood as an ephemeral product of the overlap of species populations acting independently in response to disturbance. Loucks also calculated that the energy production of the community is equally dependent on the random disturbance and rejuvenation of the forest community.

Finally, there have also been a number of tests of the diversity-stability concept using theoretical and mathematical approaches. Among these investigators Robert May, Professor at Princeton, has presented several important analyses of the concept. In his study of a complex theoretical system, Paper 21, May shows that increased complexity tends to reduce stability. May emphasizes that a relationship between diversity and stability would be the result of special strategies in that given system and not a general mathematical consequence.

These various papers on community stability are obviously applicable to successional problems, as was discussed by Henry Horn (1974) in his recent review of secondary succession. Hopefully, they will not only lead to a more rigorous way to treat these data and concepts but will also provide the framework for a dynamic theory of community stability and recovery. Clearly, there is a large gap between the relative clarity of the mathematical systems approach and the almost bewildering complexity of populations and individuals in most natural communities. Systems concepts may form a link between these extremes by identifying the key components and dynamic processes in a system and providing simplifying assumptions to explain the dynamics of these components. Further, the system approach may lead to a stress on the dynamic properties of succession rather than on the pattern. Thus, we might predict that the response of the community to disturbance through succession is more significant than the possible orderly sequence of communities in space or time. The essence of succession may well be *response* not pattern!

REFERENCE

Horn, H. S. 1974. The Ecology of Secondary Succession. In *Annual Review of Ecology and Systematics, Vol. 5,* eds. R. F. Johnson, D. W. Frank, and C. D. Michener, pp. 25–37. Palo Alto, Calif.: Annual Reviews, Inc.

17

Reprinted from pp. 13–24 of *Diversity and Stability—Ecological Systems,* Brookhaven Symp. Biol. No. 22, 1969, 000 pp.

The Meaning of Stability

RICHARD C. LEWONTIN
Committee on Mathematical Biology, University of Chicago, Chicago, Illinois 60637

To many ecologists their science has seemed to undergo a major transformation in the last 10 years, from a qualitative and descriptive science to a quantitative and theoretical one. Yet the seminal works of mathematical ecology were written by Lotka and by Volterra more than 40 years ago, and we have barely progressed beyond them. To others, the change has seemed to be from a static to an evolutionary view of ecology, yet it is 60 years since Clements began constructing the theory of succession, which is nothing if it is not an evolutionary theory of the community. Yet there *has* been a change in ecological theory in recent years, as a glance at the pages of *The American Naturalist,* or even of *Ecology,* will show. In my view, the change has come about through a union of these two strains – the mathematical and the evolutionary – to produce the beginning, and only the bare beginning, of an exact theory of the evolution of communities of organisms. Such an exact theory must "explain" in some sense the present state of the biosphere, but must also contain statements about the past history of living communities and about their future as well. Indeed, their present state can be regarded as only a snapshot, catching at one instant of time the instantaneous configuration of an ever-shifting system.

To carry out such an ambitious program as the modern ecologist has planned requires a general abstract frame on which an exact theory can be constructed, and it is not surprising that this frame turns out to be, for ecology in general, the same as that for the specialized branch of ecology called population genetics, that is, *the concept of the vector field in* n*-dimensional space.* This concept is the most fundamental one we have for dealing with the transformations of complicated dynamical systems in time. It comes to ecology and population genetics in large part from physics, by analogizing the changes through time of a population or community with the changes in the position of a particle in space, or, on a more generalized level, with the position of a population of particles in a phase space.

THE DYNAMICAL SPACE

For any physical system we can construct a description that is a list of the variables necessary to specify the system completely plus a list of the values of those variables taken for some specific case. For example, for the purposes of guiding a space capsule to the moon it is necessary and sufficient to know, at any instant, the three coordinates of the capsule in physical space, the three components of velocity, and the three components of acceleration along these (or other orthogonal) axes. The components of acceleration result partly from the rocket power given to the capsule and partly from the combined gravitational field of all celestial bodies, but that does not complicate the space. Thus the space capsule moves or "evolves" in a "space" of 9 dimensions. The 9 dimensions may be x, y, and z coordinates of a Cartesian system with velocity and acceleration along the same axes, or some other set of orthogonal axes as, for example, distance from the center of the earth, azimuth, and altitude angles, together with velocity and acceleration values.

The total history of such an object can be described as the succession of positions of the object in the n-dimensional hyperspace of its description. The n-dimensional description we will call a *position vector*. A complete theory of evolution of the position vector would then be a set of rules relating the position of the point in hyperspace at one instant to its position at some other instant. It is the set of rules that determines the dimensionality of the space. For example, it is impossible to specify the position of a space capsule at some future time only from a knowledge of its present position. The components of velocity and acceleration are both also required because the laws of motion contain them. The rules of transformation specify, for every point in the hyperspace, the direction in which an object in the space would move at that point and the magnitude of the motion. We may imagine that at each point in the space there is an arrow pointing in some direction and having a length proportional to the magnitude of change. Such an arrow is a *transformation vector,* and the space of these vectors is a *vector field.* If we could visualize such a vector field and if it were nicely behaved, we could trace out the path that a particle would take in the space. When we sprinkle iron filings on a sheet of paper over a magnet we are visualizing the vector field for a small dipole moving under the influence of the magnet.

If the dimensions of the dynamical space have been properly chosen, there will be one and only one vector at each point and there will be a proper vector field. This is equivalent to saying that trajectories traced out by the vector field never cross. For if they crossed, then at the point of crossing there would be an ambiguity. Such an ambiguity means that all the information necessary for specifying the direction and rate of change is not contained in the coordinates of the point. Thus the dimensions used to construct the hyperspace are not sufficient to describe completely changes in the dynamical system. We will say that a set of dimensions is *sufficient* if paths in a hyperspace constructed from that set of dimensions never cross. We will then define the dimensionality of a dynamical system as the smallest number of dimensions making a sufficient set.

The rules, then, make the vector field of transformation and thus determine the configuration of the space in which the particle moves. If we are to see clearly what is meant by stability in ecology, it is essential that we differentiate between this vector field and its properties, on the one hand, and the objects – individuals, populations, species – that move in it, on the other. That is, we must distinguish between the configuration of the hyperspace in which the community is moving and the actual trajectory of the community. This distinction has not always been made. As we will see, constancy of the position vector does not imply stability in the field of transformation vectors.

STATIONARY, TRANSIENT, AND CYCLIC POINTS

In the hyperspace of the dynamical system we must distinguish three sets of points. The first is the set of points at which the magnitude of the transformation vector is zero. This is the set of *stationary points* of the system. If the position vector takes one of these values, the system does not change. *Transient points* of the space are those for which the transformation vector lengths are non-zero so that the system will continue to evolve through them. We must distinguish between cyclic and noncyclic transient points. Cyclic transient points are those to which the object may return over and over again because the configuration of the vector field causes the position vector to follow a closed path through the space. Finally, a noncyclic transient point is one through which the system would pass less than infinitely often. Any particular dynamical hyperspace may contain no, one, or more than one point falling into each of these classes, and it is the locations of these points and their numbers that determine the qualitative nature of the evolutionary system.

NEIGHBORHOOD STABILITY

Consider the nature of the vector field very near a stationary point. Do all the vectors point toward the stationary point? If they do, then it is a *stable* point, since a small perturbation of the system will result in the system returning to that point. It is necessary to specify that the perturbation is *small* because there may be several such stable or attractive points in the space and each will have its own *basin of attraction*. If the perturbation is sufficiently large to carry the system out of one basin of attraction into another, the original point is still a stable point even though the system did not return to it. Here again we must distinguish between the behavior of the position vector and the configuration of the vector field of transformation vectors. We may put this in a different way. Let $[v^*]$ be the position vector at the stationary point and $[v]$ be the position vector at some other point. Let T be the transformation that transforms $[v]$ in some unit of time. Then

$$[v(t+1)] = T[v(t)] \tag{1}$$

and

$$[v(t+n)] = T^{(n)}[v(t)] \,. \tag{2}$$

The question is, does

$$T^{(n)}[v(t)] \rightarrow [v^*] \tag{3}$$

as n goes to infinity, and, if so, for what set of $[v]$? The set of $[v]$ for which (3) is true defines the basin of attraction of $[v^*]$. In general, this is a very difficult mathematical problem. The usual approach is that of Lyapunov, which is to examine $[v]$ only arbitrarily close to $[v^*]$, that is, in the neighborhood of the equilibrium. Then, for a large class of transformations, those that converge when expanded in a Taylor series, the transformation, T, is arbitrarily close to a linear transformation **L**. If we now set

$$[x] = [v] - [v^*] \tag{4}$$

so that $[x]$ is the vector of deviations of the system from the equilibrium point $[v^*]$, we can write the dynamical equation (1) as

$$[x(t+1)] = \mathbf{L}[x(t)] \tag{5}$$

where $\mathbf{L}-\mathbf{I}$ is the matrix of partial derivatives of the system T with respect to the components of the vector $[v]$, evaluated at $[v^*]$. The problem of stability is, then, does

$$[x(t+n)]\rightarrow 0 \qquad (6)$$

as n goes to infinity?

The behavior of $[x(t+n)]$ in the limit is determined by the characteristics of the matrix **L** and in particular by the eigenvalues of that matrix. If all the eigenvalues, λ_i, have their real parts less than unity, the equilibrium is stable. If any of the eigenvalues have real parts greater than unity, then the equilibrium is unstable to perturbations in some directions. That is, the transformation vectors point toward the equilibrium from some directions and away from it in others. Presumably no natural system can remain at such an unstable point, since perturbations in the wrong direction are certain to occur by chance. Finally, if the roots are complex with real parts equal to unity, the point in question is a cyclic point, part of a cyclic path. The classical prey-predator model of Lotka and Volterra is an example of a system in which the eigenvalues of **L** are of this form so that the prey and predator species both undergo a cyclic oscillation in numbers. These oscillations can be represented in our general framework as a closed path in a plane whose two dimensions are $\mathcal{N}_1$ and $\mathcal{N}_2$, the number of prey and the number of predators.

Neighborhood stability or "Lyapunov stability" is the one best understood and easiest to test for, because it is mathematically most tractable. By assuming that we will examine the vector field *arbitrarily close* to the equilibrium point, we are allowed to make nonlinear processes into linear ones that have a unique solution. Simple rules for testing stability can be derived by using the eigenvalues of the matrix **L**. Many authors have considered this kind of stability for population problems, for example, Lotka[1,2] and Leslie[3,4] for population growth. Especially in population genetics, where stability has been a preoccupation for a long while, Lyapunov stability is part of the standard repertoire of all theoreticians and appears explicitly or implicitly in every paper on the subject of equilibria.

GLOBAL STABILITY

The trouble with neighborhood stability is precisely that it is a simple yes or no test for the behavior of a system *at a point,* i.e., for arbitrarily small perturbations. But how will the system behave for large perturbations? If the system is far away from a stable equilibrium point, will it go toward that point or toward some quite different one? This is the question of the size and configuration of the domain of attraction of an equilibrium. It is a difficult and in large part unsolved problem, and no neat solution such as the size of the largest eigenvalue of some matrix can be invoked. Wider considerations of stability require the introduction of other concepts, some of which are as yet unexplored mathematically, but if a realistic view of the stability of ecological systems is to be obtained such new concepts and techniques are needed.

The first concept to be considered is the well-known notion of *global* stability. A locally stable point is said to be globally stable if the system converges to that point from all other points in the space. This global stability is a negative condition in that a point is globally stable only if no other point in the space is a stable point or a cyclic point. To test for the global stability of a point then involves showing that there are no other stable points, and this in turn may be trivially easy or impossibly difficult.

If the system of equations describing the transformation is linear, there can be only one equilibrium point for which all the components of the position vector are non-zero. Thus, if there is a locally stable point of this type, it must also be globally stable. However, there may also be other stable equilibria in which one or more of the components is zero, and each of these would be globally stable within the restriction that only that set of components could be zero. This points out that global stability must be defined in terms of the universe of interest. If we are interested only in that part of the space not including the margins where some of the position vector components are zero, then a linear system can have only one stable point and it will be globally stable. But if we are interested also in missing components, then even a linear system is not restricted to a single stable point.

One of the most interesting problems of community ecology comes under this rubric. Can there be more than one stable community composition in a given habitat? In our terms we are asking whether, in an n-dimensional space of the numbers of n different species, there can be more than one locally stable point. If the system of equations governing the species composition of the community is linear, then only one stable composition is possible with *all the species represented.* However, there may be other stable points with some of the species missing. Thus a given set of species abundances might be globally stable given the restriction of a certain taxonomic diversity, but not globally stable if lower taxonomic diversity is allowed. Moreover, we may state a converse proposition. If a given set of species has no locally stable point, those species cannot be made to coexist stably by adding another species. That is, if there is no locally stable point in n dimensions (not including the boundaries) there is no locally stable point in any $n+k$ dimensions within which those n dimensions are embedded.

If the system of equations describing the transformation of state is nonlinear, nothing of the foregoing holds. There may be multiple stable points with all species present so that local stability does not imply global stability. Moreover species that cannot coexist can be made stable by a third species. Man is the third species in many cases. He stabilizes unstable species composition by making nonlinear adjustments in the species abundances. In general, a study of local stability does not give information about global stability, and an analysis involving exclusion of other stable points is necessary.

DISSIPATION AND STABILITY

In the physical systems, the change of state is often described in terms of the change in the level of energy. In classical mechanics a system will move to a point of (locally) minimum potential energy, and the evolution of the system can be described in terms of the dissipation of energy. In general, at time t there is an energy function $W(t)$ such that

$$W(t+k) \leqslant W(t) \tag{7}$$

for all t and k. The equality in (7) holds only at equilibrium, and for all transient states of the system the function W is always decreasing in time. This energy function is sometimes called a Lyapunov function, and, at a point showing Lyapunov stability, the Lyapunov function is at a local minimum. We must emphasize a *local* minimum however, since, if there are other stable points, the Lyapunov function will also obey relation (7) in the neighborhood of these points. The Lyapunov function is like a topographic surface with peaks, valleys, pits, and saddles. The stable points are at the pits, which are local minima of W. Unstable points are peaks of the Lyapunov surface, and the system will move away from these peaks toward lower values of W.

For ecology the problem of Lyapunov functions is to what extent they represent interesting biological statements. To say that a physical system moves to a state of minimum poential energy gives an insight into the workings of the physical universe. It somehow increases one's "understanding," by providing a unifying concept for a diversity of systems. What is the equivalent ecological law? What is minimized during succession? Will it be a useful and illuminating quantity or simply a mathematical formalism with no intuitive content? MacArthur[5] has recently defined a quantity that is minimized in a purely competitive situation. No general solution in ecological terms has yet been given. An example of a mathematical form is

$$W(n)=\sum_i(1-|\lambda_i|^2)|y_i(n)|^2 \tag{8}$$

where the λ_i are the eigenvalues of $\mathbf{L}$ [see Eq. (5)] and the $y_i(n)$ are the associated orthogonal deviations from equilibrium, calculated from the x_i.

BOUNDED SYSTEMS

In the discussion of global stability it was pointed out that whether or not a locally stable point was also globally stable depended in part on whether points for which some components are zero are considered. In a biological context, the value zero (or infinity) has a special meaning. The presence or absence of species is sometimes the point of interest regardless of some variation in their numbers or relative abundance from time to time. In the n-dimensional space used for describing the state of a system we will distinguish between *interior points* and *boundary points*. The boundary set will include all points for which one or more of the vector components is zero, but may also include some other points. For example, if the number of tree holes suitable for tree swallow nests in an area is K, we may wish to consider K as a boundary for the number of nesting pairs of the swallows and ask whether the number of adult females is always less than K. The boundary set is defined by the problem. We will say that a system is *dynamically bounded* in some interior set S if at all points in the neighborhood of the boundary set B the transformation vectors point into the interior set S. That is, *no point in the boundary set is stable.* For example, let us take the problem of invasion. Zero abundance for any species is an equilibrium, since once the number gets to zero it will not change. Yet the structure of the community may be such that if one pair of individuals of this species were introduced the numbers would then rise. Thus the boundary value, although an equilibrium, is not a *stable* equilibrium, and we will say that the community is *dynamically bounded* away from zero for this species. Only some random external force could eliminate the species, and if it were reintroduced it would increase. It is important to note that to be dynamically bounded does *not* imply that some interior point is stable! It is entirely possible for no particular interior point to be stable, yet for the system to be always kept away from the edges. For example, if the environment varies randomly in such a way that one species is favored on the average but a second species is very strongly favored in occasional years, there will be no stable equilibrium abundance of the two species relative to each other, but neither species will ever be eliminated.

RELATIVE STABILITY

The concepts of local and global stability are absolute. Either a point is locally or globally stable or it is not. This does not distinguish *degrees of stability*, although such relative stability has important biological content. To understand relative stability we must return to the Lyapunov function and the basin of attraction of a stable point. If there are multiple stable points for a system, they may differ in the conformation of their domains of attraction. One stable point might have a very small basin of attraction, but the gradient of the W function might be very steep in the neighborhood of the point. Indeed the basin of attraction might itself lie at the top of a steep *upward* gradient of W. Such a region would resemble a volcano with a crater at its summit. The center of the crater is a stable point but an unlikely one for any system because its domain of attraction is small and it is surrounded by regions of large W. However, if the system should by chance get into the basin of attraction, it would very rapidly evolve into the potential well with its steep sides and, once in that basin, would be held very strongly.

By contrast we may consider a stable equilibrium point with a very large domain of attraction but with a very shallow gradient in its W function. Thus the system is very likely to fall into this domain, but will move very slowly toward the equilibrium and can be easily perturbed from it. The depth of the potential well can be easily quantified as the difference between the values of W at the equilibrium and at the edges of the basin. Indeed a general index of shape is the ratio of the depth to the radius of the basin. Setting the subscript 0 to stand for equilibrium and the subscript B to stand for a boundary point of the basin of attraction, a shape index would be

$$I = \sqrt{\sum_i (v_{i0} - v_{iB})^2} \Big/ (W_B - W_0) \,. \qquad (9)$$

Small values of I would indicate small deep regions of stability, large values denoting large shallow ones.

A third aspect of relative stability is in the shape of the basin of attraction along the various dimensions of the state space. The basin need not be circular (hyperspherical) in cross section. There may be great stability to perturbation in some directions but little stability in others. The eigenvalues of the matrix **L** give a measure of this shape in the neighborhood of the equilibrium. The rate of change of position along each dimension of the hyperspace is a linear function of the eigenvalues. By a suitable transformation of axes the space may be represented along a set of orthogonal axes each associated with one of the eigenvalues. The stability of the equilibrium with respect to each of these axes is then directly measured by its associated eigenvalue. Eigenvalues close to unity correspond to weak stability in the sense that a perturbation along an associated orthogonal dimension will result in a very slow return to equilibrium along that axis. An eigenvalue close to zero on the other hand means a high stability and rapid return.

A fourth measure of relative stability arises when we consider larger deviations from equilibrium. Since the rate of change of position along each dimension is a function of all the eigenvalues even in a linear system, it may be strongly dominated by the imaginary parts of some complex eigenvalues when the position is far from equilibrium. This means that there will be oscillation, with the system taking a spirally winding path inward toward the equilibrium, with very long time periods and an apparently erratic behavior. Oscillatory phenomena have received a great deal of attention in ecology, but usually the search is for some driver of the oscillation rather than for internal mechanisms intrinsic to the dynamical system itself. Part of the problem of relative stability is whether the complex eigenvalues play an important role far from equilibrium. Somehow we judge a system to be more stable if the oscillatory component is weaker, and this, in turn, is a question of the relative sizes of the eigenvalues.

Once again, for nonlinear systems the eigenvalue approach breaks down when the system is far from equilibrium, and stability theory of nonlinear processes is poorly developed.

STATISTICAL ENSEMBLES

The problem of relative stability discussed above is important because we believe the world of real objects to be subject to a host of random perturbations, random at least with respect to the dynamical system we have specified. I do not wish to go here into the question of the ontological status of random events. It is sufficient to say that any predictive system that can be managed practically will have variables that must be considered as extrinsic to the system and random with respect to it. Since the state of the system is being perturbed, it will never be exactly at any equilibrium point or in any equilibrium cycle. Rather, its position will be determined by a combination of random perturbations on the one hand and a restoring force, expressed by the Lyapunov function, on the other. Through time the system will occupy a series of points in the hyperspace, and the ensemble of those points will form a cloud around the stable equilibrium position. This time ensemble is really an expression of the probability density of the system in n-space. The density of the cloud will follow the shape of the relative stability measures previously discussed. If I is large, the cloud will be very dense around the stable point but will be of small radius. The density can be decreased and the radius increased for this very same equilibrium point by increasing the size of the random perturbations. When I is small, the cloud will be large and diffuse, but again it may be less so if the perturbations are smaller. It is useful to think of the "temperature" of the ensemble measuring the mean square distance of points in the ensemble from the average (not necessarily the deterministic equilibrium point).

Let $[x]$ again be the vector of deviations from equilibrium for the system. Then, if the transformation of the system is a set of linear equations F, the probability density $p(x)$ at point x is the solution of the Fokker-Planck equation

$$\sum_{i=1}^{n} \frac{\partial}{\partial x_i}(F_i p) - \frac{1}{2}\sum_{i,j=1}^{n} \frac{\partial^2}{\partial x_i \partial x_j}(d_{ij} p) = 0 \tag{10}$$

where d_{ij} is an element in the variance-covariance matrix of the perturbations. This equation has the general solution of the form

$$p(x) = K^{-W} \tag{11}$$

where W again is the Lyapunov function of the system. As we expect, Eq. (11) tells us that the probability density is highest where the Lyapunov function is at a minimum. In fact $-\log p(x)$ has exactly the shape of the Lyapunov function itself. The constant K is a function of the variance-covariance structure of the random perturbations.

This general theory has been applied by Kerner[7] to the specific case of a prey-predator community, and he derived an expression of the general form of (11). Kerner calls attention to the parallel with the Gibbs ensemble of statistical mechanics and especially to the notion of "temperature." If we define the "temperature" of the ensemble as the mean square distance of points in the ensemble from their mean position in the hyperspace, then we have a descriptive parameter expressing the cloud density and the tendency to remain near the equilibrium point. As "temperature" is defined, it is, in fact, the variance of the probability density function.

CONSTANCY AND STABILITY

The notion of the equilibrium ensemble of position vectors points up the important difference between constancy and stability. Constancy is a property of the actual system of state variables. If the point representing the system is at a fixed position, the system is *constant*. Stability, on the other hand, is a property of the dynamical space in which the system is evolving. Thus, if there is a stable equilibrium point in the vector field, it does not follow that the community must be constant at that point. Because of the random perturbations of the environment or of numbers of organisms in it, the actual state of the system is in constant flux. Yet there is *stability* in the dynamical sense because the ensemble of state points of the system maintains a steady state distribution with its center near a stable point of the deterministic dynamical system and with a constant "temperature" determined by the statistical properties of the random perturbations.

STRUCTURAL STABILITY

The vector field of transformation vectors is determined by the laws of transformation of the quantities defining the dynamical space. There will be a certain number of stable equilibria, unstable points, and perhaps limit cycles. We may now ask what would happen to the vector field if a very small change were made in the parameters of the equations determining the transformation. If the set of equations defining the system is "well behaved" in some sense, a very small change in parameters will make a very small change in the configuration of the vector field. The locations of the various stationary points or limit cycles may change slightly and the values of the Lyapunov function very slightly at each point of the space. But nothing catastrophic will occur, and the difference will hardly be noticed. If we regard the equations T as mapping the parameter space into the space of equilibrium points, then neighborhoods map into neighborhoods, infinitesimal changes in one space causing only infinitesimal changes in the other. On the other hand the laws of transformation may be very badly behaved so that a small change in parameters completely alters the kind of vector space produced. Neighborhoods of the parameter space do not map into neighborhoods of equilibrium space. Such a system is said to be *structurally unstable.* The Lotka-Volterra model of prey-predator interaction is an example of a structurally unstable system. In the simplest model there is no self-damping term, and the numbers of prey (N_1) and predator (N_2) change in time according to the equations

$$dN_1/dt = r_1N_1 - kN_1N_2; \quad dN_2/dt = KN_1N_2 - d_2N_2, \tag{12}$$

where r and d are the birth and death rates of the prey and predator, respectively, and the k's are species interaction constants. As is well known, the solution to these equations is set of closed curves in the N_1N_2 plane so that there is a perpetual undamped oscillation of prey and predator. If, however, second-order terms are added in N_1 and N_2 which introduce self-limitation, the closed curves in the N_1N_2 plane break open and become spirally winding paths leading to a stable point. Thus an infinitesimal change in the coefficients of the higher-order terms (from zero to nonzero) causes a radical alteration in the outcome of the species interaction. Several cases are known in population genetics.

The existence of such structural instabilities is very disquieting, since it means that predictions are very sensitive to assumptions. This makes an adequate predictive theory of ecology very difficult, and it must be hoped that such structural instabilities will prove to be the exception.

STABILITY AND HISTORY

A stable point or a stable limit cycle is one toward which all the neighboring vectors point. It is then a point that will be reached from any direction. If a community is at such a stable point, it is impossible to say from which direction in the hyperspace it came. Thus, historical information is destroyed at equilibrium. All is not lost, however, since at least we know that the community began in the basin of attraction of the particular equilibrium point. On the other hand, if the system is at a transient point, it is possible to reconstruct both its past and its future, since one and only one trajectory passes through each transient point in a space of sufficient dimension.

In population biology there are often two opposing modes of explanation. The first states that history is relevant to the present state of populations, species, and communities, and that their present state cannot be adequately explained without reference to specific historical events. This is equivalent to saying that multiple stable points exist and that the interesting problem is why the system is at a *particular* stable point, or else to claiming that the system is at a transient point. The second mode of explanation attempts to explain the present state of natural populations or communities without any historical information, but as a result of certain fixed forces.

This mode is equivalent to saying that the system under study is at a point of global stability, or else is one point in a stationary ensemble of points, removed from a globally stable point only because of small random perturbations. The first mode of explanation concentrates on the particular community and its trajectory in the dynamical space, while the second concentrates on properties of the dynamical space alone. However, only its equilibrium properties can be deduced in this way. If the general configuration of the dynamical space is to be discovered, that is, if the dynamical laws governing the transformation and evolution of communities are to be discovered, it is essential to work with non-equilibrium systems through time either by a search for evolving communities or by a deliberate perturbation analysis of communities at present at stable points.

REFERENCES

1. Lotka, A. J., *Proc. Natl. Acad. Sci. U.S.* **8**, 399 (1922).
2. Lotka, A. J., *Elements of Physical Biology,* Williams & Wilkins, Baltimore, 1925; republished as *Elements of Mathematical Biology,* Dover Publications, New York, 1956.
3. Leslie, P.H., *Biometrika* **33**, 183-212 (1945).
4. Leslie, P.H., *Biometrika* **35**, 213-45 (1948).
5. MacArthur, R., *Proc. Natl. Acad. Sci. U.S.,* in press (1969).
6. Falk, H. and Falk, C.T., *Biometrics* **22**, 27-37 (1969).
7. Kerner, E.H., *Ann. N.Y. Acad. Sci.* **96**, 975-84 (1962).

DISCUSSION

Watt: Richard Bellman, one of the world's leading experts on the stability theory of systems of differential equations, has recently written a book arguing that this body of theory is inappropriate for dynamically controlled systems, as opposed to systems where "control" is merely built into the design at the outset. I wonder if you would agree or disagree with him.

Lewontin: I tried to argue that the classical theory of neighborhood stability is, indeed, inappropriate, if that is what Bellman is arguing. I say that the classical theory of neighborhood stability, the theory expounded in Bellman's book – the conditions of eigenvalues of the linearized transformation – is inappropriate for two reasons. One is that it reveals nothing about global stability, which is of interest for ecological purposes. But more important, it is not a theory that leads directly to the description of a probability density cloud, the sort of Gibbs ensemble which is much more interesting in ecology. We need a theory much more closely allied to statistical mechanics, with some dissipation. We are really dealing with a kinetic theory of gases but with imperfectly distributed gases, and I agree entirely that the description of the local stable points is only the beginning.

Watt: You are saying what Bellman is saying, but I wonder where he is leading us, and whether statistical mechanics is the answer. It seems that statistical mechanics is based on assumptions that don't match the realities.

Lewontin: I am not talking about statistical mechanics which assumes a conserved quantity, which assumes perfect elastic collision. I am talking about a system in which we describe the dissipation of the population, its going toward some stable point, and compensate for it by perturbations – in fact, theory of diffusion process. I have written a Fokker-Planck equation that contains this dissipation in it and which, when solved, will give a description of the probability distribution of points in n-dimensional space. If you ask me how probable it is that communities will have 7 of one species, 14 of another, 209 of a third, and so on, I can answer that, if you tell me two things: (a) what is the configuration of the dynamical space as far as its deterministic elements are concerned, and (b) how much random perturbation goes on. That is what is done in genetics when we determine the probability of gene frequencies over an ensemble of populations from the rules of deterministic transformation by which gene frequencies are being forced by natural selection and the rules of random variation. One puts these together in a diffusion equation or some kind of a more exact description of a Markov process, and comes up with a probability distribution. I think that is the appropriate mathematics for community ecology.

WATT: Doesn't the Fokker-Planck equation involve implicit assumptions about the statistical character of exogenous perturbations operating on the system that don't correspond to what happens in nature?

LEWONTIN: I don't know whether they correspond or not, and the difference between you and me is that you are used to thinking very exactly about very exact situations. I don't consider the Fokker-Planck to be a perfect equation. Statistical perturbation theory is the appropriate theory, and, if the Fokker-Planck equation is inappropriate because the process is not Markovian, or because it does not satisfy the proper rules about the higher moments, then we have to set up some other equation. The Fokker-Planck equation is derived by neglecting the higher moments. If we can't neglect them, we need a bigger equation, which is more difficult to solve.

WATT: I am still trying to find out what the theory has to be and I mentioned the Fokker-Planck as an example. I am sure the Fokker-Planck equation itself is a wrong equation.

FOSBERG: You discussed three types of points. Which of them is represented by a dynamic equilibrium that, although it revolves about a stable point, does not necessarily return to that original point? The situation you described seemed to involve a return to the original point.

LEWONTIN: No, it doesn't. In the second half of my talk I tried to show that any real biological system does not get back to the equilibrium point because of the perturbations. I am trying to arrive at a description of the system by building up from a deterministic system which has such stable points. If there were no further perturbations, it would finally go back to such points The real biological system, the community, is moving in such a potential field, but it is always being knocked out by some other forces; or, if you like, you can regard this particular vector field with its stable point as only one of a series of vector fields, each of which is applicable at a given instant of time. So, you never return to any particular stationary point. The stationary point is the center of the cloud of probability. It is exactly like taking a pingpong ball in a bowl, and wiggling the bowl constantly – the ball is always moving around yet you can still describe the system as a potential well plus random perturbations.

18

Reprinted from *Ecology* **36**(3):533–536 (1955)

FLUCTUATIONS OF ANIMAL POPULATIONS, AND A MEASURE OF COMMUNITY STABILITY[1]

Robert MacArthur

FLUCTUATIONS

Consider a food web as in Figure 1. This is interpreted to mean that S_1 eats S_2 and S_3, S_2 eats S_3 and S_4, and S_3 and S_4 rely upon a food supply, *i.e.* a source of energy, not shown.

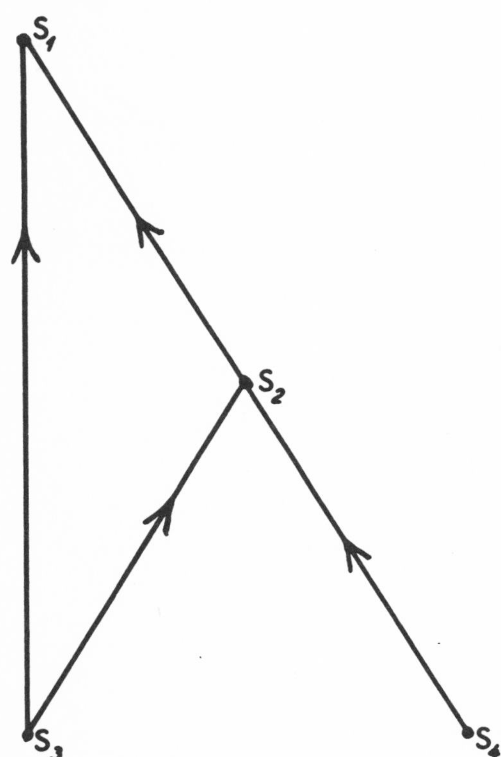

FIG. 1. A sample food web. S_1, S_2, S_3, S_4 are species, and arrows indicate direction in which energy flows.

[1] Contribution from the Osborn Zoological Laboratory, Yale University, New Haven, Connecticut.

Three assumptions will be made and a conclusion will be deduced from these. Since the conclusion is not always correct, it will be justifiable to conclude that one or more of the assumptions is responsible. First, temporarily assume that the amount of energy entering the community (at the lowest trophic level, of course) does not vary with time. Second, assume that the length of time that energy is retained by a species before being passed on to the next doesn't change from time to time. For example, if the animals die young in some year, this would violate the assumption. Third, assume that the population of each species varies directly with the food energy available. If a species first overeats and then starves (essentially a predator-prey reaction) or follows changes in available energy only after a time lag, then this assumption is violated.

These three assumptions imply that the population of each species tends to a specific constant, independent of the initial populations of the species. Proof: Since no species is perfectly efficient at transferring energy from its prey to its predators (Lindeman 1942), energy leaves the web from each species, and in view of the assumptions the amount of energy leaving equals that entering. This is then equivalent to saying that the energy leaving the community is the energy which reenters. That is, the consequences of this will be identical. This energy can now be entered on the food web of Figure 1 as S_5 and it now looks like Figure 2. Let p_{ij} be the proportion of the energy of S_i which goes to S_j. Since all of the energy transferences are shown,

$\Sigma_j p_{ij} = 1.$

This equation shows that the food web considered as an energy transformer is what is known in probability theory as a Markov chain (Feller 1950). Furthermore, if the number of links in a complete cycle of the energy is called the length of the cycle, the greatest common divisor of these lengths is 1 and so the conditions set forth in Shannon and Weaver (1949) for the Markov chain to be "ergodic" are satisfied. Therefore, as shown in Feller (1950), the amount of energy at each point tends toward a constant, independent of the initial conditions. In view of the assumptions, the species populations also approach constant values. This completes the proof.

Since populations of species often fluctuate in nature, it can be concluded that one or more of the three assumptions is failing to hold. Furthermore, it is clear from the theorem that the structure of the food web cannot

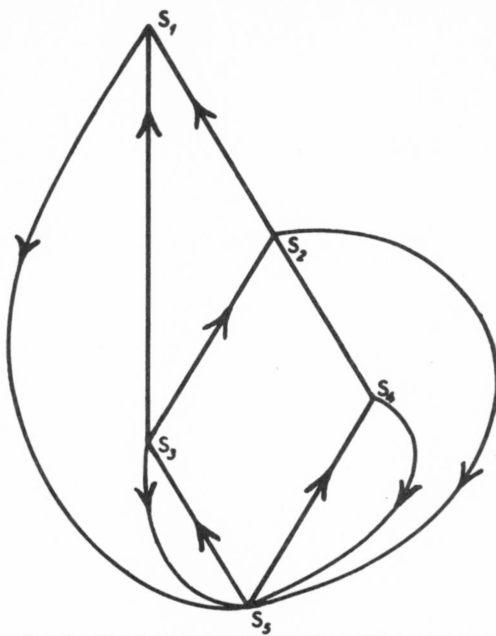

FIG. 2. The food web of Figure 1 modified to include lost energy.

be responsible for fluctuations. Therefore, an explanation of the fluctuations must lie in predator-prey relations, time lag phenomena, or other violations of assumption three; in environmental variations in available energy, violating assumption one; or in variations in age at death, violating assumption two.

COMMUNITY STABILITY

In some communities the abundance of species tends to stay quite constant, while in others the abundances vary greatly. We are inclined to call the first stable and the second unstable. This concept can be made more precise, however. Suppose, for some reason, that one species has an abnormal abundance. Then we shall say the community is unstable if the other species change markedly in abundance as a result of the first. The less effect this abnormal abundance has on the other species, the more stable the community.

This stability can arise in two ways. First, it can be due to patterns of interaction between the species forming the community; second, it can be intrinsic to the individual species. While the second is a problem requiring knowledge of the physiology of the particular species, the first can be at least partially understood in the general case. Only this kind of stability will be considered here.

The interactions between species are represented graphically in a "food web" of the community, in which "*a* eats *b*" is represented by

a
|
b

with the predator above the prey. Energy goes from *b* to *a* in this process. More generally it passes from the sun to the lowest trophic level and thence up through the web to the highest level of carnivore. A qualitative condition for stability can now be stated. The amount of choice which the energy has in following the paths up through the food web is a measure of the stability of the community (Odum 1953). To see this, consider first a community in which one species is abnormally common. For this to have a small effect upon the rest of the community there should be a large number of predators among which to distribute the excess energy, and there should be a large number of prey species of the given species in order that none should be reduced too much in population. To sum up, a large number of paths through each species is necessary to reduce the effect of overpopulation of one species. Second, suppose one species is abnormally uncommon. For this to have minimum effect upon the rest of the community, each predator of the species should have a large number of alternate foods to reduce the pressure on the scarce species and yet maintain their own abundance at very nearly the original level. There is good evidence for this in Lack (1954). We see that in either case the amount of choice of the energy in going through the web measures the stability.

A difficulty arises in making this definition quantitative, because our intuition of what stability means is ambiguous. For the concept "effect upon the other species of the community" required in the definition can be interpreted in many different ways. It could be the average (over all species) maximum change in abundance, or the relative changes, or mean square changes, or different species could be weighted differently, etc. The intuitive requirements seem to be that if each species has just one predator and one prey the stability should be minimum, say zero, and that as the number of links in the food web increases the stability should increase.

One further (arbitrary) requirement will specify the function, however. For proof, see Shannon and Weaver (1949). The most convenient requirement is that the stability of two webs such as in Figure 3 and Figure 4 should be equal. Here the ½'s between *ab* and *ae* signify that each of *b* and *c* is an equally possible food of *a* (*i.e.* $\frac{1}{2} + \frac{1}{2} = 1$). Now energy has complete freedom of choice to go to *a* from section *bcd* or from section *efg*. Hence if *S(abe)* refers to the stability of *abc* considered as a separate community, etc., it is reasonable to define the stability of the whole web in Figure 3 as $S(abc) + \frac{1}{2} S(bcd) + \frac{1}{2} S(efg)$. If we say that this should be equal to $S(a'c'd'f'g')$ in Figure 4, the appropriate stability function is

$$S = -\Sigma p_i \log p_i$$

where the p_i are the numbers in Figure 4. (*E.g.*, for either the web of Figure 3 or of Figure 4 the stability is $-4(\frac{1}{4} \log \frac{1}{4}) = \log 4$). To find the stability of any web, construct an equivalent one by making a separate line from top to bottom for each different path the energy can take. The p_i in the definition are the products of the q_i of the original web along each path. It may be significant that this is identical in form to the functions measuring entropy in Maxwell-Boltzmann statistics and information in information theory (Shannon & Weaver 1949). Entropy is also analogous in the sense that maximum entropy is the condition for chemical equilibrium while maximum stability may be a condition for community equilibrium (Hinshelwood 1951). It should be pointed out that choice among various functions to define stability precisely rests only upon usefulness of the definition (provided of course that the intuitive conditions mentioned above are satisfied).

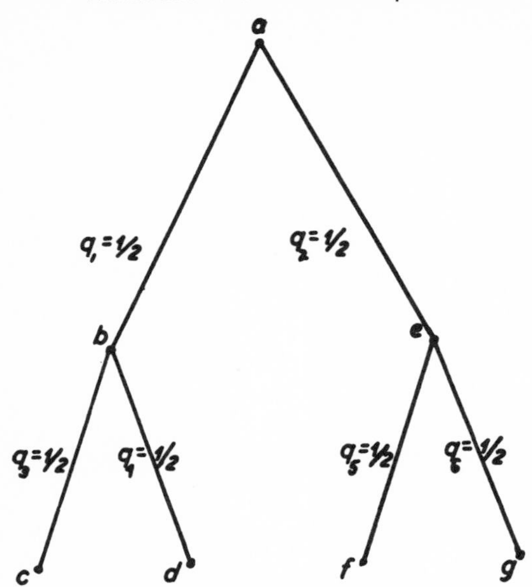

FIG. 3. A food web. *a, b, c, d, e, f* and *g* are species, and *q* on the line joining predator and prey signifies that fraction of the total number of prey species formed by the prey species in consideration.

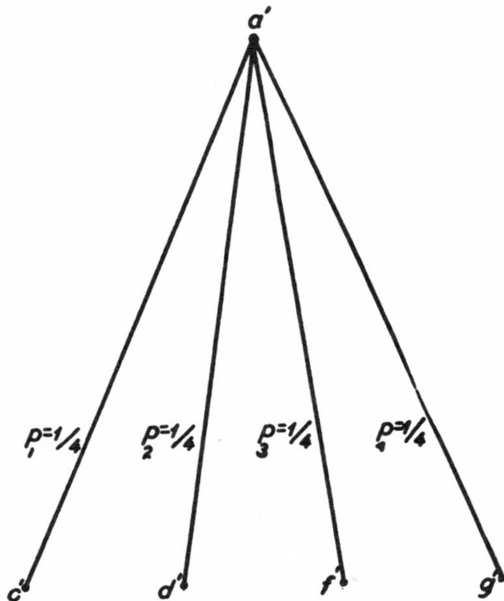

FIG. 4. A food web with stability equal to that of the food web in Figure 3.

There are several properties of this stability which are interesting.

1. Stability increases as the number of links increases.

2. If the number of prey species for each species remains constant, an increase in number of species in the community will increase the stability.

3. Combining 1 and 2, it follows that a given stability can be achieved either by large number of species each with a fairly restricted diet, or by a smaller number of species each eating a wide variety of other species.

4. The maximum stability possible for *m* species would arise when there are *m* trophic levels with one species on each, eating all species below. Similarly, the minimum stability would arise with one species eating all the others—these others being all on the same trophic level. (These are plotted in Figure 5.)

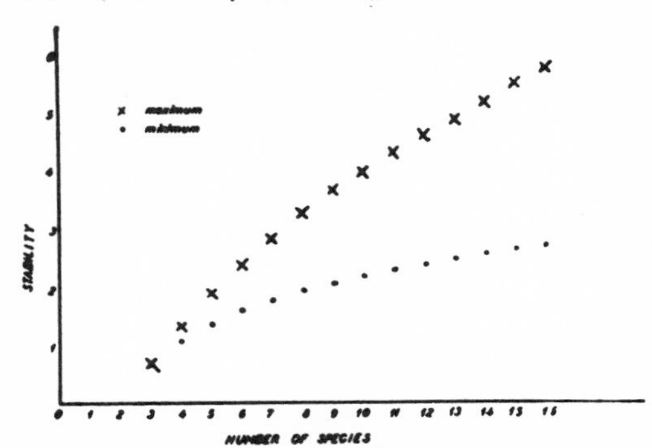

FIG. 5. Maximum and minimum stabilities for a given number of species. All communities must have stabilities within these limits.

Some interesting conclusions may be drawn from these properties. First, (1) implies that restricted diet lowers stability. But restricted diet is what is essential for efficiency. Furthermore, efficiency and stability are the two features required for survival under natural selection. Efficiency enables individual animals to outcompete others, but stability allows individual communities to outsurvive less stable ones. From this it seems reasonable that natural selection operates for maximum efficiency subject to a certain necessary stability. Combining this with the properties listed above the following seem plausible.

A. Where there is a small number of species (*e.g.* in arctic regions) the stability condition is hard or impossible to achieve; species have to eat a wide diet and a large number of trophic levels (compared to number of species) is expected. If the number of species is too small, even this will not assure stability, and, as in the Arctic, populations will vary considerably.

B. Where there is a large number of species (*e.g.* in tropical regions) the required stability can be achieved along with a fairly restricted diet; species can specialize along particular lines and a relatively small number of trophic levels (compared to number of species) is possible. (The possibility that the large number of species in tropical communities may be a result of, rather than the cause of, the restricted diets would not alter the above argument.)

REFERENCES

Feller, W. 1950. An introduction to probability theory and its applications. New York: Wiley.

Hinshelwood, C. H. 1951. The structure of physical chemistry. London: Oxford.

Lack, David. 1954. The natural regulation of animal numbers. London: Oxford.

Lindeman, R. L. 1942. The trophic-dynamic aspect of ecology. Ecology, **23**: 399-418.

Odum, E. P. 1953. Fundamentals of ecology. Philadelphia: Saunders.

Shannon, C., and W. Weaver. 1949. The mathematical theory of communication. Urbana: Univ. Illinois Press.

19

Reprinted from *Science* **173:**1134–1136 (Sept. 17, 1971)

STABILITY AND DIVERSITY AT THREE TROPHIC LEVELS IN TERRESTRIAL SUCCESSIONAL ECOSYSTEMS

L. E. Hurd, M. V. Mellinger, L. L. Wolf, and S. J. McNaughton

Department of Biology, Syracuse University

Ecosystem stability and its relationship to other ecosystem properties are among the most widely discussed ecological concepts (*1*). The origin of this discussion can be traced at least to MacArthur's (*2*) assertion that food-web diversity should stabilize a community by providing alternate channels for energy flow. Definitions of stability have ranged from convergent points in an *n*-dimensional field (*3*) to turnover time, the ratio of biomass to net productivity (*4*). We here use a definition derived from thermodynamics: stability is the ability of a system to maintain or return to its ground state after an external perturbation (*5*). The degree of stability is characterized by (i) the amplitude of the deflection from the ground state, (ii) the rapidity of response to the perturbation, and (iii) the rate at which any deflection is damped. We report here experiments designed to determine the stability of net productivity and species richness of successional abandoned hayfields after an experimental nutrient input to the primary producers. We believe this is the first experimental treatment of current ideas that relate stability to diversity and ecosystem age (*6*).

We studied two adjacent abandoned hayfields in central New York, one 1.15 hectares (2.84 acres) and the other 1.11 hectares (2.75 acres), which had not been agriculturally employed for 6 and 17 years, respectively. Macroclimate and soil series are identical (*7*). Each field was divided into two control and two treatment plots. A 10-10-10 (N, P, K) pebble fertilizer was applied to the treatment plots with a commercial lawn spreader at a rate of 560 kg/ha (555 lb/acre) on 1 May 1970 (*8*). Since enrichment has a destabilizing effect upon competitive systems (*9*), this would seem to be a particularly efficacious test of the relationship between stability and other ecosystem properties.

Producers were sampled monthly by the harvest method (*10*) from two quadrants (1.0 × 0.5 m) per plot randomized on a grid system (*11*). Samples were dried at 90°C for 20 hours and weighed. Producer biomass increased linearly with time to peak biomass in September. The best fit regression lines were determined by least squares (*12*), and net productivity, exclusive of herbivore consumption, was expressed in grams per square meter per day from the slopes of these lines. Species diversity for the producers and consumers was assessed as the number of species recorded at peak biomass (*13*). Arthropod consumers were sampled at regular intervals with a power vacuum sampler covering a 0.093-m^2 (1-square-foot) circular area

Table 1. Net productivity and diversity in control and treatment plots of two successional fields. Values are means in each case. Net productivity for the producers is in grams per square meter per day, and for the herbivores and carnivores, in milligrams per square meter per day. Diversity for the producers is number of species per 0.5-m² area; for the herbivores and carnivores, per 0.093-m² area. *F* values were derived from a 2 × 2 factorial design analysis of variance. F_A, age effect; F_T, treatment effect; F_{AT}, interaction effect. Producer *F* values were with 1,12 degrees of freedom; herbivore and carnivore *F* values were with 1,76 degrees of freedom.

	Producers		Herbivores				Carnivores			
			Early		Late		Early		Late	
	6-year field	17-year field	6-year field	17-year field	6-year field	17-year field	6-year field	17-year field	6-year field	17-year field
					Net productivity					
Control	4.46	2.68	6.06	3.56	4.17	3.77	4.75	2.43	1.61	1.01
Treatment	8.76	4.56	7.92	10.71	4.77	11.85	4.45	5.05	1.17	1.36
F_A	1549.7*		0.2		100.9*		3.6		0.6	
F_T	1662.0*		93.6*		172.0*		6.8†		0.0	
F_{AT}	255.2*		31.8*		127.8*		10.6*		2.4	
					Diversity					
Control	10.25	17.50	3.40	3.65	3.35	4.95	1.85	1.75	1.50	2.50
Treatment	9.50	18.00	4.20	5.50	5.45	4.45	2.35	3.05	1.30	2.05
F_A	46.0*		17.7*		1.4		3.4		17.3*	
F_T	0.0		51.8*		10.2*		31.0*		0.4	
F_{AT}	0.3		8.1*		26.9*		6.1†		0.4	

* $P < .01$. † $P < .05$.

(*14*). These sample positions were also randomized on the grid system. Samples were dried at 70°C for 48 hours and weighed. There were two arthropod biomass peaks, one in late May and the other in late July or early August. Net productivity of the arthropod herbivores and carnivores, in milligrams per square meter per day, was determined by summing the positive slopes of the biomass curves.

Net producer productivity in both control and treatment plots was significantly greater in the young field (6 years from last cultivation) than in the old field (17 years from last cultivation) (Table 1). Producer productivity was significantly greater in treatment plots than in control plots, in both fields. There was a significant interaction between age of the field and response to treatment. Simple effects analysis (*15*) indicated that the old field ($F_{1,12} = 307.4$, $P < .01$) responded less than the young field ($F_{1,12} = 1610.0$, $P < .01$). Although the number of species was significantly greater on the old field, there was no diversity response to fertilization. It might be argued that one growing season is insufficient time for additional plant species to invade. Catastrophic perturbations which open multiple niches, however, are followed by rapid appearance of several previously absent species over periods of just a few weeks (*16*), and it is well known that soils typically contain a reservoir of viable seeds of many species that do not occur in the community itself (*17*).

Arthropod consumers responded to the perturbation somewhat differently than the producers did. Herbivore productivity at the first biomass peak was significantly greater in the fertilized plots (Table 1). However, simple effects analysis indicated that nutrient input to the producer level affected herbivore productivity in the old field significantly more ($F_{1,76} = 117.3$, $P < .01$) than in the young field ($F_{1,76} = 8.1$, $P < .01$). In agreement with the plants, the first peak herbivore diversity was greater in the old field. In addition, however, species diversity was greater in the treatment plots, and the magnitude of the diversity increase was greater in the old field ($F_{1,76} = 50.5$, $P < .01$) than in the young field ($F_{1,76} = 9.4$, $P < .01$). In the control plots, the young field herbivore productivity at the first peak was greater, but species diversity was less, indicating higher productivity per species in the young field.

Herbivore productivity response at the second peak was similar to the first peak (Table 1). Simple effects analysis indicated that the major treatment effect was contributed again by the old field ($F_{1,76} = 298.1$, $P < .01$), not the young field ($F_{1,76} = 1.6$, $P > .10$). While herbivore diversity at the second peak increased in the control plots with successional age, it decreased on the treatment plots with age. Simple effects analysis indicated a significant increase in species diversity in the treatment part of the young field ($F_{1,76} = 35.1$, $P < .01$) but no significant difference between treatment and control plots in the old field ($F_{1,76} = 2.0$, $P > .10$). Herbivore productivity per species decreased seasonally in the control and treatment plots in the young field, while in the old field herbivore productivity per species decreased in the control and increased in the treatment plots.

Carnivore productivity at the first peak (Table 1) was enhanced more in the old field ($F_{1,76} = 7.3$, $P < .01$) than in the young field ($F_{1,76} = 0.2$, $P > .10$). Fertilizer treatment increased species diversity in both fields. Productivity per species in the young field decreased with treatment but increased slightly with treatment in the old field. There were no significant treatment or age differences in carnivore productivity at the late season peak (Table 1). Species diversity was greater in the old field, but there was no treatment effect upon late season carnivore diversity.

Current theory (*6*) states that stability and diversity increase while productivity decreases during succession. Our results support this theory completely only at the producer trophic level. For the producers the old field is more diverse, less productive, and more stable than the young field. Since producer diversity does not increase upon perturbation and productivity does, additional productivity is not a result of adding more plant species. In fact, in successional time, net productivity within both control and treatment plots is an inverse function of species richness at the producer level. Patten's (*18*) studies of plankton populations indicated that environmental effects were damped temporally in the plankton. Our studies indicate that this damping property of the ecosystem's biotic component increases with successional time. The average fertilizer stimulation of net primary productivity was 97 percent in the 6-year-old field and 71 percent in the 17-year-old field ($t = 4.17$; $P < .01$; d.f. $= 6$).

Productivity of both herbivores and carnivores in the control plots declined with successional time. Consumer diversity increased in the control plots with successional time except for the first carnivore peak. Both of these observations also support current successional theory.

The primary departure from current theory is that stability decreased with successional time, and with species diversity, at the consumer trophic levels. This is particularly apparent at the herbivore level. Old field herbivore populations were much more responsive, either proportionately or absolutely, to the experimental treatment than young field populations at both biomass peaks. Since the species composition is mainly annual arthropods, it seems likely that reduced mortality, rather than increased fecundity, generated the effect. In addition, there may have been immigration in reponse to the increased food supply. The ability of the fertilizer to enhance herbivore productivity contradicts the hypothesis of Hairston *et al.* (*19*) and the data of Pulliam *et al.* (*20*), both indicating that herbiovores are not food limited. In fact, only a relatively small proportion of total plant biomass actually may be available to the herbivores (*21*), so that enhancement of producer productivity may have a substantial effect in increasing the amount of food available to the consumers.

Similarly, the carnivores were much less stable in the old field during the first biomass peak. However, the perturbation effect was seasonally damped in this trophic level, and there was little response during the second biomass peak.

An important property of ecosystem succession is the attenuation, with time, of ecosystem output, that is, net productivity. This attenuation occurs at all three trophic levels in the fields we examined. It is clear, however, that this attenuation is accompanied by increased stability only at the producer level. Although arthropod consumer productivity declines with successional age, the disruptive effect of the perturbation increases with age. Perturbation effects are undoubtedly a function of the kind of perturbation employed and the species composition of the communities affected (*22*). The most responsive species will be those specialized for, and limited by, the niche parameters influenced by the environmental modification (*13*). Species with different critical niche dimensions will be affected less by the same modification.

Although MacArthur (*2*) argued that stability of a population should increase as the number of food species available to it increases, our results suggest that increased diversity at the plant and herbivore levels generates decreasing stability at the next higher trophic level. This is similar to Watt's (*23*) results from examining herbivorous insects. In our experiment, the only reasonable explanation of the lower stability in the old field consumers appears to be the possibility that fertilization preferentially enriched palatable food classes at the primary level. Documentation of such an enrichment would require careful determination of food preferences at the herbivore level, and the availability of these food classes within the plant communities.

The third aspect of stability, the long-term damping effect, is not reported here because such evidence will require several years to obtain and, at any rate, will not affect the interpretation of the first two components of the stability definition.

References and Notes

1. G. M. Woodwell and H. H. Smith, Eds., *Brookhaven Symp. Biol.* **22** (1969).
2. R. H. MacArthur, *Ecology* **36**, 533 (1955).
3. R. C. Lewontin, *Brookhaven Symp. Biol.* **22**, 13 (1969).
4. R. Margalef, *Amer. Natur.* **97**, 357 (1963); S. J. McNaughton, *Ecology* **49**, 962 (1968).
5. H. von Foerster, *Brookhaven Symp. Biol.* **10**, 216 (1957).
6. E. P. Odum, *Science* **164**, 262 (1969).
7. Both fields are on Honeoye-Lima soils [see G. W. Olson, J. E. Witty, R. L. Marshall, *Cornell Misc. Bull.* **80**, 45 (1969)].
8. Fertilization information obtained from Cooperative Extension, U.S. Department of Agriculture, Cornell University, Ithaca, N.Y.
9. M. L. Rosenzweig, *Science* **171**, 385 (1971).
10. E. P. Odum, *Ecology* **41**, 34 (1960); C. R. Malone, *Amer. Midland Natur.* **79**, 429 (1958).
11. Principal plant species on the 6-year-old field were *Phleum pratense, Solidago canadensis,* and *Picris hieracoides*; on the 17-year-old field the principal plant species were *Phleum pratense, Solidago canadensis, Picris hieracoides, Poa compressa, Fragaria virginiana,* and *Hieracium pratense.*
12. All slopes of the best fit regression lines were significantly different from zero (t range: 4.83 to 17.00; $P < .02$; d.f. = 3).
13. S. J. McNaughton and L. L. Wolf, *Science* **167**, 131 (1970).
14. E. Harrell and R. Davis, *J. Econ. Entomol.* **58**, 791 (1965); G. Barrett, *Ecology* **49**, 1019 (1968).
15. R. G. D. Steel and J. R. Torie, *Principles and Procedures of Statistics* (McGraw-Hill, New York, 1960), p. 202.
16. A. W. Sampson, *Bull. Calif. Agric. Sta.* **685** (1944); J. S. Horton and C. J. Kraebel, *Ecology* **36**, 244 (1955); J. R. Sweeney, *Univ. Calif. Publ. Bot.* **28**, 243 (1956); I. F. Ahlgren and C. F. Ahlgren, *Bot. Rev.* **26**, 483 (1960).
17. H. J. Oosting and M. E. Humphreys, *Bull. Torr. Bot. Club* **67**, 253 (1940); J. Major and W. T. Pyott, *Vegetatio* **13**, 253 (1966); M. C. Kellman, *Can. J. Bot.* **48**, 1383 (1970).
18. B. C. Patten, *Science* **134**, 1010 (1961).
19. N. Hairston, F. Smith, L. Slobodkin, *Amer. Natur.* **94**, 421 (1960).
20. H. R. Pulliam, E. P. Odum, G. W. Barrett, *Ecology* **49**, 772 (1968).
21. P. Feeney, *ibid.* **51**, 565 (1970).
22. D. Garfinkel, *J. Theoret. Biol.* **14**, 46 (1967).
23. K. E. F. Watt, *Can. Entomol.* **96**, 1434 (1964).

24. We thank Mr. R. Amidon for the use of his land for this research. Supported by NDEA Title IV Fellowship (M.V.M.), NSF Traineeship (L.E.H.), NSF grants GB-8099 (S.J.M.) and GB-7611 (L.L.W.), and Syracuse University funds.

26 April 1971

20

Reprinted from *Am. Zool* **10:**17 5 (1970)

Evolution of Diversity, Efficiency, and Community Stability

ORIE L. LOUCKS

Department of Botany, University of Wisconsin, Madison, Wisconsin 53706

SYNOPSIS. The response in species diversity associated with successional change in vegetation, or in a more general sense, species diversity as a function of time in any system of primary producers, has been the subject of much speculation but little direct study. All evidence available shows that pioneer communities are low in diversity, that in mesic environments the peak in diversity in forest communities can be expected 100-200 years after the initiation of a secondary successional sequence (when elements of both the pioneer and the stable communities are present), and that a downturn in both diversity and primary production takes place when the entire community is made up of the shade-tolerant climax species.

The natural tendency in forest systems toward periodic perturbation (at intervals of 50-200 years) recycles the system and maintains a periodic wave of peak diversity. This wave is associated with a corresponding wave in peak primary production. Specialization for the habitats in the early, middle, and later phases of the cycle has figured prominently in species-isolating mechanisms, giving rise to the diversity in each stage of the forest succession. It is concluded that any modifications of the system that preclude periodic, random perturbation and recycling would be detrimental to the system in the long run.

Ever since the papers by MacArthur (1955, 1957), Margalef (1957), and Hutchinson (1959), research in community biology has focused increasingly on the significance to be attached to the diversity of organisms making up a community. In particular, much speculation and some generalization have been directed to the role that species diversity plays in community stability and total community productivity, as well as the reverse, the dependence of diversity and stability of the community on its productivity.

The research results for the past decade can be summarized as (1) papers contributing information on diversity for a wide range of communities, (2) theoretical expositions on the probable mechanics of evolutionary build-up in diversity, and (3) a large number of papers concerned with measures and techniques for estimating and interpreting the number of species in relation to their abundance. Most of the research has dealt with limited segments of regional community gradients, usually with emphasis on the animal component. Very few have dealt with the primary producers, particularly in terrestrial communities. As a result, there have been few opportunities to date to view diversity, and its associated ideas—stability and productivity—as time-based functions, potentially dependent on the stage of development of the native communities.

The objectives of this paper, therefore, are: (1) to add somewhat to the data on diversity and productivity in Wisconsin forest communities; (2) to show how both of these vary sharply, rising and then falling, over relatively short periods of time in the development of a forest; and (3) to offer an improved model for describing the evolution of the biological materials in ecosystems, one which accepts time-separation selection mechanisms, as well as spatial or behavioral segregating mechanisms, in the ecological and genetic isolation of species.

RECENT INTERPRETATION OF DIVERSITY

Consider the summary of community diversity, and related properties (Fig. 1), an interpretation based on recent literature. Authors have referred in general terms to the characteristic relationships be-

This research was supported by grant No. GB-3548 from the National Science Foundation. The contributions of Professor Donald Watts and of Messrs. David Parkhurst and Burton Schnur are gratefully acknowledged.

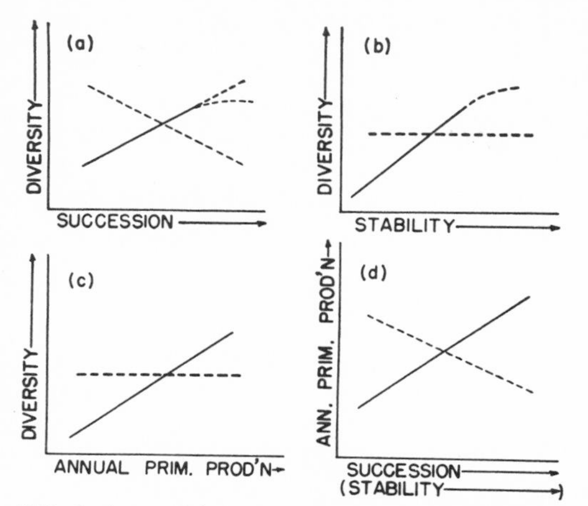

FIG. 1. A graphic summary of the relationships reported between diversity of species and other properties of biotic communities.

tween diversity, succession, stability, and primary production shown here as trend lines. Solid lines show the relationships where there seems to be some consensus with respect to the trends; dotted lines indicate trends suggested by some authors but debated by others.

Whittaker (1965) and Monk (1967) have suggested the increase in diversity with the successional sequence [Fig. 1 (a)]; others such as Margalef (1957, 1968) and Odum (1963) have indicated that this curve probably levels off and may drop near the latter portion of a successional sequence. Whittaker (1965) and Pielou (1966) have suggested that under certain conditions diversity may decrease with succession, as shown by the dotted line. The relationship between diversity and stability [Fig. 1 (b)] has many interpretations, largely dependent on one's view of stability. The work of MacArthur (1955) and the discussion of diversity under very stable environments (Connell and Orias, 1964) suggest that a high stability is associated with a high level of diversity, with no clear indication of which is dependent on the other. Over a shorter time, however, the stability of late stages in a secondary successional sequence is acknowledged by Odum (1963) to lead to leveling or possibly a decline in diversity, and there may be no relationship between diversity and stability. There is considerable hedging on the definition of stability for several very good reasons that I will re-examine in the latter part of this paper.

Figure 1 (c) expresses the consensus that with higher levels of annual primary production there will be higher levels of community diversity (Margalef 1968), although Odum (1963) and Whittaker (1965) have questioned this relationship. Their position is that there may be no consistent relationship, as shown by the horizontal dashed line. The axiomatic results of the three previous figures, summarized in Figure 1 (d), show that with more advanced succession and an increased stability, one might expect higher levels of production by the primary producers, as well as in other levels of the system. On the other hand, Odum (1963) notes a net increase in "community productivity" with advancing successions.

These figures serve both as a mechanism for presenting some of the published views on diversity and related phenomena, several of which I must seriously challenge, and as a means of introducing definitions for each of the objects under consideration. First of all, diversity will be defined here as a mean entropy per individual, simply by my choice of the Shannon-Weiner diversity measure

$$H' = -\Sigma P_i \log_2 P_i$$

where P_i is the probability of sampling the *i*th species.

The sequence we know as succession is shown equivalent to a progression in stability in Figure 1 (d) to show the view among plant ecologists that as long as change in composition (*i.e.*, succession) is evident, the system is only in the process of approaching a stable community, and is, therefore, by implication still unstable. Many of the expressions for stability used in other studies are acknowledged to be a much longer-term phenomenon, bearing little relationship to the stability of terminal communities just described. Neverthe-

TABLE 1. *Diversity expressed as the mean entropy per individual (H′) ± one standard error, for six community types (after Monk, 1967).*

Community	Mean H′ ± 1 Standard Error		
	Trees	Saplings	Seedlings
Pioneer Communities			
(1) Sand Pine Scrub	1.55 ± .02	1.56 ± .07	1.81 ± .03
(2) Sand Post Oak Phase	1.54 ± .34	1.95 ± .27	1.79 ± .29
(3) Slash Pine Phase	1.00 ± .20	1.76 ± .14	1.88 ± .16
(4) Longleaf Pine Phase	0.82 ± .17	2.03 ± .21	2.17 ± .15
Climax Communities			
(5) Mesic (non-calcareous phase)	2.50 ± .09	2.68 ± .08	1.78 ± .12
(6) Mesic (calcareous phase)	3.06 ± .09	2.74 ± .14	2.73 ± .15

less, a successional view of the approach toward stability in communities is very widespread and must be considered in this analysis. Thus, for the time being I wish to accept both a short-term (successional) and a long-term (evolutionary) view of stability.

The annual primary production of plants is defined in units of dry weight fixed per unit area per year, and when expressed as a ratio of the total annual energy input in southern Wisconsin, converts by a scaling factor to energy efficiency of the primary producer trophic level.

To examine some of the relationships between diversity, stablility, and efficiency of primary production, I have used data for forest communities in two areas: the results published by Monk (1967), and unpublished data obtained for southern Wisconsin during 1967 and 1968. These two studies represent more complete sampling of the community than in most previous studies. Both the seedling and sapling component in the forests have been sampled to obtain estimates with as great a confidence as that of the overstory trees. The primary advantage of the larger sample lies in the opportunity it affords to divide each forest on the basis of internal structural strata (overstory, mid-canopy, saplings, and seedlings) and to examine diversity and other characters in each stratum.

The results of Monk's study (1967) of diversity in tree species in deciduous forests in north central Florida (Table 1) are significant primarily because they distinguish diversity of plant species for each of the major layers the tree species occupy: seedlings, saplings, and trees. The significant result is that the communities dominated by what we would call pioneer species show a low diversity for the overstory and a higher diversity of seedlings. The reverse is true of the communities dominated by mesic hardwood species which Monk refers to as climax. In each of the latter, the greatest diversity is in the overstory tree layer, with a lesser diversity in the seedling layer. No significance is attached to the small ups and downs of species diversity in the sapling layer. Although Monk views the higher diversity of the seedling layer in pioneer forests as evidence of an increase in diversity with succession, he draws no conclusion regarding the low diversity in the understory of the so-called climax communities. The analysis of Wisconsin data was undertaken in an attempt to confirm Monk's results, and to probe the processes in the climax situation.

DIVERSITY IN WISCONSIN FORESTS

The results of an analysis similar to that of Monk for 30 stands in Wisconsin are illustrated in Figure 2. They are presented here as average diversity for three stands in each decimal segment of the Wisconsin upland compositional gradient (Curtis and McIntosh, 1951). The diversity of the overstory trees is low in the droughty, extreme environments of the low compositional index forests, ranging upward with

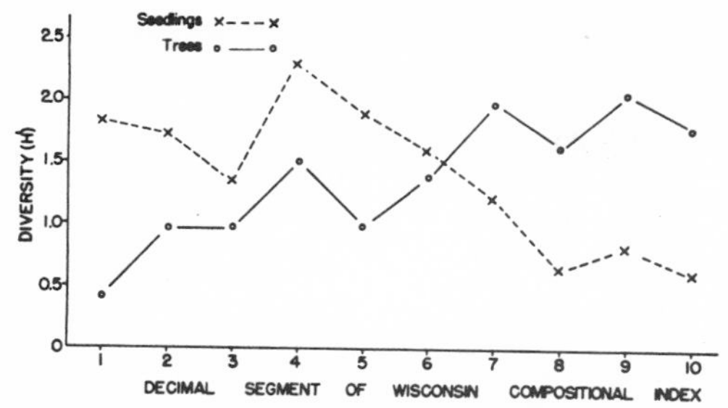

FIG. 2. The inverse relationship between diversity of species of the tree and seedling strata across the range in environments and composition represented in the southern Wisconsin compositional gradient.

small random fluctuations to a plateau level across the four right-hand segments of the gradient. On the other hand, seedling diversity begins relatively high, and after some fluctuations which are in part due to environmental differences, declines to a very low level across the three highest segments of the gradient. The break in the low end of the gradient results from the fact that the first three decimal segments are all characterized by coarse-textured soils with bedrock frequently close to the surface. The fourth segment is not greatly different in forest composition, but is quite different in texture of substrate, depth of soil, and total water storage available to seedling and sapling layers. In most respects the environment of this segment is like that of the remaining segments toward the mesic end of the gradient. The result of the more favorable environment is, as one might expect, a greatly increased diversity of both seedling and tree layers, both reflecting in part the high light conditions of these relatively pioneer communities as well as the higher supply of moisture.

While the inverse relationship between the diversity of the understory compared with the overstory is interesting, and of considerable consequence to diversity relations across the landscape in southern Wisconsin, the great differences in environment across the compositional gradient mean that these results mix the two important elements influencing diversity: (1) the restrictions imposed by a severe environment; and (2) the variations that may be imposed as a direct result of community development over time within a restricted range of environment. Total "available water capacity" has been estimated for each of the 30 stands to depths of both 36 and 60 inches. The mesic environments represented by the four highest decimal segments of the gradient are all characterized by an available water capacity of seven or more inches of water in the surface three feet of soil. Therefore, all stands, regardless of position along the compositional index, with an available water capacity of seven inches or more have been used for the following analyses of diversity in mesic environments.

To establish a time base for the analysis of diversity, age determinations were carried out on the 18 stands with high levels of substrate water supply. The estimates of age for large old-growth hardwoods utilized both historical records dating to the surveyors' descriptions in the 1830's, and correlation between diameter and age. The latter are good for the shade-tolerant species whose widths of annual ring are consistent over extended periods. The aging was sufficiently precise to allow recognition of five age-classes among the stands sampled, ranging from one whose midpoint is just over 100 years, to the oldest with a midpoint of 220 years.

The mean diversity of the seedling layer in stands across the age range from 100 to 220 years is shown in Figure 3. On theoretical grounds alone one expects the model for seedling diversity over time to show a high level immediately after disturbances, then a very low level under the young stand of saplings. One younger set of stands from a northern Wisconsin study has a seedling layer diversity of 1.4 at 30 years, giving a further indication of the shape. From the evidence here, and our knowledge of processes in the deciduous forest, we are able to say the response in seedling diversity with time is a curve which drops briefly after the initial establishment of a young forest, then rises

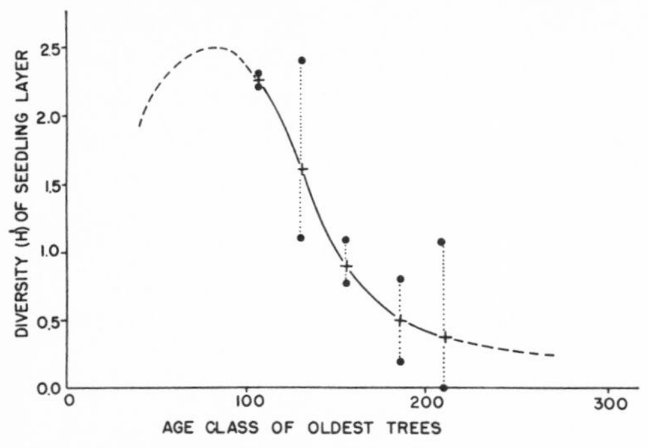

FIG. 3. Diversity of species in the seedling layer, plotted by age-class of the oldest overstory trees.

steeply to peak in the oak communities around 100 years of age, and then falls off sharply. The low diversity in the oldest age-classes sampled represents domination of the seedling layer by one very shade-tolerant species, sugar maple. At 100 years of age, the seedling layer is made up of scattered individuals of the overstory trees which survive briefly as seedlings, but which will not enter the sapling layer. In addition, there are large numbers of individuals of the more shade-tolerant species cited by Monk (1967) entering the community at this point in time. One hundred years later, there is no longer a canopy with any of the pioneer species and therefore no seedlings of species which owe their origin to the physical presence of a source of seed in the canopy.

These results imply major change in composition over this time sequence, estimated both by comparison of stands of different ages in similar environments, and by comparison of layers within the strata that may be expected to replace one another over time. It is essential, therefore, that some measures of the apparent rate of change from one stratum to the next be considered for each of the age-classes available. One potentially useful measure has been to utilize the compositional index (C. I.) for each of the layers, trees, saplings, and seedlings. An increase in C. I. from the overstory to the understory would indicate actual change in composition, not simply change in diversity. Such a measure of change works well in either pioneer or high C. I. stands, but for a number of reasons it is not satisfactory elsewhere along the gradient. The central reason lies in the fact that the compositional index is based on an "adaptation number" (Curtis and McIntosh, 1951) for all of the species in a compositional gradient, an index of the species' capability to reproduce and replace an overstory species. The results of this study show that many of the intermediate stands do not have any differences in the mean adaptation number from understory to overstory, although each stratum is in fact made up of substantially different species.

Because of the inadequacy of the compositional index for stands midway along the southern Wisconsin gradient, a new expression, the *coefficient of change,* has been developed to describe what the progression in compositional index through the canopy attempts to do. The coefficient of change is based on the index of similarity

$$c = 2w / a + b$$

where w is the quantity (relative basal area and relative density) of a species in common between any pair of forest strata (here, the tree and seedling layers), and a and b are the quantities of the species in each of the layers. The index of similarity is a decimal or percentile function used by Gleason, Bray, Curtis, and others, and usually transformed by subtraction from unity to yield an index of difference. In this example, the index gives percentage differences between the overstory and understory strata. The magnitude of the difference will be in part influenced by the total magnitude of difference in diameter between the overstory and understory layers, so that to achieve equalization across stands of different mean diameters, the magnitude of difference has been divided by the total difference in diameter, yielding units of percent change per inch of difference in diameter.

The values of the coefficient of change (Fig. 4) show another wave form similar

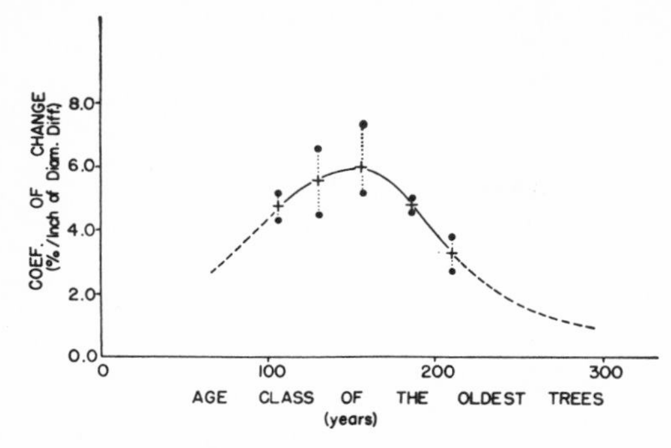

FIG. 4. The coefficient of change, a measure of the difference in composition between the tree and seedling layers, plotted as a function of age-class of the oldest trees.

to that of the seedling diversity, but displaced approximately 50 years later in the sequence of time represented by these stands. The functional dependency of the wave pattern represented by the coefficient of change on the wave pattern in seedling diversity is evident from the fact that the peak input of diversity in the seedling layer at 100 years will be followed by a corresponding change in composition some years later as the shade-tolerant seedling species gradually dominate the seedling layer.

The production of dry-matter in these forests has been estimated by common forest mensurational techniques. The total basal area of living stems in sq. ft. / acre is accepted as a good estimator of total plant biomass. A plot of basal area of stand over age of stand shows that the youngest stands support among the lowest basal areas, the middle age range is highest, and the oldest stands support a low basal area. This is a well-documented phenomenon in forestry in which the wave of high basal area developed by the initial wave of pioneer species is reached at some age determined by the longevity of the pioneer stand itself. After that point, only a lower level of basal area will be supported in perpetuity. Tables of forest yield for hardwood species also indicate that the annual increment of wood in the forest can be estimated very closely by assuming a 2% per annum increase in basal area. The total volume of increase in biomass can be estimated by multiplying the increase of basal area by the mean height of each stand, and the total increase in dry weight is estimated by a product of the volume times the mean specific gravity for the dominant species of each stand.

Figure 5 summarizes the estimates of production of dry-matter exclusive of tree foliage and herbs for all 30 forests distributed across the 10 segments of the southern Wisconsin upland gradient. The range in values among the 3 stands in each decimal segment is shown. The 2% per annum increase probably underestimates increments slightly in young stands, and overestimates it slightly in the older stands. The bars in segments 1, 3, and 6 are the estimates of production of dry matter obtained by dimension-analysis methods in three stands of the same compositional index as those adjacent to each bar. The combined results show the profound influence of the environmental extremes at the low end of the compositional index. Segment 4, with generally favorable environments, was sampled by somewhat younger stands and the estimate of primary production appears low for that reason. The other six higher segments represent the peak levels of productivity, but there is some indication of a drop in primary production among the highest CI values.

The results for the 18 stands of maximum available water capacity are shown as a function of age (Fig. 6). A wave

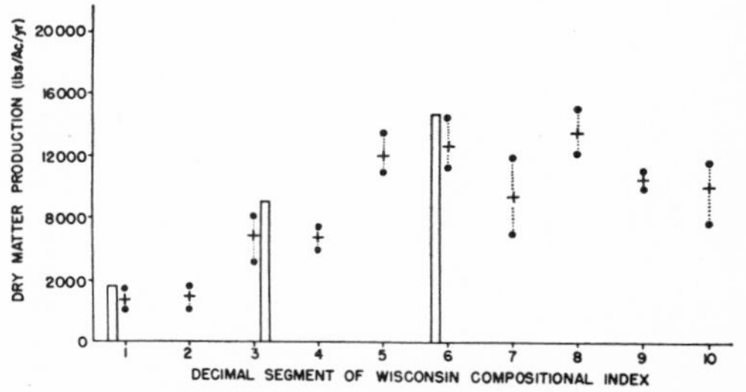

FIG. 5. Primary production fixed annually by the tree vegetation (exclusive of foliage), plotted by decimal segment along the southern Wisconsin forest gradient, shows the impact of the severe environment at the low end of the gradient.

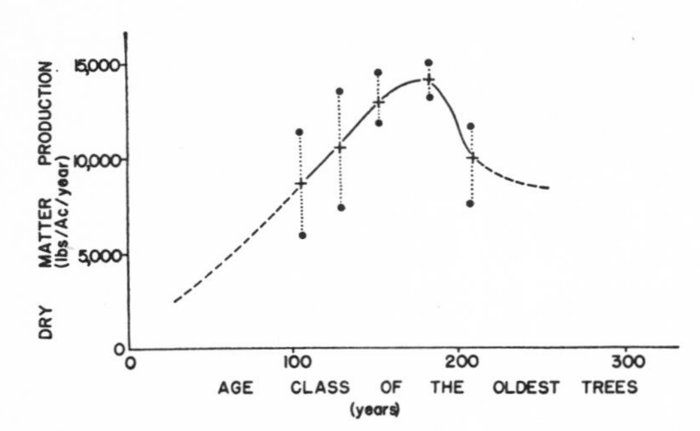

FIG. 6. Primary production fixed annually by the tree stems, plotted by age-class of the oldest trees in each stand.

response not unlike those for seedling diversity and the coefficient of change is observed again, but with the peak being reached just under 200 years of age. Again the functional dependency of this response on the two previous wave patterns can be recognized, in that the decrease in coefficient of change is attributable to the gradual dominance of both overstory and understory layers by the shade-tolerant species, particularly sugar maple. As long as there is a significant component of the pioneer species, particularly red oak, in the overstory canopy, the annual production of dry matter can be maintained at a high level. As the stand is purified to almost all sugar maple, the coefficient of change is drastically reduced, and production of dry matter ultimately is also reduced.

The wave-form during the early years after establishment of the stand is known to be the first part of a gently rising sigmoid curve. The three wave-phenomena can be viewed as taking place simultaneously in each stand, or in a composite of stands across a landscape. Taken together, they represent a step toward a more quantitative statement of what we have always known as secondary succession from white oak through red oak to sugar maple. However, the wave patterns shown here as functions of time are not once-only phenomena. They have taken place many times before, and must be thought of as repeating phenomena in a long-term sequence.

INTERPRETATION

The longer period of time over which we should consider community development is shown schematically in Figure 7. As a working hypothesis, I suggest that community characteristics be viewed as repeating wave-form phenomena triggered by random perturbations with intervals of 30 to 200 years or more. The intervals can be longer or shorter in other types of vegetation. The response property plotted in Figure 7 is the diversity of the understory layer, but any of the other transient phenomena discussed as functions of time in the previous figures could be used. Each one goes through a characteristic wave-form response every time the community sequence is restarted.

The sequence can best be described as a *stationary process,* with random perturbation. In this view, the notion of "instability" applied to a forest in which change in composition is taking place is inappropriate. The changes that take place are in fact part of a characteristic series of transient phenomena which collectively make up the "stable system" capable of repeating itself every time a perturbation starts the sequence over.

The evolutionary implications of the "stationary process", and the recurring stable system of transient phenomena, are significant because they allow recognition of the potential for selecting species to carry out specialized functions at each of

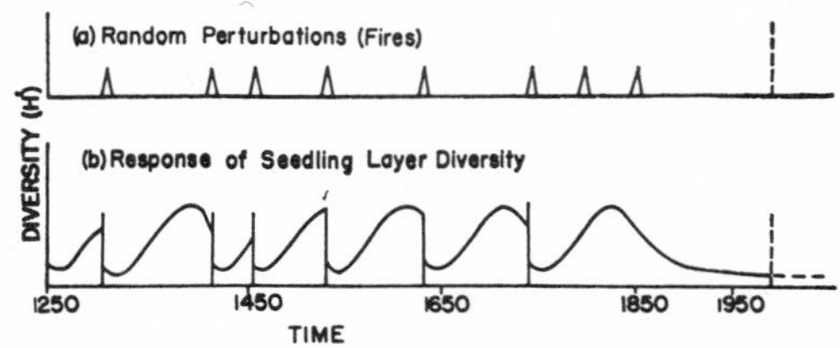

FIG. 7. A model showing the pattern of response in seedling diversity across an extended period of random perturbations in the southern Wisconsin forests. The whole series of transient response phenomena making up the stable system is restarted by each perturbation, but with control of the environment by man, the recycling process may never recur.

several phases (time-differentiated) in the responding ecosystem. We are accustomed to thinking of geographic or environmental properties as contributing to the isolating mechanisms in speciation, but I believe that this is the first indication of long-term wave-phenomena as an isolating mechanism, and I believe its role in the evolution and maintenance of diversity in ecosystems has been greatly underestimated.

To understand the selection mechanisms leading to this wave-isolation, we should consider the mechanisms by which there is a substitution of species over the period of community development. The species which make up the first wave of seedlings in this time sequence, and which later make up the early tree canopy and contribute the greatest primary production, have long been recognized as species with unique adaptations for occupying open habitat. The replacing species, on the other hand, have been thought of as distinct primarily in the physiological adaptation of leaf tissue to low light conditions.

An alternate way of looking at the differential success of these species at each age in the time sequence is to examine the mortality of the annual input of seedlings, somewhat in the form of a life-table analysis. The negative exponential has been investigated as a model capable of describing seedling depletions for three species in Ontario (Hett and Loucks 1968). The depletions of seedlings of sugar maple and other species have been investigated in Wisconsin and we find the best measure distinguishing the "replaced" and the "replacing" species is the difference in the slope of the depletion curve. A gentle slope is typical of the shade-tolerant populations in these forests, while a steep depletion characterizes the pioneer species in the shaded understory environment. The slope of the seedling depletion curves is therefore suggested as one of the characteristic properties which functionally describe the mechanisms for differentiation of diversity, the coefficient of change, and primary production along the time gradient in Wisconsin forest communities. The slope of the depletion curve can be viewed as a property of the understory populations across this time scale, significant because of the extent to which it expresses the qualitative adaptations identified with pioneer, intermediate, and shade-tolerant species.

The implications of the selection mechanism may be grasped more clearly when the annual production of dry matter in any one forest is schematically partitioned into three general categories of species: pioneer, intermediate, and shade-tolerant. Each is potentially distinguishable by a specified range in the species depletion slope. From this viewpoint, one of the primary mechanisms of selection among the pioneer species seems to have been toward maximizing the fixing of energy while occupying an open environment —a competition in which the plants which reach the largest size most quickly survive to reproduce. They have done so at the expense of survival in their own shade. The species at the other end of the time scale are specialized primarily in self-perpetuation under extreme shade. They have remarkable adaptations for saturating the environment with progeny, and appear to have done so at the expense of the capability for maximizing the conversion of energy. The answer to the question posed by Frank (1968) appears to be that in forest communities the specializations for self perpetuation are real and not tautological.

The difference in selection pressure is best illustrated by certain genera in which two species have been differentiated, primarily by the mechanisms just described, and which now play time-isolated functional roles in the system. Examples in the Wisconsin forests are the shagbark and yellowbud hickories (*Carya ovata* and *C. cordiformis*), the American and slippery elms, (*Ulmus americana* and *U. rubra*) and the red and the silver maples (*Acer rubrum* and *A. saccharum*).

CONCLUSIONS

In our most complex ecosystems, we now have the basis for predicting with some confidence the diversity of tree species to be expected of each layer within the community, and for expressing it as a function of the periods of time through which the community is developing. We should view diversity as bearing some non-linear rather than a linear relationship with time. These techniques are suggested as one of the methods by which we can extrapolate the prospective effects of man's impact on the landscape into the future.

We now have evidence for stating that the peak in diversity of species as well as the peak levels in production of dry matter characteristic of many forest communities east of the Great Plains cannot be maintained into the future if the present man-controlled environments are continued. The results reported here show that to achieve a continued high level of species diversity as well as high productivity, man should restore the mechanisms which triggered the re-cycling and rejuvenation of the biotic systems. He should restore the mechanisms by which the wave-isolated species work together over time, to produce a characteristic period of peak production and peak diversity at intervals of one to three centuries.

I offer the hypothesis that evolution in ecosystems has brought about not only adaptation to heterogeneous environments, but adaptation to a repeating pattern of changing environments, a stationary process that represents a composite of time intervals over which replacement of species is repeated over and over again. The periodicity ranges from several centuries in the northern lake forest, where white pine seems to have been re-cycled at intervals of 300-400 years, through the shorter intervals in the deciduous forests of the Mississippi Valley, to periods of a few decades and even annually at the edge of the grasslands.

On the basis of the evidence presented, and hypotheses formulated from it, I conclude that diversity of plants and animals must be understood as an ephemeral product of the partial overlap of wave-isolated species, most strongly expressed in rich as opposed to extreme environments. The peak in conversion of energy is likewise to be viewed as an ephemeral phenomenon, dependent on the random rejuvenation of the system, a process which modern man has abruptly halted. In my opinion, the interruption of the natural stationary process will, in all likelihood, be the greatest upset of the ecosystem of all time, and one that must be a central focus of the forthcoming International Biological program. It is an upset which is moving us unalterably toward decreased diversity and decreased productivity at a time when we can least afford it, and least expect it.

REFERENCES

Connell, J. H., and E. Orias. 1964. The ecological regulation of species diversity. Amer. Naturalist 98:299-414.

Curtis, J. T., and R. P. McIntosh. 1951. An upland forest continuum in the prairie-forest border region of Wisconsin. Ecology 32:476-496.

Frank, P. W. 1968. Life histories and community stability. Ecology 49:355-357.

Hett, J. M., and O. L. Loucks. 1968. Application of life-table analyses to tree seedlings in Quetico Provincial Park, Ontario. Forestry Chronicle 44:29-32.

Hutchinson, G. E. 1959. Homage to Santa Rosalie or why are there so many kinds of animals? Amer. Naturalist 93:145-159.

Margalef, D. R. 1957. La teoria de la informacion en ecologia. Men. Real Acad. Cienc. Art. Barcelona 32:373-449 (Trans. Soc. Gen. Syst. Res. 3:36-71).

Margalef, D. R. 1968. Perspective in ecological theory. Univ. of Chicago Press.

Monk, C. D. 1967. Tree species diversity in the eastern deciduous forest with particular reference to north central Florida. Amer. Naturalist 101:173-187.

Odum, E. P. 1963. Ecology. Holt, Rinehart and Winston, New York.

Pielou, E. C. 1966. Species-diversity and pattern-diversity in the study of ecological succession. J. Theor. Biol. 10:370-383.

Whittaker, R. H. 1965. Dominance and diversity in land plant communities. Science 147:250-260.

21

Reprinted from *Nature* **238**:413–414 (Aug. 18, 1972)

WILL A LARGE COMPLEX SYSTEM BE STABLE?

Robert M. May

Institute for Advanced Study, Princeton, New Jersey

Gardner and Ashby[1] have suggested that large complex systems which are assembled (connected) at random may be expected to be stable up to a certain critical level of connectance, and then, as this increases, to suddenly become unstable. Their conclusions were based on the trend of computer studies of systems with 4, 7 and 10 variables.

Here I complement Gardner and Ashby's work with an analytical investigation of such systems in the limit when the number of variables is large. The sharp transition from stability to instability which was the essential feature of their paper is confirmed, and I go further to see how this critical transition point scales with the number of variables n in the system, and with the average connectance C and interaction magnitude α between the various variables. The object is to clarify the relation between stability and complexity in ecological systems with many interacting species, and some conclusions bearing on this question are drawn from the model. But, just as in Gardner and Ashby's work, the formal development of the problem is a general one, and thus applies to the wide range of contexts spelled out by these authors.

Specifically, consider a system with n variables (in an ecological application these are the populations of the n interacting species) which in general may obey some quite nonlinear set of first-order differential equations. The stability of the possible equilibrium or time-independent configurations of such a system may be studied by Taylor-expanding in the neighbourhood of the equilibrium point, so that the stability of the possible equilibrium is characterized by the equation

$$d\mathbf{x}/dt = \mathbf{A}\mathbf{x} \qquad (1)$$

Here in an ecological context $\mathbf{x}$ is the $n \times 1$ column vector of the disturbed populations x_j, and the $n \times n$ interaction matrix A has elements a_{jk} which characterize the effect of species k on species j near equilibrium[2,3]. A diagram of the trophic web immediately determines which a_{jk} are zero (no web link), and the type of interaction determines the sign and magnitude of a_{jk}.

Following Gardner and Ashby, suppose that each of the n species would by itself have a density dependent or otherwise stabilized form, so that if disturbed from equilibrium it would return with some characteristic damping time. To set a time-scale, these damping times are all chosen to be unity: $a_{jj} = -1$. Next the interactions are "switched on", and it is assumed that each such interaction element is equally likely to be positive or negative, having an absolute magnitude chosen from some statistical distribution. That is, each of these matrix elements is assigned from a distribution of random numbers, and this distribution has mean value zero and mean square value α. (For a fuller account of such a formulation, see refs. 2 and 3.) α may be thought of as expressing the average interaction "strength", which average is for simplicity common to all interactions. In short,

$$\mathbf{A}=\mathbf{B}-\mathbf{I} \tag{2}$$

where **B** is a random matrix, and **I** the unit matrix. Thus we have an unbounded ensemble of models, one for each specific choice of the interaction matrix elements drawn individually from the random number distribution.

It is important to note that randomness only enters in the initial choice of the coefficients a_{jk}, which then define a particular model. Once the dice have been rolled to get a specific system, the subsequent analysis is purely deterministic.

The system (1) is stable if, and only if, all the eigenvalues of **A** have negative real parts. For a specified system size n and average interaction strength α, it may be asked what is the probability $P(n,\alpha)$ that a particular matrix drawn from the ensemble will correspond to a stable system. For large n, analytic techniques developed for treating large random matrices may be used to show* that such a matrix will be almost certainly stable ($P\rightarrow 1$) if

$$\alpha<(n)^{-1/2} \tag{3}$$

and almost certainly unstable ($P\rightarrow 0$) if

$$\alpha>(n)^{-1/2} \tag{4}$$

The transition from stability to instability as α increases from the regime (3) into the regime (4) is very sharp for $n\gg 1$; indeed the relative width of the transition region scales as $n^{-2/3}$.

* From equation (2) it is obvious that the eigenvalues of **A** are λ-1, where λ are those of B. The "semi-circle law" distribution for the eigenvalues of a particular random matrix ensemble was first obtained by Wigner[4], and subsequently generalized by him to a very wide class of random matrices whose elements all have the same mean square value[5]. Although the matrix **B** does not in general possess the hermiticity property required for most of these results to be directly applicable, the present results for the largest eigenvalue and its neighbourhood can be obtained by using Wigner's[4] original style of argument on $(\mathbf{B})^N(\mathbf{B}^T)^N$ where N is very large. Indirectly relevant is Mehta[5] and Ginibre[6].

Such a precise answer for any model in the ensemble in the limit $n \gg 1$ is a consequence of the familiar statistical fact that, although individual matrix elements are liable to have any value, by the time one has an $n \times n$ matrix with n^2 such statistical elements, the total system has relatively well defined properties.

Next we introduce Gardner and Ashby's connectance, C, which expresses the probability that any pair of species will interact. It is measured as the percentage of non-zero elements in the matrix, or as the ratio of actual links to topologically possible links in the trophic web. The matrix elements in **B** now either, with probability C, are drawn from the previous random number distribution, or, with probability $1 - C$, are zero. Thus each member of the ensemble of matrices A corresponds to a system of individually stable parts, connected so that each part is affected directly by a fraction C of the other parts. For large n, $\alpha^2 C$ plays the role previously played by α^2, and we find the system (1) is almost certainly stable ($P(n, \alpha, C) \to 1$) if

$$\alpha < (nC)^{-1/2} \tag{5}$$

and almost certainly unstable ($P \to 0$) if

$$\alpha > (nC)^{-1/2} \tag{6}$$

It is interesting to compare the analytical results with Gardner and Ashby's computer results for smallish n. (Their choice of **A** differs slightly from ours, but in essence they have the fixed value $\alpha^2 = 1/3$, and diagonal elements intrinsically -0.55 rather than -1.) Although our methods are based on the assumption that n is large, and are therefore only approximations when applied to $n = 4$, 7, 10, the two approaches in fact agree well when compared, being not more than 30% discrepant even for $n = 4$.

The central feature of the above results for large systems is the very sharp transition from stable to unstable behaviour as the complexity (as measured by the connectance and the average interaction strength) exceeds a critical value. This accords with Gardner and Ashby's conjecture.

Applied in an ecological context, this ensemble of very general mathematical models of multi-species communities, in which the population of each species would by itself be stable, displays the property that too rich a web connectance (too large a C) or too large an average interaction strength (too large an α) leads to instability. The larger the number of species, the more pronounced the effect.

Two corollaries are worth noting, although they should not be taken to have more than qualitative significance.

First, notice that two different systems of this kind, with average interaction strengths and connectances α_1, C_1 and α_2, C_2 respectively, have similar stability character if

$$\alpha_1^2 C_1 \simeq \alpha_2^2 C_2 \tag{7}$$

Roughly speaking, this suggests that within a web species which interact with many others (large C) should do so weakly (small α), and conversely those which interact strongly should do so with but a few species. This is indeed a tendency in many natural ecosystems, as noted, for example, by Margalef[7]: "From empirical evidence it seems that species that interact feebly with others do so with a great number of other species. Conversely, species with strong interactions are often part of a system with a small number of species. . . ."

A second feature of the models may be illustrated by using Gardner and Ashby's computations (which are for a particular α) to see, for example, that 12-species communities with 15% connectance have probability essentially zero of being stable, whereas if the interactions be organized into three separate 4×4 blocks of 4-species communities, each with a consequent 45% connectance, the "organized" 12-species models will be stable with probability 35%. That is, of the infinite ensemble of these particular 12-species models, essentially none of the general ones are stable, whereas 35% of those arranged into three "blocks" are stable. Such examples suggest that our model multi-species communities, for given average interaction strength and web connectance, will do better if the interactions tend to be arranged in "blocks"—again a feature observed in many natural ecosystems.

This work was sponsored by the US National Science Foundation.

REFERENCES

[1] Gardner, M. R., and Ashby, W. R., *Nature*, **228**, 784 (1970).
[2] Margalef, R., *Perspectives in Ecological Theory* (University of Chicago, 1968).
[3] May, R. M., *Math. Biosci.*, **12**, 59 (1971).
[4] Wigner, E. P., *Proc. Fourth Canad. Math. Cong., Toronto*, 174 (1959).
[5] Mehta, M. L., *Random Matrices*, 12 (Academic Press, New York, 1967).
[6] Ginibre, J., *J. Math. Phys.*, **6**, 44 (1965).
[7] Margalef, R., *Perspectives in Ecological Theory*, 7 (University of Chicago, 1968).

AUTHOR CITATION INDEX

SUBJECT INDEX

About the Editor

FRANK B. GOLLEY is executive director of the Institute of Ecology and Professor of Zoology at the University of Georgia. After receiving the B.S. degree in animal science from Purdue University in 1952, he continued studies at Washington State University in wildlife management (M.S., 1954) and at Michigan State University in zoology (Ph.D., 1958). Most of his career has been at the University of Georgia where he organized and directed the Savannah River Ecology Laboratory during 1961–1966 and later administered the Institute of Ecology. His research interests have been concerned with the energy flow through populations and ecosystems, population dynamics of mammals, and radiation ecology. Current interests include the theory of mineral cycling in tropical forest ecosystems and applied ecology. He currently serves as president of the Ecological Society of America and the president of the International Society of Tropical Ecology.